RICHARD BOWDLER SHARPE

CATALOGUE

OF THE

PASSERIFORMES

OR

PERCHING BIRDS

IN THE

COLLECTION OF THE BRITISH MUSEUM

COLIOMORPHÆ

Elibron Classics
www.elibron.com

Elibron Classics series.

ISBN 1-4021-5877-7 (paperback)
ISBN 1-4021-2208-X (hardcover)

This Elibron Classics Replica Edition is an unabridged facsimile of the edition published in 1877 by the British Museum, London.

CATALOGUE

OF THE

BIRDS

IN THE

BRITISH MUSEUM.

VOLUME III.

LONDON:
PRINTED BY ORDER OF THE TRUSTEES.
1877.

CATALOGUE

OF THE

PASSERIFORMES,

OR

PERCHING BIRDS,

IN THE

COLLECTION

OF THE

BRITISH MUSEUM.

COLIOMORPHÆ,

CONTAINING THE FAMILIES

CORVIDÆ, PARADISEIDÆ, ORIOLIDÆ, DICRURIDÆ, AND PRIONOPIDÆ.

BY

R. BOWDLER SHARPE.

LONDON:

PRINTED BY ORDER OF THE TRUSTEES.

1877.

PRINTED BY TAYLOR AND FRANCIS,
RED LION COURT, FLEET STREET.

PREFACE.

The Third Volume of the 'Catalogue of Birds in the British Museum' has been prepared by Mr. R. B. Sharpe, one of the Senior Assistants in the Zoological Department, and author of the two preceding volumes. It contains the commencement of the account of all the species of Passerine Birds known at present; and as no similar monograph of these Birds has been published since the year 1850, it may be hoped that the present work will not only assist the numerous students of this portion of the British-Museum collections, but also prove useful to ornithologists, travellers, and collectors.

ALBERT GÜNTHER,
Keeper of the Department of Zoology.

British Museum,
June 1877.

INTRODUCTION.

The present volume treats of the first group of the Passeriformes or Perching Birds, the *Coliomorphæ*. The total number of species enumerated is 367, of which the British Museum contains 315; so that 52 are still wanting to complete the collection: of these missing species quite half the number are of doubtful specific value, and many of them will be eventually found to be referable to species which are duly described in the present work. The number of specimens enumerated is 2014.

It is not without some satisfaction that the author is able to record the increasing assistance rendered to him by fellow labourers, by presentations to the collection and by the loan of specimens. Especial mention must be made of a donation of 1303 skins of Indian and Malayan birds made to the Museum by Capt. Stackhouse Pinwill. The series in the British Museum includes the collection of *Corvidæ* formed by Mr. John Gould, the African birds formerly belonging to the author, Mr. Wallace's fine series of Malayan Crows, and specimens collected by Mr. W. T. Blanford in Persia and by Dr. J. Anderson during the Yunan Expeditions. The Museum is further indebted for presents to the following gentlemen—Lord Lilford, Major Hayes Lloyd, Captain G. E. Shelley, Dr. B. Hinde, Captain Vincent Legge, A. O. Hume, Esq., C.B., Messrs. Woodward (of Natal), and especially to the authorities of the Smithsonian Institution, who have contributed several species which were wanting to the collection.

The special thanks of the author for valuable assistance and loan of specimens are due to The Marquis of Tweeddale, Captain Shelley, Messrs. Salvin and Godman, Howard Saunders, R. Swinhoe, Dr. Otto Finsch, Prof. J. B. Steere—but particularly to Dr. A. B. Meyer, who has sent him for description his interesting series of the genus *Rectes*, procured by him during his voyage to the Moluccas.

As in the former volumes, the affixes to the names of the persons from whom the specimens were obtained may be explained as follows :—"[P.]" means "Presented by ;" "[C.]"="Collected by ;" "[E.]"="Obtained by exchange." Where none of the above signs are used, the specimens were purchased.

R. BOWDLER SHARPE.

British Museum,
June 6, 1877.

SYSTEMATIC INDEX.

Order PASSERIFORMES.

Suborder PASSERES.

I. *ACROMYODI.*

a. Passeres normales.

SECTION TURDIFORMES.

Group I. COLIOMORPHÆ.

Fam. CORVIDÆ.

Subfam. CORVINÆ.

* Printed as *P. alpinus* in the body of the work (see App. p. 327).

CATALOGUE

OF

BIRDS.

Order II. PASSERIFORMES.

Perching birds, with a nude oil-gland and colic cæca.

Cf. Garrod, P. Z. S. 1874, p. 119.

Suborder I. PASSERES.

Anomalogonatous birds with the second, third, and fourth toes directed forwards, and the hallux backwards; the *flexor longus hallucis* muscle independent of the *flexor perforans digitorum*; the cæca coli short, but at the same time of characteristic shape; oil-gland nude; palate ægithognathous *; *tensor patagii brevis* specialized and quite peculiar.

Cf. Garrod, P. Z. S. 1876, p. 508.

The system of classification for the higher groups of Passerine birds to be followed in the present work will be that of Professor A. H. Garrod, who, after an exhaustive consideration of many points in their anatomy, has established his classification not on one, but on several characters. I am informed by him that he has devoted great attention to the characters given by Prof. Sundevall

* Cf. Huxley, P. Z. S. 1867, p. 456.

(Meth. Av. Tent. p. 1) and by Dr. Elliott Coues in the Introduction to the great work on North-American Birds, and that there are too many exceptions to be found to the characters laid down to render their diagnoses of the order comprehensive. At the same time it may be useful to give some characters propounded by the above-named ornithologists, as follows:—

Hallux stout, furnished with a larger claw than the other toes; greater coverts arranged in a single row, not reaching beyond the middle of the secondaries; sternum simple, with a single notch in the posterior margin*.

In the present work, for the lower groups, it is proposed to follow, as far as possible, the divisions of Professor Sundevall's 'Tentamen,' without adopting their exact order. However modified the arrangement may be, this useful treatise will form the basis of my classification. Several alterations proposed by Dr. Coues will be adopted by me; and at the same time the primary arrangement of the suborder as put forward by Prof. Garrod (*l. c.*) and by Mr. Wallace (in his essay entitled 'An Arrangement of the Families constituting the order Passeres') will be followed, though the sequence of the families will be somewhat changed. For American Passeres the labours of Professor Baird and Mr. Ridgway, in their great work on North-American Birds, will be consulted; and it is to be regretted that the details of the system of classification proposed by Messrs. Sclater and Salvin in their 'Nomenclator,' which seem to contain some original and highly natural combinations, have not yet been given to the world.

Dr. Coues's full characterization of the Passeres is as follows:—"Hallux invariably present, completely incumbent, separately movable by specialization of the *flexor hallucis longus*, with enlarged base and its claw larger than that of the middle digit. Neither second nor fourth toe versatile; joints of toes always 2, 3, 4, 5 from first to fourth. Wing-coverts comparatively short and few, with the exceptions of the least coverts upon the *plica alaris*, arranged in only two series, the greater of which does not reach beyond the middle of the secondary remiges. Rectrices 12 (with rare anomalous exceptions). Musical apparatus present in greater or less development

* The above characters, as a rule, suffice to separate the order *Passeres* from the so-called *Picariæ*, and they are chiefly those employed by Professor Sundevall (Meth. Av. Tent. p. 1) for his first order of birds (*Oscines*, Pallas). It may, however, be well to reproduce here the characters assigned to the order by Dr. Coues, one of the most painstaking of modern systematic ornithologists, who, in avowedly confessing that his classification is based upon that of Sundevall, modestly omits reference to the very useful compendium of additional and secondary characters which he has himself furnished. It is a fact admitted by all ornithologists who have paid attention to the classification of the class *Aves*, that a really satisfactory systematic arrangement of the families does not at present exist; but during the last ten years the labours of Professors Huxley and Parker, Professor Garrod, Dr. Murie, Dr. Elliott Coues, and other well-known anatomists have produced a wonderful change in the classification of birds; and it is not too much to expect that in another equal period a satisfactory arrangement of the class will have been propounded.

and complexity. Palate ægithognathous. Sternum of one particular mould, single-notched. Carotid single (sinistra). Nature highly altricial and psilopædic."

"As remarked by Sundevall, exceptions to the diagnostic pertinence of these two characters of hind claw and wing-coverts are scarcely found. For, in such non-Passerine birds as *Raptores* and some *Herodiones*, in which the claw is enlarged, the wing-coverts are otherwise disposed; and similarly when, as in many *Pici* and elsewhere, the coverts are of a Passerine character, the feet are highly diverse." (*Cf.* Coues, in Baird, Brewer, and Ridgway's 'North-American Birds,' p. xv.)

Division I. **ACROMYODI.**

Subdivision I. PASSERES NORMALES.

Cf. Garrod, P. Z. S. 1876, p. 518.

Section A. *TURDIFORMES.*

(TYPICAL OR THRUSH-LIKE PASSERES.)

Wing with *ten* primaries, the first more or less markedly reduced in size.

Cf. Wallace, Ibis, 1874, p. 409.

Synopsis of Groups.

a. Angle of chin produced very far forwards, so as to extend beyond the line of the anterior margin of the nasal suture.... **COLIOMORPHÆ.**
b. Angle of the chin reaching as far as the line of the anterior margin of the nasal suture, but not beyond it.
  a'. Tongue not extensile.
    a''. Bill notched near the tip of the upper mandible .... **CICHLOMORPHÆ.**
    b''. Bill entire, not notched .... **CERTHIOMORPHÆ.**
  b'. Tongue extensile.................. **CINNYRIMORPHÆ.**

Group I. COLIOMORPHÆ.

Bill stout, generally of large size, not deflected at all, or very little so; chin-angle produced before the line of the nostrils *; cutting edge of lower mandibles simple. Tongue not extensile, generally thick and fleshy, the tip horny, slight, divided, or split up into threads in various ways. Feet in most strong, large, with the claw of the middle toe oblique.

Cf. Sundevall, Av. Tent. p. 37.

Synopsis of Families.

a. Bill without a distinct subterminal notch in the upper mandible.
  a′. Toes normal, the hallux very strong, but, with its claw, not as long as the middle toe and claw .... **Corvidæ**, p. 4.
  b′. Toes abnormal: outer toe a little shorter than the middle one, longer than the inner one; hallux very large, with its claw equal to or longer than the middle toe with its claw .... **Paradiseidæ**, p. 153.
b. Bill with a notch in the upper mandible.
  c′. Nostrils placed well in front of the base of the bill and quite bare .... **Oriolidæ**, p. 188.
  d′. Nostrils more or less covered with bristles, generally completely so.
    a″. Tail forked, consisting of ten feathers. **Dicruridæ**, p. 228.
    b″. Tail moderate, rounded or square, consisting of twelve feathers.......... **Prionopidæ**, p. 270.

Family CORVIDÆ.

a. Nostrils placed high in the maxilla, as near to or nearer the culmen than the edge of the maxilla.......... *Subfam. CORVINÆ*, p. 5.
b. Nostrils placed lower in the maxilla, nearer to its lower edge than to the culmen.. .... *Subfam. FREGILINÆ*, p. 146.

* In some of the Jays the nostril and the chin-angle are nearly in a line; but in most of them, when the skeleton has been examined (as I have been able to do in nearly all Corvine genera), it gives the balance in favour of the produced angle of the chin. Some Creepers have also the last-named character; but then the general form of the bill and feet is so different as to prohibit their being confounded with Crows.

Subfamily I. CORVINÆ.

Key to the Genera.

a. Wings long, falling short of the tip of the tail by very little, the difference never being more than the length of the tarsus.
a′. Nostrils concealed with bristles.
a″. First primary long, equal to or exceeding the innermost secondaries in length.
a‴. Face bare, as well as the nostrils; forehead also devoid of plumes .... 1. TRYPANOCORAX, p. 8.
b‴. Face feathered, the nostrils hidden by bristles.
a^4. Nasal bristles short, the part of the bill uncovered by them being longer than the space between the eye and the tips of the nasal plumes; bill thin and straight, gently curved towards the tip .. 2. HETEROCORAX, p. 11.
b^4. Nasal bristles long, the space between the eye and the tips of the nasal plumes less than the remaining portion of the bill.
a^5. Tarsus equal to or less than the culmen, the latter equal to or longer than the head.
a^6. Nostrils placed in a more or less defined groove; the bill stout, but not equalling in its depth the length of the outer toe .. 3. CORVUS, p. 13.
b^6. Nostrils placed in a very deep groove; depth of bill enormous, greater than length of outer toe .... 4. CORVULTUR, p. 24.
b^5. Tarsus longer than culmen, the latter more or less shorter than the head .... 5. COLŒUS, p. 25.
b″. First primary longer than the ordinary secondaries, but not as long as the innermost of the latter.
c‴. Nasal bristles directed horizontally over the nostrils .... 6. CORONE, p. 30.
d‴. Nasal bristles erect and directed forwards and upwards .... 7. RHINOCORAX, p. 45.
c″. First primary shorter, not equal to the secondaries in length.
e‴. Nasal bristles very long, the distance between their tips and the tip of the upper mandible about equal to the distance between the eye and the tips of the nasal bristles.
c^4. Bill curved, the angle of the culmen about equal to that of the gonys towards the tip.

c⁵. Angle of mouth feathered...... 8. GAZZOLA, p. 47.
d⁵. Angle of mouth bare.......... 9. MICROCORAX, p. 48.
d⁴. Bill anomalous, the gonys ascending at a much greater angle towards the tip than the apical curve of the culmen .... 10. PHYSOCORAX, p. 50.
f'''. Nasal bristles shorter, the distance between their tips and the tip of the upper mandible much greater than the distance between the eye and the tips of the nasal bristles.
e⁴. Face and region of the eye bare .. 11. GYMNOCORAX, p. 50.
f⁴. Face and region of the eye feathered.
e⁵. Bill compressed, strongly curved; wing having the innermost secondaries longer than the ordinary or outer secondary quills . 12. MACROCORAX, p. 51.
f⁵. Bill conical and nearly straight; wing having the innermost secondaries gradually shorter than the outer ones .... 13. NUCIFRAGA, p. 52.
b'. Nostrils bare .... 14. STREPERA, p. 57.
b. Wings short, the distance between the tips of the quills and the tips of the tail-feathers always much more than the length of the tarsus.
c'. Tail very long, and strongly graduated, the two centre feathers exceeding the outer one by more than half, generally by two thirds of the length of the wing.
d''. Head not ornamented with a long and distinct crest.
g'''. Eye without a wattled ring.
g⁴. First primary sinuated and ending in a long pointed plume .... 15. PICA, p. 62.
h⁴. First primary normal.
g⁵. Tail-feathers not inclined to be spatulate at tips, being about the same breadth for their whole length.
c⁶. Bill small and nearly straight, its depth at nostrils not exceeding the length of the hind claw; nostrils densely feathered .... 16. CYANOPOLIUS, p. 67.
d⁶. Bill very stout and much curved, much deeper at the nostrils than the hind claw; nostrils not very densely feathered .... 17. UROCISSA, p. 69.
h⁵. Middle tail-feathers gradually narrowing towards the tip, where they are acuminate; bill stout, its length at base far exceeding the length of the hind claw; nostrils not very thickly plumed 18. CRYPTORHINA, p. 74.

i^6. Middle tail-feathers inclined to be spatulate, much broader at the ends than in the middle.

e^8. Nostrils thinly covered with bristles, leaving the base of the culmen unfeathered .... 19. DENDROCITTA, p. 75.

f^8. Nostrils densely covered with plumes which form a defined crest of velvety feathers hiding the base of the culmen...... 20. CRYPSIRHINA, p. 83.

h'''. Eye surrounded by a fleshy wattle.. 21. CISSA, p. 84.

e''. Head ornamented with a long crest springing from the forehead and extending beyond the nape, the plumes long and widening at the tip .... 22. CALOCITTA, p. 88.

d'. Tail moderately long and graduated, the two centre tail-feathers only slightly longer than the others.

f''. Nostrils hidden by a dense tuft of bristles.

i'''. Bill deep, compressed, the height at the nostrils equal to the hind toe (without claw).

i^4. Tail graduated .... 23. PLATYSMURUS, p. 90.

k^4. Tail-feathers square at tip ...... 24. TEMNURUS *, p. 91.

k'''. Bill moderately deep and compressed, the height at the nostrils less than the hind toe.

l^4. Tail and wing about equal in length, the difference between them not being more than the length of the hind toe (without claw).

k^5. Bill at nostrils about as high as it is broad.

g^6. Secondaries moderately long, the distance between their tips and the ends of the primaries about equal to the culmen; second primary about as long as the secondaries.

a^7. Culmen decurved .... 25. GARRULUS, p. 91.

b^7. Culmen nearly straight.... 26. PERISOREUS, p. 103.

h^6. Secondaries very long, the distance between their tips and the ends of the primaries not equal to the culmen; second primary not equalling the secondaries in length.

c^7. Head with a long crest (equalling or exceeding the culmen in length) ...... 27. CYANOCITTA, p. 106.

d^7. Head without a crest...... 28. APHELOCOMA, p. 112.

l^5. Bill at nostrils higher than it is broad; frontal plumes com-

---

* Never having examined a skin of *Temnurus*, I place it next to *Platysmurus*, with which it has always been associated.

pressed and directed forwards, forming a close-set crest...... 29. CYANOCORAX, p. 119.
m^4. Tail much longer than the wing, exceeding it by at least the length of the culmen, sometimes by more. 30. XANTHURA, p. 128.
n^4. Wing much longer than the tail, exceeding it by more than double the length of the hind toe; frontal crest long and erect, elegantly recurved .... 31. UROLEUCA, p. 137.
g''. Nostrils always more or less exposed.
l'''. Bill nearly straight, longer than hind toe and claw.................... 32. GYMNOKITTA, p. 138.
m'''. Bill stout, conical, not longer than hind toe and claw.
o^4. Angle of mouth without pendent wattle.
m^5. Head feathered.
i^6. Bill stout, but rather long, measured from gape equal to about twice and a half its height at the nostrils ...... 33. PSILORHINUS, p. 139.
k^6. Bill conical, measured from gape double its height at nostrils .... 34. STRUTHIDEA, p. 140.
n^5. Head bare .... 35. PICATHARTES, p. 141.
p^4. Angle of mouth with a pendent wattle .. .. 36. GLAUCOPIS, p. 142.
n'''. Bill long and slender, longer than the hinder toe and claw.
q^4. With a conspicuous wattle at the gape.
o^5. Sexes differing: the male with a stout straight bill; the female with a long curved one ...... 37. HETERALOCHA, p. 143.
p^5. Sexes alike.................. 38. CREADION, p. 144.
r^4. With no wattle at gape .... 39. FALCULIA, p. 145.

1. TRYPANOCORAX*.

Trypanocorax, *Bp. Ann. Sci. Nat.* i. p. 133 (1854), typo haud indicato; *Sundevall, Meth. Av. Tent.* p. 43 (1872)...................... Type. T. frugilegus.

Range. The whole of Europe and the countries round the Mediterranean; Northern Asia and China, ranging into North-western India.

---

* Bonaparte incidentally mentions this genus as "*Trypanocorax* (Kaup);" and he also alludes to a genus *Frugilegus* of Lesson. Neither of these generic names have I been able to find in the published works of these authors, although the latter is adopted by Mr. G. R. Gray in the 'Handlist' as a section of the genus *Corvus*. Sundevall also quotes Kaup as the author of the generic name; but until the exact place of publication is discovered, the Swedish professor must have the credit of establishing the genus, to which he was the first to assign peculiar characters.

Key to the Species.

a. Forehead, lores, and throat bare .... *frugilegus*, p. 9.
b. Forehead and lores only bare, throat feathered .... *pastinator*, p. 10.

1. Trypanocorax frugilegus.

La Corneille moissoneuse, *Briss. Orn.* ii. p. 16.
Corvus frugilegus, *Linn. S. N.* i. p. 156; *Temm. Man. d'Orn.* i. p. 110; *Naum. Vög. Deutschl.* ii. pl. 55; *Roux, Orn. Prov.* pl. 132; *Werner, Atlas,* Omnivores, pl. 5; *Gould, B. Eur.* iii. pl. 24; *Macgill. Brit. B.* i. p. 535; *Yarr. Brit. B.* ii. p. 91; *Schlegel & Susem. Vögel Eur.* ii. t. 3. fig. 2; *Kjærb. Orn. Dan.* pl. 11. fig. 2; *Gray, Gen. B.* ii. p. 315; *Nordm. in Démid. Voy. Russ. Mérid.* iii. p. 121; *Rüpp. Syst. Uebers.* p. 74; *Bp. Consp.* i. p. 384; *Horsf. & Moore, Cat. B. Mus. E.I. Co.* ii. p. 557; *Schlegel, Vog. Nederl.* pl. 139; *id. Dier. Nederl.* pl. xiii. figs. 4, 5; *Fritsch, Vög. Eur.* tab. 28. figs. 7, 8; *Sundev. Svensk. Fogl.* pl. 20. fig. 3; *Schlegel, Bijdr. Dierk.* 4to, ii. *art.* Corvus, p. 11, pl. 1. fig. 16; *Jaub. & Barth. Lapomm. Rich. Orn.* p. 94; *Linderm. Vög. Griechenl.* p. 71: *Filippi, Viagg. Pers.* p. 350; *Jerd. B. Ind.* ii. p. 302; *Gould, B. G. Brit.* iii. pl. 60; *Schlegel, Mus. P.-B.* Coraces, p. 26; *Degl. & Gerbe, Orn. Eur.* i. p. 201; *Doderl. Avif. Sicil.* p. 53; *Heugl. Orn. N.O.-Afr.* p. 498; *Gray, Hand-l. B.* ii. p. 12; *Salvad. Faun. Ital. Ucc.* p. 171; *Hume, Ibis,* 1871, p. 404; *Shelley, B. Egypt,* p. 159; *Severtz. Turkest. Jevotn.* p. 63; *Dresser, B. Eur.* pt. xxxv.; *Irby, B. Gibr.* p. 127; *Scully, S. F.* 1876, p. 156; *Blanf. East. Persia,* p. 263.
Le Freux, *Montb. Pl. Enl.* iii. pl. 484.
La Corneille, *Montb. t. c.* pl. 483.
Colœus frugilegus, *Kaup, Natürl. Syst.* p. 114.
Corvus agrorum, *Brehm, Vög. Deutschl.* p. 170.
Corvus granorum, *Brehm, t. c.* p. 171.
Corvus advena, *Brehm, t. c.* p. 171.
Corvus agricola, *Tristr. P. Z. S.* 1864, p. 444; *Gray, Hand-l. B.* ii. p. 12.
Trypanocorax frugilegus, *Loche, Expl. Sci. Algér. Ois.* i. p. 113; *Sundev. Meth. Av. Tent.* p. 43.

Adult male. (Sweden, May 9, 1870.) Above glossy purplish black, the head and neck rich purple, with a slight greenish shade over the eyes, on the sides of the crown, and on the ear-coverts; forehead, lores, fore part of cheeks, and throat bare; wings dull brown, the least wing-coverts purple like the back, the remainder of the coverts and the quills with a slight gloss of dull green, and remains of purple on the secondaries; tail dull greenish black, glossed with purple near the base; throat purplish black, with greyish bases to the feathers, rather more blue in shade than the sides of the neck; rest of under surface dull purplish black, not so glossy as the back; bill and feet black; iris blackish brown. Total length 17 inches, culmen 2·25, wing 12·4, tail 7·5, tarsus 2·2.

Adult female. Similar to the male. Total length 16 inches, culmen 2·2, wing 12·2, tail 7·4, tarsus 2·1.

Winter plumage. Very much more brilliant than the breeding-dress, which quickly loses much of its gloss.

Young. Much more dusky than the adults, without the purple gloss showing to any great extent; wings black, shining, with purple on the secondaries and with green near the edge of the wings; lores, nasal bristles, sides of face, and throat glossy black; under surface of body dusky black; bill black, flesh-coloured at base.

Hab. The whole of Europe, ranging into Central Asia and North-western India.

a. Ad. sk.	England.	Purchased.
b, *c*. Juv. sk.	England.	Purchased.
d. ♀ ad. sk.	Greatham, Durham, April 17, 1866.	J. H. Gurney, Jun., Esq. [P.].
e, *f*. Pull. st.	Cambridgeshire.	Mr. J. Baker [P.].
g. Juv. st.	Cambridgeshire.	Purchased.
h. Var. st.	Samborne Park, Hants.	Maj.-Gen. Warell [P.].
i. Ad. sk.	Avington Park, Hants, Nov. 25, 1872.	E. Shelley, Esq. [P.].
k. Jun. sk.	Avington Park, Hants, Nov. 25, 1872.	E. Shelley, Esq. [P.].
l. Var. st.	Devonshire.	Col. Montagu [P.].
m. ♂ ad. sk.	Dunkeld, N.B., May 2, 1876.	A. B. Brooke, Esq. [C.].
n, *o*. ♂ ♀ ad. sk.	Skara, Sweden, May 9, 1870.	Purchased.
p, *q*. ♂ ♀ juv. sk.	Skåne, Sweden, June 22, 1874.	Purchased.
r. ♀ juv. sk.	Skåne, Sweden, April 8, 1874.	Purchased.
s. Ad. sk.	Belgium, Dec. 8, 1870.	Brussels Museum [P.].
t. Juv. sk.	Belgium, Dec. 8, 1870.	Brussels Museum [P.].
u. Ad. st.	France.	R. B. Sharpe, Esq. [P.].
v, *w*. ♀ ad. sk.	Buyukdere, Turkey, Feb. 20, 1874 (*T. Robson*).	R. B. Sharpe, Esq. [P.].
x. Ad. sk.	Crimea.	J. Gould, Esq.
y. Ad. sk.	Erzeroom.	J. Gould, Esq.
z. Ad. sk.	Nazareth.	Canon Tristram [C.].
a'. Imm. sk.	Jerusalem.	Canon Tristram [C.].
b'. ♂ ad. sk.	Mesopotamia (*Commander Jones*).	Secretary of State for India [P.].
c'. Juv. sk.	Mesopotamia.	Euphrates Exped. [C.].
d'. Juv. sk.	Gangaotri, Himalayas.	
e', *f'*. Ad. sk.	[China.]	Capt. Blackiston [C.]*.
g'. Skeleton.	England.	
h'. Skull.	Würtemberg.	Dr. Günther [C.].

2. Trypanocorax pastinator.

Corvus pastinator, *Gould, P. Z. S.* 1845, p. 1; *Blyth, J. A. S. B.* xv. p. 25; *Gray, Gen. B.* ii. p. 315; *Bp. Consp.* i. p. 384; *Horsf. & Moore, Cat. B. Mus. E.I. Co.* ii. p. 558; *Gray, Hand-l. B.* ii. p. 12; *David, N. Arch. Mus.* vii. *Bull.* p. 9.
Corvus frugilegus, *Schl. Faun. Jap.* p. 80.
Frugilegus pastinator, *Swinh. P. Z. S.* 1871, p. 383.

Adult (type of species). General colour above and below purplish black, rather deeper and blacker on the sides of the head; quills black, glossed with purple on the secondaries and with steel-green

---

* I am assured by Mr. Whitely, from whom we obtained this and many other birds of Captain Blakiston's, that there can be no doubt of its having been received in a consignment of the latter gentleman's, though whether it was actually killed in China of course it is difficult to say.

on the primaries; tail greenish black, shaded with purple towards the base; frontal plumes and space between the eye and the nostril bare, as well as a small patch at the base of the lower mandible; bill and feet black. Total length 18 inches, culmen 1·85, wing 11·75, tail 7·2, tarsus 2·05.

Obs. This Rook differs from *C. frugilegus* in having only the frontal space and the base of the lower mandible bare, the throat being feathered. In the plumage there is no contrast between the head and the rest of the upper surface, the general tone of the plumage being uniform purple from head to tail; on only one specimen can I detect a faint shade of greenish on the sides of the neck. The specimen mentioned below from Irkutsk is a much blacker purple than any of the Chinese birds. The bill varies a good deal in size, the Japanese birds having, as a rule, rather a smaller bill than those of China.

Hab. China and Japan, ranging into Eastern Siberia.

a. Ad. sk.	Chusan, China, Jan. 1843.	J. Gould, Esq. (Type of species.)
b. Juv. sk.	Shanghai, Dec. 10, 1850.	J. Webb, Esq. [P.].
c, d. Ad. sk.	Shanghai.	Shanghai Museum [P.].
e. Ad. sk.	China.	Capt. Blakiston [C.].
f. Juv. sk.	China.	Capt. Blakiston [C.].
g. Juv. sk.	Japan.	Purchased.
h. Ad. sk.	Japan.	Secretary of State for India [P.]. (Spec. *a* of Horsf. & Moore, Cat. p. 558.)
i. Ad. sk.	Nagasaki, Japan.	Capt. St. John [C.].
k. ♂ ad. sk.	Irkutsk, April 23.	Dr. Dybowski [C.].

2. HETEROCORAX.

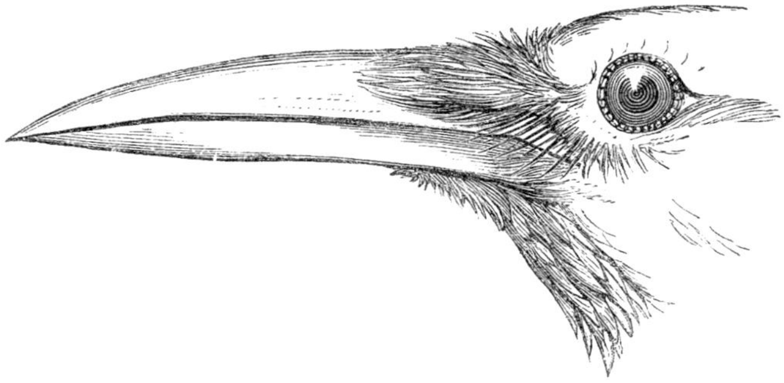

Head of *H. capensis*.

Range. Southern and North-eastern Africa.

1. Heterocorax capensis.

La Corneille du Cap, *Levaill. Ois. d'Afr.* ii. pl. 52.
Corvus capensis, *Licht. Verz. Doubl.* p. 20; *Rüpp. Faun. Abyss.* Taf. 10. fig. 3; *id. Syst. Uebers.* p. 75; *Gray, Gen. B.* ii. p. 315; *Bp. Consp.* i. p. 385; *Cab. Mus. Hein.* Th. i. p. 230; *Horsf. & Moore, Cat. B. Mus. E.I. Co.* ii. p. 557; *Grill, Zool. Anteckn.* p. 38; *Sundev. Crit. om Levaill.* p. 29; *Schl. Bijdr. Dierk. Amsterd.* fol. *art.* Corvus, p. 12, pl. 1. fig. 19; *id. Mus. P.-B.* Coraces, p. 27; *Gray, Hand-l. B.* ii. p. 12; *Sharpe, Cat. Afr. B.* p. 57; *Gurney, in Anderss. B. Dam. Ld.* p. 155.
Corvus macropterus, *Wagler, Syst. Av.* Corvus, sp. 10.
Corvus segetum, *Temm. texte art.* Corvus *in Pl. Col.*; *Layard, B. S. Afr.* p. 163.
Corvus levaillanti, *Des Murs in Lefebvr. Voy. Abyss. Zool.* p. 104 (*nec Less.*).
Trypanocorax capensis, *Bp. C. R.* xxxvii. p. 829.
Corvus minor, *Heugl. Syst. Uebers.* p. 35.
Corvus capensis minor, *Heugl. Orn. N.O.-Afr.* p. 409.

Adult male. General colour above and below dusky black, the gloss having somewhat of a greyish brown shade, the back and wings glossed with purple, as also the under surface; the quills externally dull greenish black; "bill, legs, and toes dark horn-colour; irides dark brown" (*Andersson*). Total length 18 inches, culmen 2·5, wing 13·8, tail 7·7, tarsus 2·7.

Young. Dusky brown where the adult is glossy black, nearly all the gloss being absent in the young; the wings and tail as in the adult; bill horn-coloured at tip.

Obs. Professor Schlegel (M. P.-B. *Coraces*, p. 27) refers to a species which he names "*Corvus minor*, Heugl.," and which, he says, resembles in every respect *C. capensis*, notably in the form of the beak, feathers, and in its colours, but is very much smaller. The only example in the Leiden Museum was purchased of Verreaux, in 1858, but without an exact locality attached to it; and Dr. von Heuglin (Orn. N.O.-Afr. p. 500) doubts whether it is the same species to which he assigned the name of *C. minor* in 1856. The only differences between *C. minor* and true *C. capensis* consist in the large size of the latter; and this is not sufficiently marked to enable me to separate them, as will be seen by the following measurements, which, indeed, show a preponderance in size in favour of the bird from North-eastern Africa.

	Total length.	Culmen.	Wing.	Tail.	Tarsus.
	in.	in.	in.	in.	in.
a. ♀. Angollalá (*Harris*)	19·0	2·6	14·3	8·0	2·9
b. Juv. Damara Land (*Andersson*)	16·6	2·5	12·6	7·3	2·55
c. ♂. Ondonga (*Andersson*) ..	18·0	2·5	13·8	7·7	2·7
d. Ad. Swellendam (*Verreaux*)..	20·0	2·6	13·9	7·8	2·75

It will be seen from the above that it is impossible to separate the Abyssinian species on the score of larger dimensions; and therefore Professor Schlegel's *C. minor* still requires identification.

Hab. South and North-east Africa.

a. ♂ ad. sk.	Ondonga, Ovampo Land, Nov. 11, 1866 (*Andersson*).	R. B. Sharpe, Esq.
b. Imm. sk.	Damara Land (*Andersson*).	R. B. Sharpe, Esq.
c. Juv. sk.	Cape of Good Hope.	
d. Ad. sk.	Cape of Good Hope.	Earl of Derby [P.].
e. ♀ juv. sk.	Karee river.	M. Jules Verreaux [C.].
f. ♂ ad. sk.	Swellendam, April 1834.	M. Jules Verreaux [C.].
g. ♀ ad. sk.	Angollalá, Shoa, Feb. 1842.	Sir W. C. Harris [C.].

3. CORVUS*.

Type.

Corvus, *Linn. S. N.* i. p. 155 (1766) .... C. corax.
Pterocorax, *Kaup, J. f. O. Suppl.* p. lv (1854) ....

Range. The whole of Europe and Asia above the line of the Himalayas, ranging into North-western India, Australia, North America, and thence into Mexico.

Key to the Species.

a. With no white collar round the hind neck.
- *a'*. Hackles of the throat very long, and extending over the fore neck.
  - *a''*. Larger: head and neck purplish or glossy steel-black like the rest of the upper surface. *corax*, p. 14.
  - *b''*. Smaller: head and neck brownish, differing from the rest of the upper surface .... *umbrinus*, p. 17.
- *b'*. Hackles of the throat moderately long, but confined to the throat.
  - *c''*. Base of plumage of neck and fore neck in adult bird dark grey.
    - *a'''*. Much larger: bill 1·05–1·1 inch deep..... *leptonyx*, p. 19.
    - *b'''*. Very much smaller: bill 0·85–0·9 inch deep .... *culminatus*, p. 20.
  - *d''*. Base of plumage of neck and fore neck in adult bird snowy white.................. *coronoides*, p. 20.

b. With a pure white collar round the hind neck.
- *c'*. Across the chest a broad white collar; throat and rest of under surface black .... *torquatus*, p. 21.
- *d'*. Entire breast white; throat and abdomen black *scapulatus*, p. 22.

CORVUS HAWAIENSIS.

Corvus tropicus, *Bloxham in Byron's Voy.* p. 249.
Corvus hawaiensis, *Peale, U.S. Expl. Exp. Orn.* 1848, p. 106; *Cass t. c.* 2nd edit. p. 119, pl. vi.; *Gray, B. Trop. Isl. Pacif.* p. 24; *Dole, Pr. Bost. Soc. N. Hist.* xii. p. 300; *Scl. Ibis*, 1871, p. 359; *Gray, Hand-l. B.* ii. p. 14.

Entire plumage fuliginous brown, with a slight tinge of cinereous; quills light reddish brown, with their shafts white on their under surfaces; all the plumage dark cinereous at the bases of the feathers; bill and legs black, the former lighter at the tip. Total length 18¾ inches, wing 12, tail 8, bill 2¼, tarsus 2½. (*Cassin, l. c.*)

Female (?). Total length 17½ inches. *Cassin, l. c.*

Hab. Hawaii, Sandwich Islands.

* The late Mr. G. R. Gray, in the 'Hand-list' (ii. p. 11), gives as a synonym of this genus, "*Corax*, Bp. 1850." In the 'Conspectus,' Bonaparte does not use the name in a generic sense, but separates the genus *Corvus* into sections, one of which he calls *Coraces* and another *Cornices*. It is therefore a mistake to credit Bonaparte with the manufacture of the genera *Corax* and *Cornix*.

1. Corvus corax.

Corvus corax, *Linn. S. N.* i. p. 155; *Temm. Man. d'Orn.* i. p. 107; *Wils. Am. Orn.* ix. p. 136, pl. lxxxvi. fig. 3; *Naum. Vög. Deutschl.* ii. Taf. 53. fig. 1; *Roux, Orn. Prov.* pl. 129; *Werner, Atlas*, Omnivores, pl. xx. fig. 2; *Audub. Orn. Biogr.* ii. p. 476; *id. B. Amer.* pl. ci.; *Macgill. Br. B.* i. p. 498; *Yarr. Br. B.* ii. p. 63; *Gould, B. Eur.* iii. pl. 220; *Gray, Gen. B.* ii. p. 315; *Nordm. in Démid. Voy. Russ. Mérid.* iii. p. 118; *Blyth, Cat. B. Mus. A. S. B.* p. 89; *Cab. Mus. Hein.* Th. i. p. 233; *Schl. u. Susem. Vög. Deutschl.* ii. Taf. 1; *Midd. Sibir. Reis. Zool.* p. 161; *Schl. Vog. Nederl.* pl. 136; *Kjærb. Orn. Dan.* pl. xi. fig. 3; *Jaub. et Barth. Lapomm. Rich. Orn.* p. 91; *Schl. Dier. Nederl. Vog.* pl. xiii. fig. 1; *id. Bijdr. Dierk. Amsterd.* fol. *art.* Corvus, p. 4, pl. 1. figs. 1, 2; *Fritsch, Vög. Eur.* tab. 28. fig. 6; *Schrenck, Reis. Amurl. Vög.* p. 326; *Linderm. Vög. Deutschl.* p. 71; *Sundev. Sv. Fogl.* pl. xx. fig. 2; *Filippi, Viagg. Pers.* p. 350; *Radde, Reis. Sibir. Vög.* p. 211; *Degl. & Gerbe, Orn. Eur.* i. p. 196; *Schl. Mus. P.-B.* Coraces, p. 11; *Doderl. Avif. Sicil.* p. 61; *Gray, Hand-l. B.* ii. p. 11; *Salvad. Faun. Ital. Ucc.* p. 171; *David, N. Arch. Mus.* vii. *Bull.* p. 9; *Gould, B. Gt. Br.* iii. pl. 57; *Severtz. Turkest. Jevotn.* p. 63; *Finsch, Abhandl. Ver. Brem.* iii. p. 40; *Dresser, B. Eur.* parts xxxvii., xxxviii.; *Irby, B. Gibr.* p. 126; *Scl. & Salv. Nomencl. Av. Neotr.* p. 40; *Blanf. East. Persia*, p. 261.

Le Corbeau, *Montb. Pl. Enl.* 495.

Corvus maximus, *Scop. Ann.* i. p. 34.

Corvus clericus, *Sparrm. Mus. Carls.* pl. ii. var.

Corvus carnivorus, *Bartr. Trav. E. Florida*, p. 290; *Baird, B. N. Amer.* p. 560, pl. xxi.; *Lord, Pr. R. A. Inst.* iv. 1864, p. 121; *Dall & Bann. Tr. Chic. Acad.* i. p. 285; *Cooper, Orn. Calif.* i. p. 282; *Gray, Hand-l. B.* ii. p. 11.

Le Grand Corbeau, *Levaill. Ois. d'Afr.* ii. p. 8, pl. 51.

Corvus leucophæus, *Vieill. N. Dict. d'Hist. Nat.* viii. p. 27; *Gray, Gen. B.* ii. p. 315; *Degl. & Gerbe, Orn. Eur.* i. p. 197.

Corvus major, *Vieill. N. Dict.* viii. p. 27 (*ex Levaill.*); *Gray, Gen. B.* ii. p. 315; *id. Hand-l. B.* ii. p. 11.

Corvus leucomelas, *Wagler, Syst. Av.* Corvus, sp. 4.

Corvus cacolotl, *Wagler, Isis*, 1831, p. 527; *Gray, Gen. B.* ii. p. 315.

Corvus sylvestris, *A. E. Brehm, Vög. Deutschl.* p. 163.

Corvus littoralis, *id. t. c.* p. 164.

Corvus peregrinus, *id. t. c.* p. 164.

Corvus montanus, *id. t. c.* p. 165.

Corvus nobilis, *Gould, P. Z. S.* 1837, p. 79; *Gray, Hand-l. B.* ii. p. 11.

Corvus littoralis, *Holb. in Kröyer's Tidschr.* iv. p. 390.

Corvus vociferus, *Cabot, Bost. Journ. N. Sci.* iv. p. 464.

Corvus lugubris, *Agass. Proc. Bost. Soc. N. H.* ii. p. 188.

Corvus thibetanus, *Hodgs. Ann. N. H.* (2) iii. p. 203; *Horsf. & Moore, Cat. B. Mus. E.I. Co.* ii. p. 553; *Jerd. B. Ind.* ii. p. 294; *Gray, Hand-l. B.* ii. p. 11; *Henders. & Hume, Lahore to Yark.* p. 234; *Scully, Str. F.* 1876, p. 155.

Corax maximus, *A. E. Brehm, Allg. deutsch. naturh. Zeit.* 1857, p. 445.

Corvus ferroensis, *Schl. Bijdr. Dierk. Amsterd.* folio, *art.* Corvus, p. 6.

Corvus corax, var. carnivorus, *Baird, Brewer, & Ridgw. N. Amer. B.* ii. p. 234.

Adult male. General colour above glossy steel-black with a purplish lustre, the base of the feathers greyish; wings browner and more bronzy than the back, the coverts and the secondaries purplish at their bases, the latter also washed with purple externally; primaries externally steel-black, with no shade of green; tail purplish black, inclining to steel-black on the two outer tail-feathers; head glossy purplish black, the sides of the face and the feathers of the neck silky black, with somewhat of a purplish-brown shade; entire under surface of body glossy blue-black, slightly shaded with purple, the feathers of the throat extremely lanceolate and long; these plumes glossy purplish brown, those of the fore neck more distinctly glossed with purple; bill and legs black; iris brown. Total length 24 inches, culmen 3·15, wing 17·5, tail 10·5, tarsus 2·85.

Obs. The following Table of measurements is arranged according

---

CORVUS LAWRENCII.

Corvus lawrencii, *Hume, Lahore to Yark.* p. 235; *id. S. F.* 1873, p. 205; *Adam, t. c.* p. 385; *Hume, Nests and Eggs Ind. B.* p. 408; *Stoliczka, S. F.* 1864, p. 474; *Ball, S. F.* 1875, p. 207; *Hume, t. c.* p. 493.

This Raven, the ordinary species of the Panjáb, was separated by Mr. Hume on account of its small size. He writes:—"In a Greenland specimen the bill measures 3·4 at front, straight from forehead to point; while, similarly measured, the bills of the Panjáb do not exceed 2·8, and those of the Tibetan birds 3·18. The height of the two mandibles closed is about 1·0 in the Panjáb birds, 1·2 in the Tibetan birds, and 1·25 in the Greenland specimen. No male of the Panjáb bird, out of numbers measured in the flesh, exceeded 24·75 inches in length, no female exceeded 23·75, and no male weighed more than 2 pounds 5 ounces." Mr. Dresser, in the 'Birds of Europe,' unhesitatingly refers *C. lawrencii* to *C. corax*; but I think it will most probably turn out to be *C. umbrinus*.

CORVUS CRYPTOLEUCUS.

Corvus cryptoleucus, *Couch, Pr. Acad. Nat. Sci. Philad.* 1854, p. 66; *Baird in B. N. Am.* p. 565, pl. xxii.; *Dresser, Ibis,* 1855, p. 494; *Cooper, Orn. Calif.* i. p. 284; *Gray, Hand-l. B.* ii. p. 13; *Coues, Key N. Amer. B.* p. 162; *Baird, Brewer, & Ridgw. N. Am. B.* ii. p. 242; *Coues, B. Northwest,* p. 206.

Glossy black, with violet reflections; feathers of neck all round, back and breast snow-white at the base; feathers of throat lanceolate; bristly feathers along the base of the bill covering it for nearly two thirds of its length. The fourth quill is longest, the third and fifth equal, the second longer than the sixth, the first about equal to seventh. Length about 21 inches, wing 14·0, tail 8·5. (*Baird & Ridgw. l. c.*)

Hab. Valley of Rio Grande and Gila; abundant on the Llano Estacado and at Eagle Pass, Texas; Colorado (*Baird, Brewer, & Ridgw. l. c.*).

Obs. The authors of 'North-American Birds' keep this species distinct from *C. corax*, with the following diognoses (ii. p. 233):—

1. *C. corax*, var. *carnivorus*. Length about 25·00, wing 17·00, tail 10·00; gradation of tail 1·60 to 2·40. Feathers of the neck and breast light grey beneath surface.
2. *C. cryptoleucus*. Length about 21·00, wing 14·00, tail 8·50; gradation of tail 1·25. Feathers of neck and breast snowy white beneath surface.

Not having seen a specimen of this Raven, I reproduce the description given in the above-mentioned work.

to the size of the birds, and not according to locality, the wing being taken as the standard of measurement:—

		Total length. in.	Culm. in.	Wing. in.	Tail. in.	Tarsus. in.
1. Ad.	Thibet (*Hodgson*)	24·0	3·35	19·6	11·2	2·8
2. Ad.	Mexico (type of *C. nobilis*)	25·0	3·45	18·9	10·5	2·9
3. ♀ ad.	Onon river (*Dybowski*) ..	..	2·9	18·45	10·5	2·5
4. Ad.	Ladakh (*Strachey*)	25·0	3·3	17·9	11·0	2·75
5. ♀ ad.	Jerusalem (*Tristram*)....	24·5	3·2	17·7	9·8	2·7
6. ♀ ad.	England	23·5	3·3	17·7	10·0	2·9
7. ♂ ad.	Sardinia (*Brooke*)	25·0	3·15	17·6	10·0	2·9
8. Ad.	Vancouver's Island (*Lord*)	24·0	3·45	17·5	10·2	2·85
9. Ad.	Mexico	22·5	2·95	17·45	10·0	2·8
10. Ad.	England	24·0	3·25	17·3	10·2	2·7
11. Ad.	Fort Simpson (*Ross*)	22·0	2·95	17·1	9·8	2·65
12. ♂ ad.	S. Spain (*Irby*)	23·5	2·95	16·9	9·8	2·8
13. ♂ ad.	Fort Simpson (*Ross*)	24·0	2·95	16·8	9·5	2·6
14. Ad.	Switzerland	25·0	3·0	16·6	10·0	2·8
15. ♀ ad.	W. side Rocky Mts. (*Lord*)	23·0	2·85	16·6	10·1	2·5
16. ♀ ad.	Skåne (*Meves*)	21·0	3·2	16·5	9·6	2·65
17. ♂ juv.	Skåne (*Meves*)	23·0	3·05	16·5	9·8	2·85
18. Ad.	Mexico	24·0	2·95	16·5	10·0	2·55
19. Ad.	England	24·0	3·15	16·5	10·0	2·95
20. Ad.	49th parallel (*Dawson*)....	23·0	3·05	16·35	10·0	2·5
21. Ad.	49th parallel (*Dawson*)....	23·0	2·85	15·95	10·0	2·6

After a careful perusal of the large series at my disposal in the Museum, I find myself entirely of the same opinion as Mr. Dresser, viz. that there is only one species of large Raven, and that *C. thibetanus*, *C. nobilis*, &c. must be considered synonymous with *C. corax*. The differences given by the authors of the great work on North-American birds for the separation of the American Raven do not hold good in the set of skins now before me; and so various is the size of bill, even in individuals from the same locality, that any separation on account of this would prove quite untenable. The dimensions given above show how these vary in the Museum series.

Hab. The whole of Europe, Northern and Central Asia, North America, ranging into Mexico.

a, b. ♂ ♀ ad. st.	England.	Purchased.
c. Jun. st.	England.	Purchased.
d. Ad. st.	Cambridgeshire.	Mr. J. Baker [C.].
e. Ad. st.	Devonshire.	Col. Montagu [P.].
f. Ad. sk.	Devonshire.	Col. Montagu [P.].
g. ♀ ad. sk.	Skåne, Sweden, Nov. 18, 1873.	Purchased.
h. ♂ juv. sk.	Skåne, Sweden, March 9, 1874.	Purchased.
i. Ad. sk.	Switzerland.	Purchased.
k. ♂ ad. sk.	Villacidio, Sardinia, May 22, 1871.	A. B. Brooke, Esq. [C.].
l. ♂ ad. sk.	Near Casa Vieja, South Spain, May 1874.	Lieut.-Col. Irby [P.].
m. ♀ ad. sk.	Jerusalem.	Canon Tristram [C.].

n. Ad. sk.	Ladakh (*Strachey*).	Secretary of State for India [P.].
o. Ad. sk.	Thibet (*Hodgson*).	Secretary of State for India [P.]. (Type of *C. thibetanus.*)
p. ♀ ad. sk.	Onon river.	Dr. Dybowski [C.].
q, *r*. Ad. sk.	Fort Simpson, N. A.	B. R. Ross, Esq. [P.].
s, *t*. Ad. sk.	49th parallel (*Dawson*).	N. A. Boundary Commission [P.].
u. Ad. sk.	West side of Rocky Mts.	J. K. Lord, Esq. [C.].
v. ♀ ad. sk.	West side of Rocky Mts.	J. K. Lord, Esq. [C.].
w. Ad. sk.	Vancouver's Island.	J. K. Lord, Esq. [C.].
x. Ad. sk.	Mexico.	Purchased.
y. Ad. sk.	Mexico.	Purchased.
z. Ad. sk.	Mexico.	J. Gould, Esq. (Typeof *C. nobilis.*)
a'. Skeleton.		
b'. Sternum.	England.	J. Ray, Esq.
c'. Skull.	England.	W. Yarrell, Esq.

2. Corvus umbrinus.

Corvus umbrinus, *Sundev. Œfv. k. Vet. Akad. Forh. Stockh.* 1838, p. 199 (*ex Hedenb. MS.*); *id. op. cit.* 1850, p. 130; *Cab. Mus. Hein.* Th. i. p. 231; *Brehm, J. f. O.* 1854, Extrah. p. 97; *Schl. Bijdr. Dierk. Amsterd.* fol. p. 8, Taf. 1. fig. 14; *id. Mus. P.-B.* Coraces, p. 19; *Gray, Hand-l. B.* ii. p. 12; *Heugl. Orn. N.O.-Afr.* pp. 505, cxxiv; *Finsch & Hartl. Vög. Ostafr.* p. 373; *Shelley, B. Egypt*, p. 158; *Dresser, B. Eur.* parts xxxiv., xxxviii.; *Blanford, E. Pers.* p. 262.

Corvus infumatus, *Wagn. Münch. gel. Anz.* 1839 (*teste Finsch & Hartl. t. c.*).

Corvus corax, *Horsf. & Moore, Cat. B. Mus. E.I. Co.* ii. p. 552 (*nec L.*); *E. C. Taylor, Ibis*, 1859, p. 49; *Adams, Ibis*, 1864, p. 22.

Corvus umbrinus, *Pr. Würt. Naum.* 1857, p. 433.

Corax umbrinus, *Brehm, Reis. Habesch*, pp. 216, 319.

Adult. Black, with a dull bronzy gloss, or with a purple shade near the base of many of the feathers; head brown, slightly glossed

---

CORVUS RUFICOLLIS.

Corvus ruficollis, *Less. Tr. d' Orn*, p. 329; *Pucher. Rev. et Mag. de Zool.* 1853, p. 548; *Gray, Hand-l. B.* ii. p. 12; *Dresser, B. Eur.* part xxxviii.; *Bocage, Jorn. Acad. Lisb.* 1867, p. 129; *id. op. cit.* 1875, p. 13.

Corvus capensis, *Bolle, J. f. O.* 1856, p. 20 (*nec Licht.*).

Corvus corone, *Keulem. N. T. D.* iii. p. 365 (*nec L.*).

Adult (of large size). Black, glossed with violet steel-colour; head, neck, and abdomen (partly) umber-brown; primary-coverts and primary quills shining with bronzy green on the outer web; bill tolerably stout, elongated, curved, black in colour; tarsi elongated, black, as well as the toes, which are of moderate length; quills scarcely reaching beyond the tip of the tail; third and fourth quills equal and longest. (*Bocage, l. c.*)

Hab. Cape-Verd Islands.

Obs. The history of this species, of which I have not as yet seen an example, is thoroughly detailed by Professor Barboza du Bocage. According to his careful studies it is closely allied to *C. umbrinus*, from which, indeed, it appears to me difficult to separate it; but it is impossible to decide without seeing the species.

with bronze, the hinder part and sides of the neck umber-brown, without any gloss at all; lores, nasal bristles, and feathers round the eye black; the ear-coverts and cheeks brown, like the sides of the neck; under surface of body brown, glossed with purple on the breast, flanks, abdomen, and under tail-coverts; feathers of the throat and fore neck long and pointed, brown, with very little blue gloss on some of the plumes; under wing-coverts and axillaries purplish black; bill and feet black; iris very dark brown. Total length 21 inches, calmen 2·75, wing 15·5, tail 8·6, tarsus 2·75.

Obs. The distinctive character of this Raven, viz. the brown head and neck, is more clearly shown in the birds of worn plumage than in newly moulted birds. The differences between it and the common Raven seem, however, to be well maintained in a series of skins. Great variation exists in the size of the bill, some specimens having a beak as large as that of *C. corax*; and one or two of the birds in the Museum collection have been labelled *C. corax* in consequence—as, for instance, two from Quetta and another from Ferozepore. both of which I refer to *C. umbrinus*. They are distinguished clearly by the brown neck, and by the greater extent and purity of the white bases to the feathers, the latter by no means a strongly marked character in any Crow. From the following measurements it would seem that there are two races of brown-necked Raven, often inhabiting exactly the same localities, as is the case with *Corvus culminatus* (*intermedius*, Adams) and *Corone levaillanti*. Should the larger form be considered by ornithologists, as mentioned above, to belong to true *C. corax*, then I contend that the whole of these birds must be treated as races only of the ordinary Raven. At present the constancy of the brown neck seems to be sufficient for the specific separation of *C. umbrinus*.

	Total length.	Culmen.	Wing.	Tail.	Tarsus.
	in.	in.	in.	in.	in.
1. Egypt (*Shelley*)	21·0	2·95	14·9	8·4	2·95
2. Dead Sea (*Tristram*)	21·0	2·75	15·5	8·6	2·75
3. ♀. Jerusalem (*Tristram*)	20·0	2·6	14·55	8·4	2·55
4. ♂. Mesopotamia	22·0	2·8	16·6	9·3	2·6
5. ♀. Gwader (*Blanford*) .	17·5	2·45	14·75	8·4	2·45
6. ♂. Dizak (*Blanford*) ..	18·5	2·65	15·9	8·7	2·4
7. Quetta (*Griffith*) ..	22·0	3·2	17·5	9·6	2·7
8. Quetta (*Griffith*)....	22·5	2·8	16·6	9·2	2·5
9. Ferozepore (*Hay*) ..	23·0	3·05	16·4	9·1	2·55

Of the above series the birds from Mesopotamia, Ferozepore, and Quetta seem to be much the largest, Capt. Shelley's Egyptian specimen being the only one approaching them. They appear to me to be *C. umbrinus* showing an approach towards the true *C. corax*. A much larger series is necessary before the whole question can be determined.

Hab. Egypt and Palestine, eastwards to Afghanistan and North-western India.

a.	Imm. sk.	Egypt, March 9, 1868.	Capt. G. E. Shelley [P.].
b.	Ad. sk.	Shore of the Dead Sea.	Canon Tristram [C.].
c.	♀ ad. sk.	Jerusalem.	Canon Tristram [C.].
d.	♂ ad. sk.	Mesopotamia.	Euphrates Exped. [C.].
e.	♀ ad. sk.	Gwader, Baluchistan, Jan. 14, 1872.	W. T. Blanford, Esq. [C.].
f.	♂ ad. sk.	Near Dizak, Baluchistan, March 25, 1872.	W. T. Blanford, Esq. [C.].
g.	Ad. sk.	Quetta, Afghanistan (*Griffith*).	Secretary of State for India [P.].
h.	Jun. sk.	Quetta, Afghanistan (*Griffith*).	Secretary of State for India [P.].
i.	Ad. sk.	Ferozepore (*Capt. Hay*).	Secretary of State for India [P.].

3. Corvus leptonyx.

Corvus corax, *Webb & Berth. Orn. Canar.* p. 9; *Harcourt, Ann. Nat. Hist.* (2) xv. p. 437; *Loche, Expl. Sci. Algér. Ois.* i. p. 108.
Corvus leptonyx, *Peale, U.S. Expl. Exp.* 1848, p. 105.
Corvus ruficollis, *Cass. op. cit.* p. 116, pl. v.
Corvus tingitanus, *Irby, Ibis*, 1874, p. 264; *Dresser, B. Eur.* parts xxxvi., xxxviii.; *Irby, B. Gibr.* p. 126.

Adult male. General colour above steel-black with a purplish gloss, especially on the scapulars and quills; the secondaries bronzy brown, probably the remains of previous plumage; tail purplish black, with a slight greenish tinge on the inner webs, this being faintly indicated on the inner webs of the secondaries also; head all round and under surface of the body black, with a distinct purplish steel gloss; the ear-coverts and feathers of the throat and thighs faintly glossed with green; bill and feet black. Total length 20 inches, culmen 2·7, wing 15·7, tail 9·1, tarsus 2·85.

There can be little doubt that, if Peale's specimen of *C. leptonyx* really came from Madeira (as seems to be unquestionably the case), his name must be employed for the species instead of *C. tingitanus* of Irby, since Mr. Godman obtained in Teneriffe a specimen which is inseparable from Maroccan examples of *C. tingitanus*, and he says that this same species is "occasionally found in Madeira." The only difficulty in accepting the identity of these two species lies in the fact that in the plate of *C. leptonyx* the bird is described and figured as brown on the head and neck, which fact induced the late Mr. Cassin to refer it to *C. ruficollis* of Lesson. Lately, however, Professor Barboza du Bocage has shown, in a most carefully written essay, that the latter species is a bird of the Cape-Verd archipelago, and is in all probability distinct from *C. leptonyx* of Peale. In this view I am disposed to agree; and I believe that the brown colour on the head and neck in the type of *C. leptonyx* is caused by the worn plumage of the specimen, as I am quite certain it is in the wings of those which Colonel Irby brought from Tangier. It is therefore not a specific character in *C. tingitanus*. I believe the North-African Raven to be a distinct species, allied to *C. corax*, of which it is a small representative, as Mr. Dresser has already pointed out (*l. c.*). The typical examples of *C. tingitanus* measure as follows:—

	Total length.	Culmen.	Wing.	Tail.	Tarsus.
	in.	in.	in.	in.	in.
1. ♂ ad. Tangier (*Irby*)	19·0	2·85	14·85	8·0	2·85
2. ♀ ad. Tangier (*Irby*)	20·0	2·65	15·7	8·8	2·8

Hab. North-eastern Africa, Canary Islands, and Madeira.

a, b. ♂ ♀ ad. sk.	Tangier, April 26, 1876.	Lieut.-Col. Irby [P.]. (Types of *C. tingitanus.*)
c. Ad. sk.	Tangier (*Olcese*).	R. B. Sharpe, Esq. [P.].

4. **Corvus coronoides***.

Corvus coronoides, *Vig. & Horsf. Tr. Linn. Soc.* xv. p. 261; *Wagler, Isis,* 1829, p. 748; *Gray, Gen. B.* ii. p. 315; *Bp. Consp.* i. p. 385; *Cab. Mus. Hein.* Th. i. p. 231.
Corvus australis, *Gray, Hand-l. B.* ii. p. 12 (pt.).

Adult male. Black, with a rich purple gloss, all the feathers of the upper surface snowy white at base; wings purplish black like the back, some of the feathers externally bronzy brown; tail purplish black, the inner webs of the outer feathers shaded with greenish; under surface of body black, slightly glossed with purplish, all the feathers snow-white at base; bill and legs black; "iris brown" (*Elsey*). Total length 17·5 inches, culmen 2·25, wing 12·9, tail 7·2, tarsus 2·45.

Obs. A second specimen from the Yarrundi river is marked by Mr. Gould as having the iris "greyish brown."

Hab. Australia and Tasmania.

a. Ad. sk.	Australia.	Linnean Society [P.]. (Type of species.)
b. Ad. sk.	Australia.	
c. Ad. st.	Van Diemen's Land.	Ronald Gunn, Esq. [P.].
d. Ad. sk.	Murray river.	Sir George Grey [P.].
e. Ad. sk.	South Australia.	J. Gould, Esq. [C.].
f. ♂ ad. sk.	Yarrundi river.	J. Gould, Esq. [C.].
g. Ad. sk.	Port Essington (*Stanley*).	J. Gould, Esq.
h. Ad. sk.	Port Essington.	Dr. Sibbald [P.].
i. Ad. sk.	N.W. Australia.	J. R. Elsey, Esq. [P.].

5. **Corvus culminatus.**

Corvus culminatus, *Sykes, P. Z. S.* 1832, p. 96.
Corvus intermedius, *Adams, P. Z. S.* 1859, p. 297; *Jerd. B. Ind.* ii. p. 297; *Hume, Lahore to Yark.* p. 237.

Adult†. Steel-black above, shaded with violet on the back, wing-

* Those references to Australian Crows wherein the colour of the iris is not mentioned must be added to the synonymy of this species; those wherein the iris is recorded as white must be added to *Corone australis* (*vide infrà*).

† The type specimen being in rather poor order, I have described a more recent skin from Madras.

coverts, and secondaries; the primary coverts, bastard wing, and outer aspect of the primaries steel-green, with which colour the other quills are shaded on the inner web; tail-feathers black, externally glossed with violet and with steel-green on the inner web; head glossy black, slightly shaded with violet on the crown, and with a decided greenish cast on the sides of the face and the neck; under surface of body blue-black, slightly shaded with violet under certain lights; the plumes of the throat slightly lanceolate, greenish black, with a purplish lustre on the lower ones; bill and legs black. Total length 18 inches, culmen 2·35, wing 11·2, tail 6·8, tarsus 2·1.

Obs. Mr. Hume has treated of the difficult question of this Crow with his usual ability in 'Lahore to Yarkand;' but I cannot agree with him entirely; and I believe that when he re-examines his series he will find that Dr. Jerdon was quite right in admitting *Corvus intermedius* as a distinct species from *C. levaillanti*, as will be seen below; the bird which I call *C. culminatus* (and with which *C. intermedius* is identical) is not confined to the Himalayas, but is spread all over India; and, further, Dr. Jerdon was very nearly right in saying that it is "no larger than *C. splendens*." The full range of the species and the value of the differences advanced in the present work will have to be left to Mr. Hume and his coadjutors to work out in the field. The following are the measurements of our series of *C. culminatus*:—

	Total length. in.	Culmen. in.	Wing. in.	Tail. in.	Tarsus. in.
1. Deccan (*Sykes*)	18·5	2·6	11·5	6·5	2·2
2. Madras	18·0	2·35	11·2	6·8	2·1
3. Himalayas	16·0	2·15	11·85	7·5	2·1
4. Behar (*Hodgson*)	16·0	2·5	11·7	7·0	2·1
5. Nepal (*Hodgson*)	16·0	2·35	11·0	7·2	2·1

Hab. Indian peninsula.

a. Ad. sk.	Nepal.	B. H. Hodgson, Esq.
b. Ad. sk.	Behar.	B. H. Hodgson, Esq.
c. Ad. sk.	Himalayas.	
d. Ad. sk.	Madras.	
e. Ad. sk.	Deccan (*Sykes*),	Secretary of State for India [P.].

6. Corvus torquatus.

Corvus torquatus, *Less. Traité*, p. 328*; *Gray, Hand-l. B.* ii. p. 13; *David, N. Arch. Mus.* vii. *Bull.* p. 9; *Swinh. P. Z. S.* 1871, p. 383.
Corvus pectoralis, *Horsf. & Moore, Cat. B. Mus. E.I. Co.* ii. p. 557; *Schl. Mus. P.-B.* Coraces, p. 5.

Adult. General colour above black with a purplish gloss, the wings and tail with a very slight shade of steel-green in certain lights; under surface of body black, very slightly glossed with

* Lesson's description shows that this is the species intended by him, notwithstanding that he gives Australia as the locality. Besides, Professor Schlegel states (*l. c.*) that the Leiden Museum received a specimen from the Paris Museum labelled *C. torquatus*.

purple; the plumes of the throat lanceolate, but lax in texture; nape, hinder neck all round, and upper part of mantle white, the lower plumes of the latter adjoining the back white, more or less shaded with black towards the tip, the latter colour somewhat predominating; chest white, forming a crescentic band, some of the feathers mesially streaked or shaded with glossy black; bill and legs black. Total length 19 inches, culmen 2·45, wing 14·1, tail 8·6, tarsus 2·5.

Obs. Very great difference in size exists in this species. None of the Museum specimens have the sexes or exact localities indicated, so that I am unable to say whether the difference is sexual or whether there are two races of the bird. That the dimensions vary considerably will be seen from the following list.

	Total length. in.	Culmen. in.	Wing. in.	Tail. in.	Tarsus. in.
1. China (type of *C. pectoralis*) ..	17·0	2·25	11·5	7·5	2·3
2. China (*Reeves*)	17·0	2·3	12·1	7·3	2·45
3. Fokien....................	19·0	2·3	13·9	8·7	2·65
4. China (*Blakiston*)	19·0	2·45	14·1	8·6	2·5
5. China (*Blakiston*)	17·0	2·35	13·0	7·7	2·55
6. China (*Blakiston*)	17·5	2·4	12·8	7·8	2·5

Young. Dusky black, without any gloss underneath; wings and tail as in adult; the white mantle and pectoral patch obscured by dusky tips to many of the feathers, some of the chest-plumes mesially washed with dusky blackish.

Hab. China.

a. Ad. sk.	China.	Zoological Society. (Type of *C. pectoralis.*)
b. Ad. sk.	China.	J. R. Reeves, Esq. [P.].
c. Juv. sk.	China.	Secretary of State for India. (*Cf.* Horsf. & Moore, Cat. p. 557, spec. *a.*)
d, e, f. Ad. sk.	China.	Capt. Blakiston [C.].
g. Ad. sk.	Fokien.	J. Gould, Esq.

7. Corvus scapulatus.

Corneille du Sénégal, *Montb. Pl. Enl.* iii. pl. 327.

Corvus cornix, *Bodd. Tabl. Pl. Enl.* p. 20 (*nec L.*).

La Corneille à scapulaire blanc, *Levaill. Ois. d'Afr.* ii. p. 14, pl. 53; *Sundev. Crit. om Levaill.* p. 20.

Corvus scapulatus, *Daud. Traité*, ii. p. 232; *Wagler, Syst. Av.* Corvus, sp. 8; *Less. Traité*, p. 328; *Rüpp. Syst. Uebers.* p. 75; *Gray, Gen. B.* ii. p. 315; *Bp. Consp.* i. p. 386; *Cab. Mus. Hein.* Th. i. p. 231; *Schl. Bijdr. Dierk. Amsterd.* fol. *art.* Corvus, p. 2; *Kirk, Ibis*, 1864, p. 321; *Layard, B. S. Afr.* p. 168; *Schl. Mus. P.-B.* Coraces, p. 3; *Gray, Hand-l. B.* ii. p. 13; *Bocage, Jorn. Lisb.* 1870, p. 13; *Finsch, Tr. Z. S.* vii. p. 256; *Blanf. Geol. & Zool. Abyss.* p. 393; *Heugl. Orn. N.O.-Afr.* p. 501; *Cab. in Von der Decken Reis.* iii. p. 33; *Finsch & Hartl. Vög. Ostafr.* p. 374; *Sharpe, Cat. Afr. B.* p. 57; *Gurney in Anderss. B. Dam. Ld.* p. 154; *Antin. & Salvad. Viagg. Bogos*, p. 128; *Sharpe & Bouvier, Bull. Soc. Zool. France*, i. p. 46.

Corvus scapularis, var. æthiops, *Hempr. & Ehr. Symb. Phys.* fol. z.

Corvus dauricus, *Desj. P. Z. S.* 1831, p. 45.

Corvus curvirostris, *Gould, P. Z. S.* 1836, p. 18; *Gray, Gen. B.* ii. p. 315; *Bp. Consp.* i. p. 386; *Cab. Mus. Hein.* Th. i. p. 232, note; *Hartl. Orn. W.-Afr.* p. 114; *Antin. Cat. descr. Ucc.* p. 59; *Gray, Hand-l. B.* ii. p. 13.
Corvus leuconotus, *Swains. Classif. B.* i. p. 133, pl. v.; *Jard. & Selby, Ill. Orn.* iv. pl. xxxii.; *Gray, Gen. B.* ii. p. 315.
Corvus phæocephalus, *Cab. Mus. Hein.* Th. i. p. 232; *Gray, Hand-l. B.* ii. p. 13.
Corvus madagascariensis, *Bp. C. R.* xxxvii. p. 829; *Hartl. Faun. Madag.* p. 52; *Gray, Hand-l. B.* ii. p. 13.
Pterocorax scapulatus, *Brehm, Illustr. Thierl.* iii. p. 350.
Corax scapulatus, *A. E. Brehm, Reis. Habesch*, p. 216.

Adult male. Above glossy purplish black, with a shade of deep steel-blue under certain lights; the abdomen and under wing-coverts not quite so glossy; feathers of the throat slightly lanceolate, but lax in texture, the plumes of the lower throat white at base; hind neck, mantle, sides of neck, and entire breast from the fore neck to the abdomen pure white; under wing-coverts black, excepting the innermost axillaries, which are white; bill and feet black; "iris hazel-brown" (*Andersson*). Total length 18 inches, culmen 2·4, wing 13·9, tail 7·9, tarsus 2·5.

Adult female. Similar to the male. Total length 17·5 inches, culmen 2·5, wing 13·9, tail 7·3, tarsus 2·4.

Young. Similar to the adults, but more dusky, especially on the throat and abdomen; the white neck and breast dull white, with dusky edgings to a great many of the feathers.

Obs. Professor Schlegel remarks that all the supposed species of African White-breasted Crows have no real existence; and in this I thoroughly agree with him. Some specimens are brighter and more glossy than others; and these are doubtless older or freshly-moulted birds. Specimens of male and female, whose sex has been ascertained by dissection, show no difference in the sexes, both being equally bright-plumaged; but the younger birds seem to be rather duller-coloured.

Hab. The whole of Africa below the Sahara; Madagascar.

a. Juv. sk.	South Africa.	Major-General Hardwicke [P.].
b. Ad. sk.	South Africa.	Earl of Derby [P.].
c, d. ♂ ad. sk.	Otjimbinque, Damara Land, July 1866 (*Andersson*).	R. B. Sharpe, Esq.
e. ♂ ad. sk.	Mossamedes (*Anchieta*).	R. B. Sharpe, Esq.
f. Ad. sk.	Benguela.	J. J. Monteiro, Esq. [C.].
g. Ad. sk.	Clarence, Fernando Po, June 1842.	L. Fraser, Esq. [C.].
h. Ad. sk.	Accra.	R. B. Sharpe, Esq. [P.].
i. Ad. sk.	Fantee.	
k. Ad. sk.	Wassaw (*Blissett*).	R. B. Sharpe, Esq. [P.].
l. Ad. st.	River Gambia.	
m. ♂ ad. sk.	Angollalá, Shoa, Feb. 1842.	Sir W. C. Harris [C.].
n. ♀ ad. sk.	Wadela Plateau, April 21, 1868.	W. T. Blanford, Esq. [C.].
o. ♂ ad. sk.	Rairo, Habab, Aug. 13, 1868.	W. T. Blanford, Esq. [C.].
p. ♂ ad. sk.	Zoulla, Annesley Bay, Jan. 16, 1868.	W. T. Blanford, Esq. [C.].

q. Ad. sk.	Mozambique.	Purchased.
r. ♂ ad. sk.	N.E. Madagascar (*Van Dam*).	R. B. Sharpe, Esq.
s. ♂ ad. sk.	Nossi-Bé.	MM. Pollen and Van Dam [C.].

4. CORVULTUR.

	Type.
Corvultur, *Less. Traité d'Orn.* p. 327 (1831)............	C. albicollis.
Archicorax, *Gloger, Hand-b. Naturg.* p. 274 (1842)	C. albicollis.

Range. Southern and North-eastern Africa.

Key to the Species.

a. Round the neck a plain white collar; feathers of throat lanceolate; round the chest a concealed white collar .... *albicollis*, p. 24.

b. Nape broadly white, as also the hind neck, connected by a narrow white band down the hind neck; feathers of throat rounded; no white collar on the chest .... *crassirostris*, p. 25.

1. Corvultur albicollis.

South-Sea Raven, *Lath. Gen. Syn. Suppl.* p. 75.
Corvus albicollis, *Lath. Ind. Orn.* p. 151; *Daud. Traité d'Orn.* ii. p. 227, pl. xiv.; *Wagler, Syst. Av.* Corvus, sp. 5; *Layard, B. S. Afr.* p. 167; *Gray, Hand-l. B.* ii. p. 14.
Le Corbivau, *Levaill. Ois. d'Afr.* ii. pl. 50; *Sundev. Crit. om Levaill.* p. 29.
Corvus cafer, *Licht. Cat. Rer. Nat. Hamb.* p. 9; *Gray, Gen. B.* ii. p. 315; *Schl. Mus. P.-B.* Coraces, p. 3.
Corvus vulturinus, *Shaw, Gen. Zool.* vii. p. 343.
Corvultur albicollis, *Less. Traité d'Orn.* p. 327; *Sharpe, Cat. Afr. B.* p. 57; *Gurney, in Anderss. B. Dam. Ld.* p. 153; *Shelley, Ibis*, 1875, p. 77.
Corvultur cafer, *Bp. Consp.* p. 387.
Archicorax cafer, *Cab. Mus. Hein.* i. p. 233.
Archicorax albicollis, *Cab. in Von der Decken's Reis.* iii. p. 33; *Finsch & Hartl. Vögel Ost-Afr.* p. 371.

Adult. General colour above and below glossy black, without any admixture of brown, excepting a slight shade on the inner secondaries; head purplish brown-black; round the hinder neck a very broad white collar; throat and fore neck dull brown, contrasting with the rest of the under surface, from which it is separated by a concealed band of white, the feathers composing which are either edged with white or are pure white, sometimes with dark brown centres; "bill very dark brown, the tips of the mandibles of a white horn-colour; legs and toes brownish black; iris hazel-brown" (*Andersson*). Total length 18 inches, culmen 2·95, wing 15·7, tail 7·1, tarsus 2·95.

Young. More dusky brown than the adult, the lower feathers of the white neck-collar mesially streaked with brown; below with a very distinct white collar across the chest, all the feathers of which, however, are much mixed with brown.

Hab. South Africa.

a. Ad. st.	South Africa.	J. Gould, Esq.
b. Ad. sk.	South Africa.	J. Rocke, Esq. [P.].
c. Juv. sk.	South Africa.	South-African Museum.
d. Ad. sk.	Knysna, Feb. 25th, 1866 (*C. J. Andersson*).	R. B. Sharpe, Esq.
e. Ad. sk.	Kanye, Matabili country (*Exton*).	R. B. Sharpe, Esq.

2. Corvultur crassirostris.

Corvus crassirostris, *Rüpp. N. Wirb.* p. 19, Taf. viii.; *Schl. Bijd. Dierk. Amst.* fol. p. 1; *id. Mus. P.-B.* Coraces, p. 2; *Gray, Hand-l. B.* ii. p. 14.

Corvultur crassirostris, *Rüpp. Syst. Uebers.* p. 75; *Bp. Consp.* i. p. 887; *Horsf. & Moore, Cat. B. Mus. E.I. C.* ii. p. 552; *Heugl. Syst. Uebers.* p. 36; *Blanf. Geol. & Zool. Abyss.* p. 394.

Corvus albicollis, *Des Murs in Lefebvr. Voy. Abyss. Zool.* p. 104 (*nec Lath.*).

Archicorax crassirostris, *Cab. Mus. Hein.* p. 233; *Finsch, Tr. Z. S.* vii. p. 256; *id. & Hartl. Vögel Ost-Afr.* p. 370; *Heugl. Orn. N.O.-Afr.* pp. 507, cxxv.

Adult male. General colour above purplish black, with somewhat of a hoary shade under certain lights; the wings more or less mixed with brown where the disarranged bases to the coverts or the inner webs of the secondaries show; head all round, sides of neck, and entire under surface purplish black, the ground-colour of the feathers being brown, the circular margins purplish black; the under wing-coverts rather browner than the breast; on the hinder neck a large patch of white, occupying the whole of the sides and back of the nape, then continued in a narrower line down the back of the neck, and widening out slightly on the hind neck; bill black, the tip white; feet black; iris very dark brown. Total length 24 inches, culmen 3·85, wing 17·7, tail 11, tarsus 3·1.

Hab. North-east Africa.

a. Ad. st.	Abyssinia.	Dr. Rüppell [C.].
b. Ad. sk.	Abyssinia.	R. B. Sharpe, Esq.
c. Ad. sk.	Shoa.	Sir W. C. Harris [C.].
d. ♂ ad. sk.	Angollala, Shoa, October 1841.	Sir W. C. Harris [C.].
e. Skull.		Purchased.

5. COLŒUS.

	Type.
Lycus, *Boie, Isis*, 1822, p. 55 (*nec Fabr.* 1787)	C. monedula.
Monedula, *Brehm, Isis*, 1828, p. 1273 (*nec Coquebert,* 1798)	C. monedula.
Colæus, *Kaup, Natürl. Syst.* p. 114 (1829)..........	C. monedula.

Range. The whole of Europe and Northern Asia, extending into Northern China and into North-western India.

Key to the Species.

a. Sides of neck more or less greyish or white; lighter than the back, and forming a collar of greater or less intensity.
  a′. Collar round hind neck silvery grey, intensifying on the sides of the neck into greyish or creamy white.
    a″. Abdomen leaden black .... { *monedula*, p. 26. / *collaris*, p. 27. }
    b″. Abdomen dingy ashy grey .... *neglectus*, p. 28.
  b′. Collar round the hind neck broad, creamy white; abdomen creamy white like hind neck .... *dauuricus*, p. 28.
b. Sides of neck glossy black like the back: no light collar .... *ossifragus*, p. 29.

1. Colœus monedula.

Le Choucas, *Briss. Orn.* ii. p. 24.
Le Choucas à collier, *Briss. t. c.* p. 27.
Le Choucas blanc, *Briss. t. c.* p. 27 (var.).
Le Choucas noir, *Briss. t. c.* p. 28; *Montb. Pl. Enl.* iii. pl. 522.
Corvus monedula, *Linn. S. N.* i. p. 156; *Temm. Man. d'Orn.* i. p. 111; *Naum. Vög. Deutschl.* ii. Taf. 56. fig. 1; *Roux, Orn. Prov.* pl. 133; *Werner, Atlas,* Omnivores, pl. vi.; *Gould, B. Eur.* iii. pl. 223; *Macgill. Br. B.* i. p. 552; *Yarr. Br. B.* ii. p. 102; *Gray, Gen. B.* ii. p. 315; *Schl. u. Susem. Vög. Eur.* ii. Taf. iv. fig. 1; *Nordm. in Démid. Voy. Russie Mérid.* iii. p. 120; *Schl. Dier. Nederl. Vög.* pl. xiii. fig. 6; *id. Vog. Nederl.* pl. 140; *Kjærb. Orn. Dan.* Taf. xi. fig. 1; *Schl. Bijdr. Dierk. Amsterd.* folio, p. 15; *Jaub. et Barth. Lapomm. Rich. Orn.* p. 95; *Schrenck, Reis. Amurl. Zool.* p. 324; *Sunder. Sv. Fogl.* pl. xx. fig. 4; *Linderm. Vög. Griechenl.* p. 69; *Gould, B. Gt. Br.* iii. pl. 61; *Filippi, Viagg. Pers.* p. 263; *Degl. et Gerbe, Orn. Eur.* i. p. 202; *Schl. Mus. P.-B.* Coraces, p. 33; *Gray, Hand-l.* ii. p. 13; *Doderl. Avif. Sicil.* p. 63; *Salvad. Faun. Ital. Ucc.* p. 173; *Heugl. Orn. N.O.-Afr.* p. 498; *Shelley, B. Egypt,* p. 160; *Hume, Nests &c. Ind. B.* p. 414; *Irby, B. Gibr.* p. 128; *Dresser, B. Eur.* part xlvi.; *Scully, Str. F.* 1876, p. 158; *Blanf. East. Persia,* p. 263.
La Grolle ou Choucas gris, *Montb. Pl. Enl.* iii. p. 523.
Corvus spermolegus, *Vieill. N. Dict. d'Hist. Nat.* viii. p. 40; *Gray, Gen. B.* ii. p. 315; *Schl. u. Susem. Vög. Eur.* ii. Taf. I. fig. 2.
Lycus monedula, *Boie, Isis,* 1822, p. 551; *Bp. Consp.* i. p. 384.
Colæus monedula, *Kaup, Natürl. Syst.* p. 114; *Horsf. & Moore, Cat. B. Mus. E.I. Co.* ii. p. 562; *Jerd. B. Ind.* ii. p. 302; *Hume, Lahore to Yark.* p. 239.
Monedula turrium, *Brehm, Vög. Deutsch.* p. 172; *Loche, Expl. Sci. Algér.* i. p. 114.
Monedula arborea, *Brehm, t. c.* p. 173.
Monedula septentrionalis, *Brehm, t. c.* p. 173.
"Monedula nigra, *Brehm,*" *Bp. Consp.* i. p. 384.
Monedula spermolegos, *Brehm, Naum.* 1855, p. 273.

Adult male (Skåne, April 5, 1874). General colour black with somewhat of a greyish shade on the back, the feathers of which have obsolete greyish margins; wings black, glossed with purple, the primaries and innermost secondaries with green; tail purplish black, with a slight greenish shade on the edges of many of the feathers; nasal bristles rusty black; crown of head glossy blue-black, forming a cap; rest of head and neck hoary bluish grey, the feathers having a hairy appearance, the grey being more silvery where it adjoins the cap, and rather whiter on the lower sides of the neck; ear-coverts rather more ashy grey than the head; feathers below the eye, cheeks, and entire under surface of body cindery grey, blacker on the throat, which is obscurely streaked with brown; under wing- and tail-coverts greenish black; bill and feet black; iris bluish white. Total length 13 inches, culmen 1·35, wing 9·5, tail 6·1, tarsus 1·7.

Adult female. Similar to the male. Total length 13 inches, culmen 1·3, wing 9·6, tail 6, tarsus 1·75.

Young female in second plumage (Nov. 10, 1873). Much duller in colour than the adults, more glossy, but not nearly so grey, the only shade of the latter colour being on the ear-coverts and sides of neck; under surface of body very dark, nearly as deep in colour as the upper plumage, and only slightly shaded with grey.

Hab. Europe and Northern Africa.

a. ♀ ad. sk.	Castle Eden, Durham, May 9, 1866.	J. H. Gurney, Esq., jun. [P.].
b. Var. st.	Yorkshire.	
c, d. Pull. st.	Cambridgeshire.	Mr. J. Baker [C.].
e. Ad. st.	Saffron Walden, Essex.	Mr. J. Clark [C.].
f. ♀ ad. sk.	Taplow, Berks, April 1865.	J. Gould, Esq. [C.].
g, h. ♂ ♀ ad. sk.	Taplow, Berks, May 1865.	J. Gould, Esq. [C.].
i. ♀ jun. sk.	Avington, Hants, Nov. 10, 1873.	R. B. Sharpe, Esq. [P.].
k. ♂ ad. sk.	Avington, Hants, May 5, 1876.	Edward Shelley, Esq. [P.].
l, m. ♀ ad. sk.	Avington, Hants, May 16, 1876.	Edward Shelley, Esq. [P.].
n. ♀ ad. sk.	Skåne, Sweden, Dec. 29, 1873.	Purchased.
o, p. ♂ ♀ ad. sk.	Skåne, March 23, 1874.	Purchased.
q. ♀ ad. sk.	Skåne, April 5, 1874.	Purchased.
r. Ad. sk.	Belgium, Dec. 15, 1851.	Brussels Museum [P.].
s. Ad. sk.	Algiers.	Purchased.
t. Skeleton.	England.	
u. Skeleton.	England.	
v. Skull.	England.	J. Rae, Esq.
w, x. Skulls.	S. Germany.	Dr. Günther.

Subsp. *a.* **Colœus collaris.**

Corvus collaris, *Drummond, Ann. N. H.* xviii. p. 11; *Gray, Hand-l. B.* ii. p. 13; *Dresser, B. Eur.* part xlvi. art. *C. monedula.*
Lycus collaris, *Bp. Consp.* i. p. 384.
Corvus monedula, *Severtz. Turkest. Jevotn.* p. 63; *Dresser, Ibis,* 1875, p. 237.

Differs from *C. monedula* in having a creamy white half-collar on each side of the neck. It is doubtful whether it should be separated from the Common Jackdaw, from which in its young stage it is undistinguishable; but it at least represents a very well-marked race. Total length 13 inches, culmen 1·15, wing 9·15, tail 5·1, tarsus 1·7.

Hab. Eastern Europe, ranging through Persia into Afghanistan and Cashmere.

a. ♂ ad. sk.	Finland.	J. Gould, Esq.
b. ♀ juv. sk.	Rabbath Ammon.	Canon Tristram [C.].
c. ♀ ad. sk.	Constantinople, Feb. 17, 1871 (*T. Robson*).	R. B. Sharpe, Esq. [P.].
d. Ad. sk.	Erzeroum (*Dickson & Ross*).	J. Gould, Esq.
e. Ad. st.	Bagdad.	K. Loftus, Esq. [P.].
f. Ad. sk.	Mesopotamia (*Jones*).	Secretary of State for India [P.].
g. Ad. sk.	Cabul (*Griffith*).	Secretary of State for India [P.].
h. Ad. sk.	Cashmere.	Capt. Stackhouse Pinwill [P.].
i. Ad. st.	Thibet.	Earl of Gifford [P.].
k. Skeleton.	Macedonia.	Purchased.

2. Colœus neglectus.

Corvus dauuricus, jun., *Schl. Faun. Jap.* pl. xl.
Corvus neglectus, *Schl. Bijdr. Dierk. Amsterd.* fol. *art.* Corvus, p. 16; *Gray, Hand-l. B.* ii. p. 14; *Dresser, B. Eur. art.* Corvus monedula, p. 3.
Monedula neglecta, *David, N. Arch. Mus.* vii. *Bull.* p. 9.
Lycus neglectus, *Swinh. P. Z. S.* 1871, p. 383.

Adult male. General colour above black, with a slight purplish gloss; wings blackish, glossed with bronzy purple, the inner webs with a greenish shade, especially on the innermost secondaries; tail purplish brown, washed with green; crown of head glossy purplish black, as also the lores and sides of face, the ear-coverts streaked with silvery grey; round the hind neck an indistinct collar of silvery-grey feathers; the sides of the neck dingy white mixed with black; throat and chest purplish black, the rest of the under surface dingy grey, becoming black on the thighs and under tail-coverts; under wing-coverts also black. Total length 14 inches, culmen 1·35, wing 9·1, tail 5·2, tarsus 1·9. (*Mus. R. Swinhoe.*)

Hab. North China and Japan.

a. ♂ ad. sk.	Shanghai, China.	Shanghai Museum [P.].

3. Colœus dauuricus.

Corvus dauuricus, *Pall. Reis. Russ. Reichs*, iii. Anhang, p. 694; *Gray, Gen. B.* ii. p. 315; *Schl. Faun. Jap.* Aves, pl. xl. B; *Midd. Sibir. Reis. Zool.* p. 159; *Schl. Bijdr. Dierk. Amsterd.* folio, *art.* Corvus, p. 16; *id. Mus. P.-B.* Coraces, p. 35.
White-breasted Crow, *Lath. Gen. Syn.* i. part 1, p. 376, pl. xv.
Corvus capitalis, *Wagler, Syst. Av.* Corvus, sp. 19.
Lycus dauricus, *Bp. Consp.* i. p. 384; *Swinh. P. Z. S.* 1871, p. 383.

Monedula daurica, *Schl. Faun. Jap.* Aves, p. 80; *Cab. Mus. Hein.* Th. i. p. 230; *Brehm, Naum.* 1855, p. 273.
Colæus daurica, *Horsf. & Moore, Cat. B. Mus. E.I. Co.* ii. p. 563.
Corvus monedula, var. daurica, *Radde, Reis. Sibir. Vög.* p. 207.

Adult male. Above purplish black, the feathers more glossy on their margins, the primary-coverts and quills inclining to greenish, the inner secondaries glossed with green or bronzy; tail purplish black, with a decided greenish gloss under certain lights; round the hind neck a broad collar of ashy white, occupying also all the sides of the neck; head glossy purplish black, including all the sides of the face, throat, and fore neck, the latter rather more bluish purple; the ear-coverts streaked with hoary grey, as also the side of the head adjoining; rest of under surface of body ashy white, the thighs, vent, and under wing- and tail-coverts glossy purplish black; bill and feet black; iris brown. Total length 13·2 inches, culmen 1·3, wing 9·4, tail 5·7, tarsus 1·65.

Adult female. Exactly resembles the male. Total length 13·5 inches, culmen 1·3, wing 9·1, tail 5·5, tarsus 1·75.

Young. Duller in plumage than the adults, the white collar and breast being shaded with dusky grey and the flanks washed with blackish; the grey on the ear-coverts and sides of the crown less distinct.

Hab. Northern Asia, from the Altai Mountains to Eastern Siberia; China and Japan.

a. Ad. st.	North China.	Mr. Fortune [C.].
b, c. ♂ ♀ ad. sk.	North China.	Capt. Blakiston [C.].
d. ♂ ad. sk.	Shanghai.	J. Gould, Esq.
e. Ad. sk.	Shanghai.	Secretary of State for India [P.]. (Spec. *a* of Horsf. & Moore, Cat. ii. p. 563.)
f, g, h. ♂ ♀ ad. sk.	Shanghai.	Shanghai Museum.
i. Jun. sk.	Amoor Land.	R. B. Sharpe, Esq.
k. ♂ ad. sk.	Kultuk, Southern Lake Baikal, May 18, 1871.	Dr. Dybowski [C.].
l. Ad. sk.	Altai Mountains.	Purchased.

4. Colœus ossifragus.

Corvus ossifragus, *Wilson, Am. Orn.* v. p. 27, pl. 37. fig. 2; *Wagler, Syst. Av.* Corvus, sp. 12; *Audub. Orn. Biogr.* ii. p. 268; *id. B. Amer.* pl. cxlvi.; *id. B. Amer.* roy. 8vo, iv. p. 94, pl. ccxxvi.; *Gray, Gen. B.* ii. p. 315; *Baird, B. N. Amer.* p. 571, pl. lxvii.

COLŒUS MINUTUS.

Corvus minutus, *Gundl. J. f. O.* 1856, p. 97; *Baird, B. N. Amer.* p. 569; *Allen, Bull. Harv. Coll.* ii. p. 297, note; *Baird, Brewer, & Ridgw. B. N. Amer.* p. 234, note.

"This species we have not seen; it seems, however, to be rather nearly related to *C. ossifragus*, and possesses more lengthened nasal plumes than the three West-Indian species (*C. nasicus, C. leucognalaphus, C. jamaicensis*)." (*Cf.* Baird, Brewer, and Ridgw. *l. c.*)

Hab. Cuba.

fig. 2; *Schl. Bijdr. Dierk. Amsterd.* folio, *art.* Corvus, p. 14, pl. 1, fig. 27; *Gray, Hand-l. B.* ii. p. 297; *Allen, Bull. Harv. Coll.* ii. p. 297; *Coues, Key N. Amer. B.* p. 163; *Scl. & Salv. Nomencl. Av. Neotr.* p. 40; *Baird, Brewer, & Ridgw. N. Amer. B.* ii. p. 251, pl. 37. fig. 7; *Coues, B. North-west,* p. 207, note.

Adult. General colour glossy blue-black, rather inclining to violet on the back and wings; head and neck slightly shaded with greenish, clearer on the ear-coverts; wings and tail bronzy black, somewhat shaded with steel-blue on the outer webs, and with green on the inner ones; under surface of body glossy steel-black, with a decided greenish cast on the breast, flanks, and abdomen; bill and legs black. Total length 15 inches, culmen 1·6, wing 10·3, tail 6·1, tarsus 1·85.

Hab. North America: Atlantic coast of the United States.

a. Ad. st.	North America.	H.R.H. Duke of Sussex [P.].
b. Ad. st.	North America.	

6. CORONE.

	Type.
Corone, *Kaup, Natürl. Syst.* p. 99 (1829)	C. corone.
Amblycorax, *Bp. C. R.* xxxvii. p. 829 (1853)	C. violaceus.
Anomalocorax, *Fitz. Sitz. Akad. Wien,* xlvi. Abth. 1, p. 209 (1863)	C. splendens.

Range. The whole of the Old World, excepting Africa south of the desert and Oceania: North America and Mexico.

Key to the Species.

a. Plumage with some shade of green in it, generally more distinct on the outer edge of the wing, and on the primaries and outer tail-feathers.

a'. **Back of neck distinctly of a different colour from the rest of the upper surface, being white or grey or drab-brown, and forming a collar.**

a''. Back grey or white from the nape to the tail, as also the entire under surface — *cornix*, p. 31. / *capellana*, p. 32.

b''. Back black, the hinder part of the neck drab-brown or grey .... — *splendens*, p. 33. / *insolens*, p. 34.

b'. Back of neck uniform with the rest of the upper surface, sometimes (in *C. macrorhynchus* and allies) with a slight shade of grey under certain lights.

c''. **Base** of feathers of mantle greyish or greyish white.

a'''. Throat-feathers blended, not lanceolate.

a^4. Tarsus longer than bill .... — *americana*, p. 34. / *floridana*, p. 35.

b^4. Tarsus shorter than bill .... — *caurina*, p. 36. / *mexicana*, p. 36.

b'''. Throat-feathers lanceolate.

c^4. Lanceolate feathers not confined to the throat, but extending to the fore neck and chest.
a^5. Feathers of hind neck pointed and producing a silky hair-like structure . . . . *corone*, p. 36.
b^5. Feathers of hind neck rounded . . *australis*, p. 37.
d^4. Lanceolate plumes of throat not extending to the fore neck . . . . { *macrorhyncha*, p. 38. *levaillantii*, p. 39. *japonensis*, p. 41. }
d''. Feathers of mantle snowy white at base.
c'''. Size large: wing over 13 inches; bill very long, the culmen measuring over 3 inches in length . . . . *validissima*, p. 42.
d'''. Size small: wing not exceeding 12 inches; culmen not exceeding 2·5 inches in length . . . . *philippina*, p. 42.
b. Plumage without green, always more or less purple in tone, the outer edge of the primaries and the outer tail-feathers steel-black or purplish black; feathers of the throat not distinctly lanceolate, and with a purplish shade; behind the eye a large triangular bare patch . . . . *enca*, p. 43.
α. Size moderate: wing 10·8–12·5 inches, tarsus 1·9–2·1. *enca*.
β. Size larger: wing 12·5–13·1 inches, tarsus 2·15–2·3. *orru*.
γ. Size very small; wing 9·5–9·8 inches, tarsus 1·65–1·75. *violacea*.

1. Corone cornix.

La Corneille mantelée, *Briss. Orn.* ii. p. 19; *Montb. Pl. Enl.* 76.
Corvus cornix, *Linn. S. N.* i. p. 156; *Daud. Traité*, ii. p. 231; *Naum. Vög. Deutschl.* ii. pl. 54; *Werner, Atlas*, Carnivores, pl. 3; *Less. Traité*, p. 328; *Gould, B. Eur.* iii. pl. 22; *Macgill. Brit. B.* i. p. 529; *Yarr. Br. B.* ii. p. 82; *Nordm. in Démid. Voy. Russ. Mér.* iii. p. 119; *Gray, Gen. B.* ii. p. 315; *Kjærb. Orn. Dan.* pl. xi. fig. 5; *Blyth, Cat. B. Mus. A. S. B.* p. 89; *Cab. Mus. Hein.* Th. i. p. 231; *Bp. Consp.* i. p. 385; *Midd. Sibir. Reis. Zool.* p. 160; *Schl. Vog. Nederl.* pl. 138; *Horsf. & Moore, Cat. B. Mus. E.I. Co.* ii. p. 553; *Schl. Dier. Nederl.* iii. pl. 13. fig. 3; *Sundev. Sv. Fogl.* pl. 20. fig. 2; *Fritsch, Vög. Eur.* tab. 28. fig. 1; *Jaub. & Barth. Lap. Rich. Orn.* p. 94; *Jerd. B. Ind.* ii. p. 298; *Borggr. Vögelf. Norddeutschl.* p. 81; *Degl. et Gerbe, Orn. Eur.* i. p. 200; *Schl. Mus. P.-B.* Coraces, p. 5; *Gould, B. Gt. Br.* part lix.; *G. R. Gray, Hand-l. B.* ii. p. 11; *Salvad. Faun. Ital. Ucc.* p. 173; *Doderl. Avif. Sicil.* p. 62; *Heugl. Orn. N.O.-Afr.* p. 503; *Shelley, B. Egypt*, p. 159; *Dresser, B. Eur.* part xxvi.; *R. Gray, B. W. Scotl.* p. 173; *Irby, B. Gibr.* p. 128; *Severtz. Turkest. Jevotn.* p. 63; *Dresser, Ibis*, 1875, p. 237; *Scully, Str. F.* 1876, p. 156; *Blanf. E. Persia*, p. 262.
Corvus cinereus, *Leach, Syst. Cat. Mamm. &c. B. M.* p. 18 (1816).
Corone cornix, *Kaup, Natürl. Syst.* p. 99.
Corvus subcornix, *Brehm, Vög. Deutschl.* p. 168.
Corvus cornix ægyptiaca, *Brehm, J. f. O.* 1853, p. 97.
Corvus tenuirostris, *Brehm, Naum.* 1855, p. 73 (*nec Moore*).

Adult female. Above drab-grey, with indistinct dusky shaft-stripes, the hind neck more decidedly grey; wings and tail entirely purplish black, slightly glossed with steel-green under certain lights; upper tail-coverts purplish brown, with lighter grey edges; head all round, fore neck, and centre of the chest glossy blue-black, the feathers lanceolate in shape; sides of the neck drab-grey, exactly like the remainder of the under surface; thighs dusky black; wing-coverts blue-black, the axillaries grey; bill and legs black; iris very dark brown. Total length 17 inches, culmen 2·4, wing 12·5, tail 7·8, tarsus 2·2.

Adult male. Similar to the female. Total length 18 inches, culmen 2·35, wing 12·2, tail 7·5, tarsus 2·25.

Young. Similar to the adult, but the colour rather more dingy, and the drab somewhat shaded with a purplish lustre; hind neck decidedly more dusky than in the adult; upper tail-coverts dull blackish, without any perceptible grey margin; head also dull black, without lustre, the lanceolate plumes on the fore neck not developed.

Hab. Great Britain, Central and Southern Europe, Eastern Siberia (? *C. capellana*).

Obs. On the range of this Crow and its remarkable habit of interbreeding with *C. hiemalis* see an important sketch by Professor Schlegel in the 'Museum des Pays-Bas' (*l. c.*).

a. Ad. st.	Great Britain.	
b, c. Juv. st.	Great Britain.	
d, e, f. Pull. st.	Cambridgeshire, June 1853.	Mr. J. Baker [C.].
g. ♀ ad. st.	Cambridgeshire.	Mr. J. Baker [C.].
h. Ad. st.	Wiltshire.	Col. Montagu [P.].
i. Ad. sk.	Hempstead, Norfolk.	J. H. Gurney, Esq., Junr. [P.].
k. ♀ ad. sk.	Stockholm, April 28, 1873.	Purchased.
l. ♀ juv. sk.	Stockholm, June 22, 1873.	Purchased.
m. Pull. sk.	Archangel, June 18, 1872.	E. R. Alston, Esq. [C.].
n. ♂ ad. sk.	Ust Zylma, Great Petchora river, May 10, 1875.	Henry Seebohm, Esq. [P.].
o, p. ♀ ad. sk.	Heligoland, Oct. 2, 1876.	Henry Seebohm, Esq. [P.].
q, r, s. ♂ ad. sk.	Heligoland, Oct. 2, 1876.	F. Nicholson, Esq. [P.].
t, u. ♂ ad. sk.	Heligoland, Oct. 4, 1876.	F. Nicholson, Esq. [P.].
v. Ad. sk.	Belgium.	Brussels Museum [P.].
w. ♀ ad. sk.	Sardinia, March 6, 1872.	A. B. Brooke, Esq. [C.].
x. ♂ ad. sk.	Bethel.	Canon Tristram [C.].
y. Skeleton.	England.	
z. Sternum.	England.	J. Rae, Esq. [P.].
a′, b′, c′. Skulls.	S. Germany.	Dr. Günther.

Subsp. *a.* **Corone capellana.**

Corvus capellanus, *Sclater, P. Z. S.* 1876.

Similar to *C. cornix*, but having the light parts of the plumage white instead of drab-grey. The species is founded on some (at present) living birds in the Zoological Gardens; and their appearance certainly justifies the separation of this form from the true *Corone cornix*; but I do not think they will ever be recognized as more than a race of the ordinary bird. My reasons for thinking thus are, that in the Museum are some birds in moulting-plumage, wherein the new feathers are grey in colour and not much lighter than in the European species, into which it seems to graduate.

Hab. Persia and Mesopotamia eastwards into India.

a. Ad. sk.	Erzeroum (*Dickson & Ross*).	J. Gould, Esq. [P.].
b. Ad. sk.	Mesopotamia (*Jones*).	Secretary of State for India [P.].
c. ♂ ad sk.	Mesopotamia.	Euphrates Expedition [P.].
d. ♂ ad. sk.	Kazerum, West of Shiraz, South Persia.	Major St. John [C.].
e. Ad. sk.	Deccan (*Day*).	Secretary of State for India [P.].

2. Corone splendens.

Corvus splendens, *Vieill. N. Dict. d'Hist. Nat.* viii. p. 44; *Temm. Pl. Col.* ii. pl. 425; *Wagler, Syst. Av.* Corvus, sp. 18; *Gray, Gen. B.* ii. p. 315; *Blyth, Cat. B. Mus. A. S. B.* p. 90; *Bp. Consp.* i. p. 384; *Gray, Cat. Mamm. &c. coll. Hodgs. Nepal,* p. 102; *Horsf. & Moore, Cat. B. Mus. E.I. Co.* ii. p. 559; *Schl. Bijdr. Dierk. Amsterd.* fol. p. 4; *Jerd. B. Ind.* ii. p. 298; *Schl. Mus. P.-B.* Coraces, p. 10; *Gray, Hand-l. B.* ii. p. 14.

Corvus impudiens, *Hodgs. in Gray's Zool. Misc.* p. 84.

Corvus dauricus, *Pears. J. A. S. B.* x. p. 652 (*nec Pall.*).

Corvus impudicus, *Gray, Hand-l. B.* ii. p. 14; *Hume, Str. F.* 1873, p. 206; *Ball, S. F.* 1874, p. 418; *Butler, S. F.* 1874, p. 493.

Adult. Above greyish black, with a shade of light purple on the wings; the back, as well as the tail and outer aspect of the primaries, glossed with green or steel-black; nasal bristles and crown of head purplish black, scarcely reaching to the occiput, and including all the region of the eye, cheeks, and throat, the feathers of which are lanceolate; nape and all the sides of the neck and mantle, fore neck, and chest drab-brown, more dingy underneath; all the rest of the under surface greyish black, with only a very slight gloss; bill and legs black; iris deep brown. Total length 16 inches, culmen 2·15, wing 11·0, tail 7·1, tarsus 1·8.

Obs. The Nepalese and Himalayan examples appear to be the most typical; and those of Central India and Bengal exhibit a gradual shading of black on the hind neck, which intensifies most in the race called by Mr. Hume *C. insolens*, from Burmah and Tenasserim. The Malayan example is not *C. insolens*, but is of the dark form of *C. splendens*. There is no doubt as to the correct locality of this Malaccan bird, as it was purchased by Mr. Boucard from a collector who shot the birds himself and sold them *en bloc* on his arrival in this country. I saw the collection myself before it was dispersed; it consisted of ordinary Malaccan species; and a few rarities, such as *Pericrocotus igneus*, *Spilornis bacha*, &c., were in the same consignment. I am particular in mentioning this, as it is the first instance known of the occurrence of the species in Malacca, where we should have expected to find *C. insolens*. A Ceylonese specimen, kindly presented by Capt. Vincent Legge, seems to afford a connecting link, being a shade darker than Bengal specimens, but not so thoroughly black as *C. insolens*.

Hab. The peninsula of India and Ceylon. (? Migrating into) Malacca.

a. Ad. sk.	Himalayas.	Secretary of State for India [P.].
b, c. Ad. sk.	North-west Himalayas.	Capt. Stackhouse Pinwill [P.].
d, e, f. Ad. sk.	Nepal.	B. H. Hodgson, Esq. [P.].
g. Ad. sk.	Darjiling.	Capt. R. C. Beavan [C.].
h. Ad. sk.	Saugor.	Secretary of State for India [P.].
i. ♀ ad. sk.	Bengal.	Purchased.
k. ♂ ad. sk.	Colombo, Ceylon, Feb. 25, 1876.	Capt. Vincent Legge, R.E. [P.].
l. Ad. sk.	Malacca.	Purchased.
m. Skeleton.	India.	
n. Skeleton.	Nepal.	B. H. Hodgson, Esq.

Subsp. *a*. Corone insolens.

Corvus culminatus. *Schomb. Ibis*, 1864, p. 252.
Corvus insolens, *Hume, Str. F.* 1875, p. 144.
Corvus splendens (pt.), *Blyth & Wald. B. Burm.* p. 86.

Adult. Above ashy black, with a steel-blue gloss, rather more purple on the wings, and inclining to green on the outer primaries and on the tail; neck all round smoky black, a little lighter than the back; the crown, sides of face, and throat purplish black, the feathers of the latter lanceolate and shot with green; rest of under surface of body ashy black, slightly glossed with purple. Total length 15·5 inches, culmen 1·85, wing 10·35, tail 5·3, tarsus 1·9.

Hab. Assam, Burmah, and Tenasserim.

a. Ad. sk.	Nazeerah, Upper Assam (*Dr. Reid*).	R. B. Sharpe, Esq. [P.].
b. Ad. sk.	Burmah.	Capt. Stackhouse Pinwill [P.].
c. Ad. sk.	Katha, Upper Burmah, January 19, 1868.	Dr. Anderson [C.].
d. Ad. sk.	Tenasserim (*Helfer*).	Secretary of State for India [P.]. (Spec. *f* of Horsf. & Moore, Cat. p. 560.)
e. Ad. sk.	Siam.	M. Mouhot [C.].

3. Corone americana.

Corvus corone (*nec L.*), *Wils. Am. Orn.* iv. p. 79, pl. xxv. fig. 3; *Swains. & Richards. Faun. Bor.-Amer. Birds*, p. 291.
Corvus americanus, *Audub. Orn. Biogr.* ii. p. 317; *id. op. cit.* v. p. 477; *id. B. Amer.* pl. clvi.; *id. Birds Amer.* roy. 8vo, v. p. 57, pl. ccxxv.; *Bp. Consp.* i. p. 385; *Baird, B. N. Amer.* p. 566, pl. xxiii.; *Schl. Bijdr. Dierk. Amsterd.* fol. *art.* Corvus, p. 10, pl. 1.

fig. 16; *id. Mus. P.-B.* Coraces, p. 24; *Gray, Hand-l. B.* ii. p. 12; *Allen, Bull. Harv. Coll.* p. 297; *Coues, Key N.-Amer. B.* p. 162; *Baird, Brewer, & Ridgw. N.-Amer. B.* ii. p. 243, pl. 37. fig. 5; *Coues, B. North-west*, p. 206.

Adult. General colour purplish black, with a more or less distinct gloss of violet under certain lights; the head and neck all round glossy black, with somewhat of a brownish cast; the feathers of the throat loose and blended, but not lanceolate; wings glossy steel-black, with a slight violet gloss, the outer wing-coverts, primary-coverts, and primaries externally glossed with green; tail purplish black, inclining to green on the inner webs of the feathers; under surface of body black, with a violet gloss; bill and legs black; iris brown. Total length 19·5 inches, culmen 2·05, wing 13·15, tail 8·1, tarsus 2·3.

Hab. North America.

a. Ad. sk.	North America.	E. Doubleday, Esq. [P.].
b. Ad. sk.	North America.	Purchased.
c. Ad. sk.	Upper Canada.	J. Gould, Esq.
d. Juv. sk.	Toronto.	J. Whitely, Esq. [C.].
e, f, g. Ad. sk.	49th parallel (*J. Dawson*).	N.-A.-Boundary Commission [C.].
h. ♀ ad. sk.	Colville, British Columbia.	J. K. Lord, Esq. [C.].
i. Ad. sk.	California.	Purchased.
k. Skeleton.	N. America.	Zoological Society.

CORONE FLORIDANA.

Corvus americanus, var. floridanus, *Baird, B. N. Amer.* p. 568, pl. lxvii. fig. 1; *Baird, Brewer, & Ridgw. N. Am. B.* ii. p. 247; *Coues, B. North-west*, p. 206.

Corvus americanus, *Allen, Bull. Harv. Coll.* ii. p. 297.

About the size of *C. americana*, but bill and feet larger; tail less rounded; third, fourth, and fifth quills nearly equal, third rather longer than fifth. Colour less violet above. Total length 19·5 inches, wing 12, tail 7·7, tarsus 2·6.

Hab. Southern peninsula of Florida.

Obs. Not having seen a specimen of the Florida Crow, I transcribe the following notes from the great work on North-American Birds (*l. c.*):—"This resident Crow of Florida differs in some marked features from that of the more northern localities in several characters. Although perhaps rather smaller, the bill and feet, especially the latter, are very considerably larger. The nasal feathers extend over the basal two fifths of the bill, instead of the half. The proportions of the bill are about the same: in the Florida bird it is rather the longer; the greatest difference is in the feet. The tarsal joint of the tibia is bare, the feathers scarcely coming below it, even anteriorly, instead of projecting some distance. The tarsus is almost a quarter of an inch longer, covered anteriorly by nine scutellæ instead of eight. The outer lateral toe is shorter, not reaching the base of the middle claw. The middle toe and claw are considerably shorter than the tarsus; the middle claw is shorter than in the northern bird. .... The colours differ somewhat from those of the Carrion-Crow. There is less violet; and the feathers of the back have almost a brassy gloss on their margins, as in *Crotophaga*." Dr. Elliott Coues (*l. c.*), however, does not consider the bird specifically distinct, and says that it "merely illustrates the usual relative increase in size of the bill and feet shown by birds of corresponding latitudes."

4. Corone caurina.

Corvus caurinus, *Baird, B. N. Amer.* p. 569, pl. xxiv.; *Dall & Bann. Tr. Chic. Acad.* i. p. 286; *Gray, Hand-l. B.* ii. p. 13; *Finsch, Abhandl. Brem.* iii. p. 41; *Cooper, Orn. Calif.* p. 285; *Baird, Brewer, & Ridgw. N.-Amer. B.* ii. p. 248.
Corvus americanus, var. β.? caurinus, *Coues, Key N.-Amer. B.* p. 163; *id. B. North-west,* p. 206.

Adult. General colour brown, with a dull violet gloss; wing-coverts like the back, the quills externally glossed with green, as also the primary-coverts, the inner webs with a greenish steel gloss; tail steel-black, the centre feathers glossed with violet like the back; head and neck duller black all round, the ear-coverts faintly shaded with green; throat-plumes slightly different in texture from those of the other parts of the under surface, but not lanceolate; rest of under surface black, glossed with violet; bill and legs black. Total length 15·5 inches, culmen 2·05, wing 11·1, tail 7·1, tarsus 1·95.

Hab. North-western coast of North America.

a. Ad. sk.	West side of Rocky Mountains.	J. K. Lord, Esq.	[P.].
b. Ad. sk.	Vancouver Island.	J. K. Lord, Esq.	[C.].

5. Corone mexicana.

La Grande Pie du Mexique, *Briss. Orn.* ii. p. 43.
Mexican Crow, *Lath. Gen. Syn.* i. p. 396.
Corvus mexicanus, *Gm. Syst. Nat.* i. p. 375 (*ex Lath.*); *Schl. Bijdr. Dierk. Amster.* fol. p. 11, pl. i. fig. 25; *id. Mus. P.-B.* Coraces, p. 25; *Gray, Hand-l. B.* ii. p. 13; *Baird, Brewer, & Ridgw. N.-Amer. B.* ii. p. 233; *Lawr. Mem. Bost. Soc. N. H.* ii. p. 283.

Adult. Plumage highly lustrous, blended; soft burnished steel-blue changing to violet on the crown, and with a greenish cast on lower parts. Wing 9 inches, tail 6·50, culmen 1·60, tarsus 1·20, middle toe 1·10. Wing-formula:—4, 3, 5; first quill very much longer than tenth. (*Baird, Brewer, and Ridgway, l. c.*)

6. Corone corone.

La Corneille, *Briss. Orn.* ii. p. 12; *Montb. Pl. Enl.* iii. pl. 483.
Corvus corone, *Linn. S. N.* i. p. 155; *Temm. Man. d'Orn.* i. p. 108; *Naum. Vög. Deutschl.* ii. Taf. 53. fig. 2; *Roux, Orn. Prov.* pl. 130; *Werner, Atlas,* Omnivores, pl. 1; *Gould, B. Eur.* iii. pl. 221; *Nordm. in Démid. Voy. Russ. Mérid.* iii. p. 120; *Macgill. Br. B.* i. p. 516; *Yarr. Br. B.* ii. p. 79; *Kjærb. Orn. Dan.* tab. xi. fig. 4; *Gray, Gen. B.* ii. p. 313; *Bp. Consp.* i. p. 386; *Schl. u. Susem. Vög. Eur.* ii. Taf. 2, Taf. 3. fig. 1; *Schl. F. J.* Aves, p. 79; *Midd. Sibir. Reis. Zool.* p. 160; *Schl. Vog. Nederl.* pl. 137; *id. Dier. Nederl. Vog.* pl. xiii. fig. 2; *Horsf. & Moore, Cat. B. Mus. E.I. Co.* ii. p. 553; *Fritsch, Vög. Eur.* tab. 28. fig. 3; *Jaub. et Barth. Lap. Rich. Orn.* p. 93; *Schl. Bijdr. Dierk. Amsterd.* fol. *art.* Corvus, p. 8, pl. 1. figs. 11, 12; *Schrenck, Reis. Amurl.* p. 325; *Linderm. Vög. Griechenl.* p. 70; *Jerd. B. Ind.* ii. p. 295; *Radde, Reis. Sibir.* p. 209; *Degl. et Gerbe, Orn. Eur.* i. p. 198; *Schl. Mus. P.-B.* Coraces, p. 17; *Doderl. Avif. Sicil.* p. 62; *Gould, B. Gt. Br.* iii.

pl. 58; *Gray, Hand-l. B.* ii. p. 11; *Salvad. Faun. Ital. Ucc.* p. 172; *Swinh. P. Z. S.* 1871, p. 383; *Severtz. Turkest. Jevotn.* p. 63; *Irby, B. Gibr.* p. 127; *Dresser, B. Eur.* pt. xxxviii.; *id. Ibis*, 1875, p. 207; *Hume, Nests and Eggs Ind. B.* p. 410; *Scully, S. F.* 1876, p. 156.

Corvus subcorone, *C. L. Brehm, Vög. Deutschl.* p. 167.

Corvus hiemalis, *id. t. c.* p. 167.

Corvus assimilis, *id. Vogelfang*, p. 57 *.

Corvus pseudocorone, *Hume, Nests and Eggs Ind. B.* p. 410.

Adult male. General colour above and below steel-black, with a slight shade of purple; the head and neck decidedly glossed with green, as also are the lanceolate feathers of the throat; the lower throat purplish, all the plumes being distinctly lanceolate; quills externally steel-black; tail purplish black; bill and legs black; iris brown. Total length 19 inches, culmen 2·15, wing 12·11, tail 8, tarsus 2·5.

Adult female. Similar to the male, but the gloss on the plumage not quite so brilliant. Total length 19·5 inches, culmen 2·25, wing 13·4, tail 7·8, tarsus 2·4.

Hab. Europe and Northern Asia, ranging into North-western India and China.

a. ♂ ad. sk.	Dunkeld, N.B., May 1876.	A. B. Brooke, Esq. [P.].
b. ♀ ad. sk.	Wolsingham, Durham, June 5, 1868.	J. H. Gurney, Esq., jun. [P.].
c, d. ♂ ♀ ad. sk.	Lilford Hall, Northants, March 10, 1876.	Lord Lilford [P.].
e. ♂ ad. st.	Bushey, Herts.	Purchased.
f. Ad. st.	Near Paris.	R. B. Sharpe, Esq. [P.]
g. Juv. sk.	Turin.	M. Bonelli [C.].
h. ♂ ad. sk.	Balryna, E. Siberia, May 18, 1872.	Dr. Dybowski [C.] †.
i, k, l, m. Skulls.	S. Germany.	Dr. Günther.

7. Corone australis.

Corvus coronoides, *Gould, B. Austr.* iv. pl. 18; *Schl. Bijdr. Dierk. Amsterd.* fol. *art.* Corvus, p. 8, pl. 1. fig. 13; *id. Mus. P.-B.* Coraces, p. 18.

Corvus australis, *Gould, Handb. B. Austr.* i. p. 475; *Gray, Hand-l. B.* ii. p. 12 (pt.).

Adult. Purplish black, the gloss not quite so brilliant on the wings, which incline to steel-black on the outer aspect of the primaries; tail purplish black; head and neck also purplish, not so bright as the back; feathers of the throat long and lanceolate in shape, of a greenish black colour, the plumes of the lower throat

* In Dresser's 'Birds of Europe' (*l. c.*) will be found nine trinominal names for this bird, given by the late Pastor Brehm to the species, and published by his son Dr. A. E. Brehm in an account of his father's collection. I decline to quote these names.

† Labelled *Corvus orientalis*, Eversm. (Addend. Pall. Zoogr. ii. p. 7). From this one specimen I am not able to separate the bird from the ordinary Carrion-Crow.

inclining to purple; rest of the under surface purplish black, as also the under wing- and tail-coverts; lower surface of the quills entirely black; bill and legs black; iris dark brown. Total length 18·5 inches, culmen 2·45, wing 13·7, tail 8, tarsus 2·6.

Obs. Believing as I do that there are two species of Crow in Australia, it remains to find a specific name for the *Corone*, as an examination of the type shows that *C. coronoides* is a true *Corvus* (*vide suprà*, p. 20). Instead of proposing a new specific name, it seems to me that we can adopt that of *C. australis* for the species, as Mr. Gould has applied it in his 'Handbook.' He there employs the name of *Corvus australis*, Gm. S. N. i. p. 365 (founded on the "South-Sea Raven" of Latham, Gen. Syn. i. pt. i. p. 369) for the white-eyed Crow, which he had previously figured as *C. coronoides* in his folio work. The description given by Latham suits either of the Australian species; but as Mr. Gould has restricted it, it seems better to follow him. The idea of there being two species of Crow in Australia is not new, since Mr. E. P. Ramsay, in 1865, noticed that in New South Wales there were two species, one of which had white irides, the other brown, and that these birds laid differently coloured eggs. The suggestion that the brown-eyed birds are the young, as suggested by Mr. Gould, does not seem to me feasible.

Hab. Australia.

a. Ad. sk.	Australia.	
b. Ad. sk.	Australia.	Purchased.

8. Corone macrorhyncha.

Corvus macrorhynchus, *Wagler*, *Syst. Av.* Corvus, sp. 3; *Bp. Consp.* i. p. 386; *Cab. Mus. Hein.* Th. i. p. 232; *Schl. Bijdr. Dierk. art.* Corvus, pl. 1. figs. 5, 6; *id. Mus. P.-B.* Coraces, p. 19; *Gray, Hand-l. B.* ii. p. 11.

Corvus timoriensis, *Bp. C. R.* xxxvii. p. 829.

Similar to *C. levaillanti* described in full below, but with white bases to the feathers. (*Vide infrà.*)

Hab. Malacca, Java, Sumatra, Borneo, Flores, Timor, Bali.

a. Ad. sk.	Malacca.	Hugh Cuming, Esq. [C.].
b. Ad. sk.	Penang.	A. R. Wallace, Esq. [C.].
c. Pull. sk.	Penang.	Dr. Cantor [C.].
d. Ad. sk.	Sumatra.	Purchased.
e. Juv. sk.	Sumatra.	Purchased.
f, g. ♀ ad. sk.	E. Java.	A. R. Wallace, Esq. [C.].
h. ♀ ad. sk.	E. Timor.	A. R. Wallace, Esq. [C.].
i. ♀ juv. sk.	Semao, Timor.	A. R. Wallace, Esq. [C.].
k, l. Ad. sk.	Flores.	A. R. Wallace, Esq. [C.].
m. Ad. sk.	Bali.	A. R. Wallace, Esq. [C.].
n. Ad. sk.	Bali (*Wallace*).	J. Gould, Esq.

Subsp. *a*. Corone levaillanti.

Corvus levaillantii, *Less. Traité*, p. 328; *Pucher. Rev. et Mag. de Zool.* 1853, p. 547; *Holdsw. P. Z. S.* 1872, p. 460; *Ball, S. F.* 1874, p. 418; *Hume, t. c.* pp. 243, 479, 1875, p. 143; *id. Nests and Eggs Ind. B.* p. 411; *Blyth & Wald. B. Burm.* p. 86.
Corvus corone, *Frankl. P. Z. S.* 1831, p. 120 (*nec L.*).
Corvus corax, *Royle, Himalaya*, p. 77 (*nec L.*).
Corvus macrorhynchus, *Gray & Hardw. Ill. Ind. Zool.* pl. 36. figs. 2*a*, 2*b* (*nec Wagl.*); *Jerd. Madr. Journ.* xi. p. 18.
Corvus enca, *Sundev. Ann. N. H.* xviii. p. 306 (*nec Horsf.*).
Corvus culminatus, *Gray, Cat. Mamm. &c. Nepal, Hodgs.* p. 102 (*nec Sykes*); *Blyth, J. A. S. B.* xv. p. 24, xvi. p. 727, xxiv. p. 479; *id. Cat. B. Mus. A. S. B.* p. 89; *Hutton, J. A. S. B.* xvii. pt. 2, p. 9; *Bp. Consp.* i. p. 385; *Horsf. & Moore, Cat. B. Mus. E.I. Co.* ii. p. 553 (pt.); *Jerd. B. Ind.* ii. p. 295; *Blyth, Ibis*, 1867, p. 34, 1868, p. 132; *Gray, Hand-l. B.* ii. p. 11; *Godw.-Aust. J. A. S. B.* 1874, p. 170.
Corvus sinensis, *Moore, Cat. B. Mus. E.I. Co.* ii. p. 556; *Gray, Hand-l. B.* ii. p. 11.
Corvus japonicus, *Swinh. Ibis*, 1861, p. 337 (*nec Bp.*).
Corvus japonensis, *Swinh. Ibis*, 1862, p. 260 (*nec Bp.*).
Corvus colonorum, *Swinh. Ibis*, 1864, p. 427.
Corvus andamanensis, *Tytler, Ibis*, 1866, p. 420; *id. & Beav. Ibis*, 1867, p. 328; *Gray, Hand-l. B.* ii. p. 11.

Adult. General colour glossy black, inclining to purple on the back; the head greener, and the neck all round having a kind of dusky greyish-black shade, somewhat washed with green; quills purplish black, the primaries externally and the innermost secondaries internally steel-black; tail purplish black; under surface of body dull blackish, with a slight greenish gloss, less distinct on the fore neck, which is dull greyish like the hind neck; sides of face and throat glossy green, inclining to purple on the lower throat, the feathers lanceolate in shape, but not extending beyond the throat; bill and legs black; iris dark brown. Total length 21 inches, culmen 2·3, wing 13·6, tail 9·5, tarsus 2·3.

Young. Not so glossy as the adult; the bill horn-brown, yellowish near the base of both mandibles.

				Total length circa in.	Culmen. in.	Wing. in.	Tail. in.	Tarsus. in.
1.	*C. macrorhyncha*	♂.	Bali	20	2·75	14·45	9·5	2·45
2.	"	♀.	"	20	3	14·2	9	2·45
3.	"		Malacca	20	2·75	12·2	8	2·3
4.	"		Penang	19	2·7	13·2	8	2·4
5.	"	♀.	E. Java	21	2·7	13	8	2·35
6.	"	♀.	"	20	2·7	13·3	8·2	2·35
7.	"		Sumatra	18	2·6	12·1	7·2	2·1
8.	"		Flores	20	2·75	13·3	8·3	2·1
9.	"		"	18	2·8	13·8	8·3	2·3
10.	"		E. Timor	20	2·95	13·2	8	2·1
11.	"	♀.	Semao, Timor	19	2·7	12·2	7·5	2·25
12.	*C. levaillanti*		India	21	2·6	13·65	9	2·45
13.	"		"	19	2·5	12·5	8	2·1

				Total length circa in.	Culmen. in.	Wing. in.	Tail. in.	Tarsus. in.
14.	*C. levaillanti*		India	21	2·5	13·4	9	2·05
15.	"		N. India	18	2·6	13·2	8·5	2·25
16.	"		Afghanistan	18	2·35	13·5	9·3	2·1
17.	"		Kooloo	19	2·4	12·85	8·6	2·15
18.	"		Nynee Tal	19·5	2·55	13·1	8·5	2·15
19.	"		Nepal	19·5	2·6	12·3	7·8	2·1
20.	"		"	18	2·8	13·65	8·5	2·4
21.	"	♀.	Andaman Islands ..	17·8	2·7	12	7·5	2·25
22.	"	♂.	" " ..	20	2·85	13·05	8·6	2·25
23.	"	♀.	" " ..	20·5	2·2	12·35	8	2·15
24.	"		Bhamau	18	2·65	11·5	7·3	2·15
25.	"		"	20	2·8	12·9	8·5	2·3
26.	"		"	19	2·6	12·5	8	2·4
27.	"		"	21	2·8	13·7	8·5	2·4
28.	"		Tapeng	18	2·55	12·8	8	2·4
29.	"		Tsitkau	18	2·6	12·6	8	2·25
30.	"		Ponsee............	18·5	2·55	12·65	8·5	2·25
31.	"		Yun-nan	18·5	2·45	13·1	8·1	2·25
32.	"		China	20	2·5	12·3	8	2·4
33.	"		"	19·5	2·75	13·95	8·5	2·5
34.	"		"	17	..	13·05	8·5	2·3
35.	"		Fokien............	20	2·7	13·4	8·8	2·35
36.	"		Hainan	18·5	2·7	13·7	9·2	2·35
37.	"	♂.	Ussuri river	20	2·85	14·35	9	2·45
38.	"	♀.	" "	20	2·7	13·65	9	2·2

Obs. The series of measurements which is given above shows pretty conclusively that size alone is not sufficient to separate as distinct *C. macrorhyncha* and *C. levaillanti.* Malayan specimens are found absolutely identical with others from different parts of India; but it must be observed that there is a tendency in *C. macrorhyncha* to exhibit pure white bases to the feathers, a character not observable in Indian and Chinese birds. Whether this is a point which is of such great importance in the study of Crows will be for some future worker at the group to determine; and it will require a longer period than I have been able to devote to the subject, along with a much larger series of birds to settle the matter; but I have not failed to notice that certain species which have pure white bases to the feathers in the adult have greyish bases to the same plumes in the young bird; and Mr. Dresser has observed the same fact. Thus *C. macrorhyncha* might be distinguished from *C. levaillanti* were it not for the occurrence of intermediate forms which render it impossible to draw an exact line for specific separation; and, no doubt, in cases where an equally good series of birds could be compared, a good many more supposed species would be suppressed. Between *C. sinensis* and *C. levaillanti* there is not any difference at all, and *C. japonensis* I only keep distinct on account of its large raven-like form. *C. culminata* I have already distinguished, and must leave it to the researches of Indian ornithologists to determine the value of the distinctions that I have drawn from the skins.

Hab. India and China, extending to Eastern Siberia.

a. Ad. sk.	India.	J. Gould, Esq.
b, *c*. Ad. sk.	India.	Purchased.
d. Ad. sk.	Afghanistan (*Griffith*).	Secretary of State for India [P.]. (Spec. *b* of Horsf. & Moore, Cat. p. 553. no. 831.)
e. Ad. sk.	Gangaotri, Himalayas.	Purchased.
f. Ad. sk.	Nynee Tal.	Capt. Stackhouse Pinwill [P.].
g. Ad. sk.	Kooloo (*Tweedie*).	J. Gould, Esq.
h. Ad. sk.	North India.	H. Drummond, Esq. [P.].
i, *k*, *l*. Ad. sk.	Nepal.	B. H. Hodgson, Esq. [P.].
m, *n*. Ad. sk.	Nepal.	B. H. Hodgson, Esq. [P.].
o, *p*. ♂ ♀ ad. sk.	Andaman Islands.	Purchased.
q. ♀ ad. sk.	South Andamans, Oct. 25, 1873 (*Wimberley*).	R. B. Sharpe, Esq. [P.].
r. Ad. sk.	Bhamau, Upper Burmah, Feb. 1868.	Dr. Anderson [C.].
s. Ad. sk.	Ponsee, Kakhyen Hills, March 13, 1868.	Dr. Anderson [C.].
t. Juv. sk.	Mangwyne, Sanda Valley, May 12, 1868.	Dr. Anderson [C.].
u. ♂ ad. sk.	Momien, Yun-nan, May 1868.	Dr. Anderson [C.].
v. ♂ ad. sk.	Poo-too, China, Aug. 12, 1860.	Secretary of State for India [P.]. (Type of *C. sinensis*.)
w. Ad. sk.	China.	J. Gould, Esq.
x. Ad. sk.	China.	Capt. Blakiston [C.].
y. Ad. sk.	Fokien, China.	Capt. Blakiston [C.].
z. Ad. sk.	Shanghai.	Shanghai Museum [P.].
a'. ♂ ad. sk.	Hainan, Feb. 2, 1868.	R. Swinhoe, Esq. [C.].
b', *c'*. ♂ ♀ ad. sk.	Ussuri river, Jan. 20, 1874.	Dr. Dybowski [C.].

Subsp. β. **Corone japonensis.**

Corvus macrorhynchus, *Schlegel, Faun. Japon.* Aves, p. 79, pl. 39.
Corvus japonensis, *Bp. Consp.* i. p. 386; *Midd. Sibir. Reis. Zool.* p. 161; *Schrenck, Reis. Amurl.* p. 325; *Radde, Reise Sibir.* p. 210; *Schl. Bijdr. Dierk. Amsterd.* 4to, *art.* Corvus, pl. i. figs. 3 & 4; *id. Mus. P.-B.* Coraces, p. 15; *Gray, Hand-l. B.* ii. p. 11; *Swinh. P. Z. S.* 1871, p. 382.

Adult. Above glossy purplish black, the wings rather more reddish or bronzy purple, the bastard wing and primary-coverts, as well as the outer aspect of the quills, steel-green, the inner webs of the quills also greenish, especially of the secondaries; tail-feathers steel-green, glossed with purple, especially near the base; head greenish black, the ear-coverts somewhat oily green, the neck all round with somewhat of a greyish cast; under surface black with a very slight gloss of purple, the feathers of the throat lanceolate with a dull purplish gloss; bill and feet black. Total length 21 inches, culmen 3·05, wing 14·2, tail 9·0, tarsus 2·4.

Another specimen measures 15·6 inches in the wing, tarsus 2·7.

This species is not a true Raven, but is a large form of *C. levaillantii*. It is larger in general appearance; but it is exceeded in length of wing and tail by some specimens of the Malayan birds, as will be seen by the above measurements. I therefore consider it only a race of that widespread form which, under the titles of *C. levaillantii* or *C. macrorhyncha*, is spread over the greater part of Asia.

Hab. Japan and North China "beyond the great wall." (*Swinhoe.*)

a. Ad. sk.	Nagasaki, Japan.	Capt. Blakiston [C.].
b. Juv. sk.	Hakodadi, Japan.	Capt. Blakiston [C.].
c. Ad. sk.	Japan.	J. Gould, Esq.
d. Ad. sk.	Japan.	Purchased.

9. **Corone validissima.**

Corvus validissimus, *Schlegel, Bijdr. Dierk. Amst.* 4to, p. 12. pl. i. figs. 1, 2, 21; *id. Mus. P.-B.* Coraces, p. 28; *Gray, Hand-l. B.* ii. p. 12.

Adult female. Black, with a very rich purple gloss, all the bases to the feathers pure white; wings rich purple, like the back, the edge of the wing, the primary-coverts, and the outer aspect of the primaries deep greenish black; tail purplish black, shaded with steel-green on the outer feather; head all round greenish black, the feathers of the throat distinctly glossed with green and somewhat lanceolate, the hinder neck somewhat greyish purple; remainder of under surface ashy black, with a slight purplish gloss; bill and feet black. Total length 21 inches, culmen 3·1, wing 13·4, tail 7·8, tarsus 2·15.

Adult male. Similar to the adult female described, but decidedly more grey, the feathers under certain lights having greyish margins both on the upper and under surface. Total length 20 inches, culmen 3·2, wing 13·3, tail 7·6, tarsus 2·15.

Obs. Professor Schlegel refers this species to the neighbourhood of *Corone valida*, *C. enca*, and its allies; but in reality it is a strong-billed representative of *C. macrorhyncha*, distinguishable from that species by its long and peculiarly shaped bill and by the purplish, not green, shade of the underparts. He also states that the Crows of Kaioa Island are referable to this species. The birds collected there by Mr. Wallace, however, belong to the race known as *C. orru.*

Hab. Batchian and Halmahera.

a. ♀ ad. sk.	Batchian, 1858.	A. R. Wallace, Esq. [C.].
b. Ad. sk.	Batchian, 1858.	A. R. Wallace, Esq. [C.].
c. ♂ ad. sk.	Gilolo, Halmahera, 1858.	A. R. Wallace, Esq. [C.].
d. Ad. sk.	Batchian (*Wallace*).	J. Gould, Esq.

10. **Corone philippina.**

Corvus philippinus, *Bp. C. R.* xxxvii. p. 380; *Gray, Hand-l. B.* ii. p. 12; *Walden, Tr. Z. S.* ix. p. 201.

Corvus solitarius, *Kittl. Reis. Russ. Amer. Mikronesien, &c.*, ii. p. 431 (1858, *desc. nullâ*).

Corvus brevipennis, *Schl. Bijdr. Dierk. Amsterd.* fol. p. 9, pl. i. fig. 8; *id. Mus. P.-B.* Coraces, p. 22.

Adult. Above blue-black, with very little shade of purple, all the bases of the feathers pure white, the lower back rather shaded with grey; wings and tail purplish black, with somewhat of a greenish gloss, especially on the inner webs; sides of face and throat greenish black, the plumes slightly lanceolate; rest of under surface ashy black, with a slight greenish gloss; bill and feet black. Total length 18 inches, culmen 2·45, wing 10·7, tail 7·3, tarsus 2·1.

Hab. Philippine Islands.

a. ♂ ad. sk.	Cataguan, Luzon.	Hugh Cuming, Esq. [C.].
b. Ad. sk.	Cataguan, Luzon.	Hugh Cuming, Esq. [C.].
c. Var. sk.	Philippine Islands.	Hugh Cuming, Esq. [C.].

11. Corone enca.

Fregilus enca, *Horsf. Tr. Linn. Soc.* xiii. p. 164.

Corvus enca, *Wagler, Syst. art.* Corvus, sp. 11; *Vig. Mem. Raffl.* p. 665; *Bp. Consp.* i. p. 385; *Cab. Mus. Hein.* Th. i. p. 231; *Horsf. & Moore, Cat. B. Mus. E.I. Co.* ii. p. 558; *Schl. Bijdr. Dierk., Amsterd.* 4to, *art.* Corvus, p. 13, pl. 1. fig. 23; *id. Mus. P.-B.* Coraces, p. 29; *Gray, Hand-l. B.* ii. p. 12; *Walden, Tr. Z. S.* viii. p. 74.

Corvus validus, *Bp. Consp.* i. p. 385; *Schl. Bijdr. Dierk.* 4to, p. 13 pl. i. fig. 22; *id. Mus. P.-B.* Coraces, p. 29; *Gray, Hand-l. B.* ii. p. 12; *Salvad. Ucc. Born.* p. 281; *Walden, Tr. Z. S.* ix. p. 202.

Corvus tenuirostris, *Moore, Cat. B. Mus. E.I. Co.* ii. p. 558; *Jerd. B. Ind.* ii. p. 297; *Gray, Hand-l. B.* ii. p. 12.

Corvus fallax, *Brügg. Abhandl. Verein Bremen,* v. p. 76, Taf. 3. fig. 4*.

Adult (type of species). Above dull purplish black all over, wings and tail included, the primaries a little deeper in colour; underneath dull purplish black, with somewhat of a greyish cast; bill and feet black. Total length 15 inches, culmen 2·2, wing 11·6, tail 6, tarsus 1·9.

Young. Similar to the adult, but less glossy, with more greyish on the under surface.

Table of measurements of C. enca, C. orru, *and* C. violacea.

				Total length circa in.	Culmen. in.	Wing. in.	Tail. in.	Tarsus. in.
1.	*C. enca.*		W. Java	15·2	2·35	11·15	5·7	1·95
2.	"		Java (type)	15·0	2·2	11·6	6·0	1·9
3.	"		(Type of *C. tenuirostris*)	17·5	2·65	12·4	6·6	2·05
4.	"	♀.	Sumatra........	19·0	2·5	12·5	7·0	2·0
5.	"	♀.	Sumatra........	16·5	2·45	11·4	6·0	2·05
6.	"		Banjermassing ..	19·0	2·5	11·2	6·3	2·1
7.	"		Borneo	17·0	2·4	11·4	6·2	1·95
8.	"	♂.	Macassar	16·5	2·2	11·2	6·3	2·1
9.	"	♂.	Tondano	17·0	2·2	11·5	6·1	2·05
10.	"		Sula Islands	15·2	2·35	11·0	5·9	2·0

* Dr. Brüggemann describes another Crow from Celebes as *C. annectens*, which may be specifically distinct (*l. c.* p. 4, Taf. 3. fig. 3).

				Total length circa in.	Culmen. in.	Wing. in.	Tail. in.	Tarsus. in.
11.	*C. enca.*		Sula Islands	14·5	2·2	10·8	5·8	2·0
12.	"		Sula Islands	16·0	2·4	11·1	6·1	2·0
13.	*C. orru.*	♂.	Dorey..........	20·0	2·5	12·9	7·0	2·2
14.	"	♀.	Dorey..........	19·0	2·4	12·95	7·0	2·25
15.	"	♀.	Dorey..........	19·5	2·45	12·5	6·8	2·25
16.	"	♀.	Salwati	17·0	2·4	12·55	7·0	2·2
17.	"	♀.	Mysol..........	19·0	2·65	13·1	7·2	2·15
18.	"	♀.	Mysol..........	17·0	2·4	12·6	6·5	2·15
19.	"		Morty Island....	18·0	2·5	12·7	7·0	2·25
20.	"	♂.	Ternate	20·0	2·75	13·1	7·5	2·3
21.	"	♀.	Ternate	18·5	2·5	moulting	7·3	2·2
22.	"		Kaioa Island....	17·0	2·7	12·6	6·5	2·25
23.	"		Kaioa Island....	18·5	2·5	12·75	6·8	2·2
24.	*C. violacea*	♀.	Ceram	14·8	1·9	9·5	6·0	1·65
25.	"		Ceram	15·5	1·9	9·6	5·5	1·75
26.	"		Ceram	14·0	1·95	9·8	6·0	1·75

Obs. If we consider *C. enca* as one prominent form of Purple Crow, we cannot recognize more than three races:—a large one, *C. orru*; a smaller one, *C. enca* (with which *C. valida* and *C. tenuirostris* are synonymous); and *C. violacea*, the smallest of all. Between *C. enca* and *C. valida* it is impossible to draw the line, as they differ only in size, and a series shows that they grade one into another. The foregoing list of measurements proves this; and the bare space behind the eye, though exaggerated in these birds, is generally not very much larger than in some of the other Malayan forms, and cannot be relied on as a character.

Hab. Java, Sumatra, Borneo, Celebes, and the Sula Islands.

a. Ad. sk.	Java (*Horsfield*).	Secretary of State for India [P.]. (Type of species.)
b. Juv. sk.	E. Java.	A. R. Wallace [C.].
c. Ad. sk.	W. Java.	A. R. Wallace [C.].
d. Ad. sk.	[Sumatra?]	Secretary of State for India [P.]. (Type of *Corvus tenuirostris*, Moore, *l.c.* p. 558.)
e.f. ♀ ad. sk.	Sumatra.	A. R. Wallace, Esq. [C.].
g. ♂ ad. sk.	Borneo.	Purchased.
h. Ad. sk.	Labuan.	Hugh Low, Esq. [P.].
i. Ad. sk.	Banjermassing, Borneo.	J. Motley, Esq. [C.].
k. ♂ ad. sk.	Macassar, Celebes.	A. R. Wallace, Esq. [C.].
l. Juv. sk.	Macassar, Celebes.	A. R. Wallace, Esq. [C.].
m. ♂ ad. sk.	Tondano.	A. R. Wallace, Esq. [C.].
n, o, p. Ad. sk.	Sula Islands.	A. R. Wallace, Esq. [C.].
q. Skeleton.	Java.	

Subsp. *a.* **Corone orru.**

Corvus orru, *Bp. Consp.* i. p. 385 (*ex Mull. MS.*); *Schl. Bijdr. Dierk.* 4to, *art.* Corvus, p. 9, pl. 1. figs. 9, 10; *id. Mus. P.-B.* Coraces, p. 20; *Gray, Hand-l. B.* ii. p. 12.

Adult male (Dorey). General colour purplish black, deeper and more of a bluish purple on the primaries, inner secondaries, and tail; under surface of body duller than the upper, the feathers of the throat and fore neck lanceolate and having somewhat of a greenish cast; bill and feet black. Total length 20 inches, culmen 2·5, wing 12·9, tail 7, tarsus 2·2.

Adult female (Dorey). Similar to the male. Total length 20 inches, culmen 2·4, wing 12·45, tail 7, tarsus 2·15.

Hab. Moluccan and Papuan Islands.

a, b. ♂ ♀ ad. sk.	Dorey, New Guinea.	A. R. Wallace, Esq. [C.].
c. ♀ ad. sk.	Dorey, New Guinea.	A. R. Wallace, Esq. [C.].
d. ♂ juv. sk.	Dorey, New Guinea.	A. R. Wallace, Esq. [C.].
e. ♀ ad. sk.	Salwati, June 29, 1868 (*Hoedt*).	J. Gould, Esq.
f. ♂ juv. sk.	Waigiou.	A. R. Wallace, Esq. [C.].
g, h. ♀ ad. sk.	Mysol.	A. R. Wallace, Esq. [C.].
i, k. ♂ ♀ ad. sk.	Ternate.	A. R. Wallace, Esq. [C.].
l. Ad. sk.	Morty Island.	A. R. Wallace, Esq. [C.].
m. Ad. sk.	Kaioa Island.	A. R. Wallace, Esq. [C.].
n. ♀ ad. sk.	Kaioa Island.	A. R. Wallace, Esq. [C.].

Subsp. β. **Corone violacea.**

Corvus violaceus, *Bp. Consp.* i. p. 384 (*ex Temm. MS.*); *Schl. Bijdr. Dierk.* 4to, p. 14, p. i. fig. 24; *Gray, P. Z. S.* 1860, p. 354; *Schl. Mus. B.-P.* Coraces, p. 31; *Gray, Hand-l. B.* ii. p. 13.

Amblycorax violaceus, *Bp. C. R.* xxxvii. p. 829.

Corvus modestus, *Brügg. Abhandl. Ver. Bremen*, v. p. 76, pl. 3. fig. 5.

Adult female. Dull purplish black above, a little less glossy below; wings and tail rather richer purple, the primaries blue-black; bill and feet black. Total length 14·8 inches, culmen 1·9, wing 9·5, tail 6, tarsus 1·65.

Hab. Ceram.

a, b. Ad. sk.	Ceram.	A. R. Wallace, Esq. [P.].
c. ♀ ad. sk.	Ceram.	A. R. Wallace, Esq. [P.].
d. ♀ ad. sk.	N. Ceram.	A. R. Wallace, Esq. [P.].

7. RHINOCORAX.

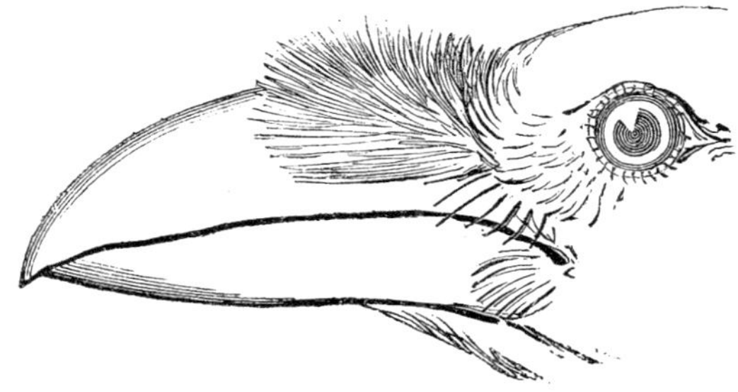

Head of *Rhinocorax affinis.*

Range. North-eastern Africa, ranging into Palestine.

1. Rhinocorax affinis.

Corvus affinis, *Rüpp. Neue Wirb.* p. 20, Taf. 10. fig. 2; *Gray, Gen. B.* ii. p. 315; *Horsf. & Moore, Cat. B. Mus. E.I. Co.* ii. p. 557; *Bp. Consp.* i. p. 385; *Blyth, J. A. S. B.* xxiv. p. 300; *Schl. Bijdr. Dierk.* 4to, p. 15, pl. 1. fig. 26; *Scl. Ibis*, 1860, p. 245; *Tristr. Ibis*, 1866, p. 72; *Schl. Mus. P.-B.* Coraces, p. 31; *Gray, Hand-l. B.* ii. p. 12; *Heugl. Orn. N.O.-Afr.* p. 502; *Finsch, Tr. Z. S.* vii. p. 256; *Blanf. Geol. & Zool. Abyss.* p. 393; *Wyatt, Ibis*, 1870, p. 16; *Finsch & Hartl. Vög. Ostafr.* p. 372; *Shelley, B. Egypt*, p. 158; *Antin. & Salvad. Viagg. Bogos*, p. 128; *Dresser, B. Eur.* pts. xxxvi. xxxviii.
Corvus brachyurus, *A. E. Brehm, J. f. O.* 1854, p. 75.
Corvus brachyrhynchos, *C. L. Brehm, Vogelfang*, p. 414.
Corvus brevicaudatus, *Müll. J. f. O.* 1855, p. 496.
Corax affinis, *A. E. Brehm, Reis. Habesch*, p. 216.
Corax brachyurus, *id. t. c.* p. 323.

Adult male. Black, glossed with purple on the back, wings, and tail; the head like the back, but duller; sides of the face and neck rather oily greenish in tint; under surface of body glossy steel-black, the lanceolate feathers confined to the whole of the throat and not extending on to the fore neck; nasal bristles black, and becoming gradually erect towards the forehead, directed forwards anteriorly, so as to hide the nostrils completely; bill and legs black; iris brown. Total length 18·5 inches, culmen 2·3, wing 14·75, tail 6·6, tarsus 2·25.

Obs. The description is taken from a Palestine skin procured by Canon Tristram. The other specimens in the Museum measure as follows:—

			Total length.	Culmen.	Wing.	Tail.	Tarsus.
			in.	in.	in.	in.	in.
a.	Ad.	Shoa (*Harris*)	17.0	2·35	15·0	6·9	2·98
b.	Ad.	Shoa (*Harris*)	17·0	2·35	15·9	7·2	2·75
c.	♂ ad.	Ankober (*Harris*)	21·0	2·7	15·8	7·3	2·8
d.	♀ ad.	Ankober (*Harris*)	19·0	2·25	15·1	7·3	2·6
e.	♂ ad.	Adigrat (*Blanford*)	19·5	2·5	15·7	6·9	2·7
f.	♀ ad.	Koomaylee (*Blanford*)	18·0	2·55	15·3	6·6	2·7
g.	♂ ad.	Ghor es Safieh (*Tristram*)	18·5	2·3	14·75	6·6	2·25

Hab. North-eastern Africa, ranging into Palestine.

a. ♂ ad. sk.	Ghor es Safieh, Palestine.	Canon Tristram [C.].
b. ♀ ad. sk.	Koomaylee, Abyssinia, Feb. 13, 1868.	W. T. Blanford, Esq. [C.].
c. ♂ ad. sk.	Adigrat, May 8, 1868.	W. T. Blanford, Esq. [C.].
d, e. ♂ ♀ ad. sk.	Ankober, May 1862.	Sir W. C. Harris [C.].
f, g. Ad. sk.	Shoa.	Sir W. C. Harris [C.].

8. GAZZOLA*.

	Type.
Gazzola, *Bp. Consp.* i. p. 75 (1850)	?
Gazzola, *Bp. C. R.* xxxvii. p. 828 (1854)	G. typica.

Head of *Gazzola typica.*

Range. Confined to Celebes.

1. Gazzola typica.

Gazzola typica, *Bp. C. R.* xxxvii. p. 828; *Sclater, Ibis,* 1859, p. 113; *Walden, Tr. Z. S.* viii. p. 74.

Corvus advena, *Schleg. Bijdr. Dierk. Amsterd.* fol. *art.* Corvus, p. 3, pl. 2 (*nec Brehm*); *id. Mus. P.-B.* Coraces, p. 6; *Wallace, Malay Archip.* i. p. 375; *Gray, Hand-l. B.* ii. p. 13.

Adult. Above purplish brown, the feathers waved with dusky brown under certain lights; the wings glossy purple, inclining to violet on the primary-coverts, primaries, and inner webs of the secondaries; tail purplish black; nasal bristles blackish; head all round glossy purplish brown, including the ear-coverts and adjoining parts, extending on to the fore neck, where the tinge is browner; nape and neck all round, as well as the mantle above and the entire under surface of the body, white; thighs and under wing- and tail-coverts blackish brown; bill and legs black. Total length 13·7 inches, culmen 1·75, wing 8·5, tail 5·2, tarsus 1·55.

Hab. Celebes.

a, b.	♂ ♀ ad. st.	Makassar.	A. R. Wallace, Esq. [C.].
c, d.	♀ ad. sk.	Makassar.	A. R. Wallace, Esq. [C.].

---

* The confusion connected with the synonymy of this genus is disentangled by Lord Walden (Tr. Z. S. viii. p. 74), whose conclusions are here adopted.

9. MICROCORAX.

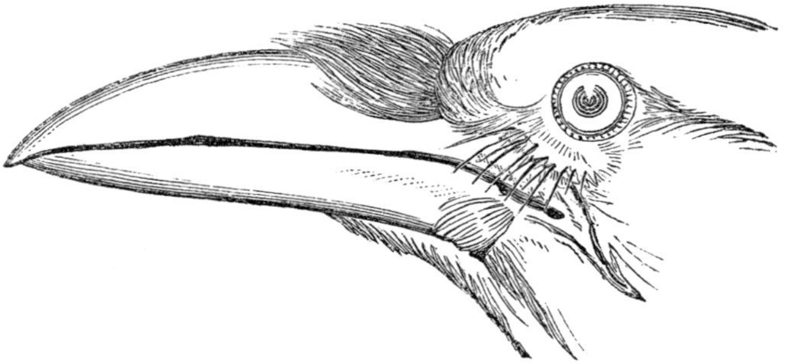

Head of *Microcorax leucognaphalus.*

Range. West-India Islands.

Key to the Species.*

a. Tarsus about equal to the culmen; feathers smoky grey at base.
  a'. Plumage cinereous or plumbeous brown; head and neck only glossy .... *jamaicensis*, p. 48.
  b'. Plumage glossy black, with a violet shade .. *solitarius*, p. 49.
b. Tarsus much shorter than the bill.
  c'. Plumage white at base .... *leucognaphalus*, p. 49.
  d'. Plumage smoky grey at base .... *nasicus*, p. 49.

1. Microcorax jamaicensis.

Chattering Crow, *Lath. Gen. Syn.* i. p. 377.
Corvus jamaicensis, *Gm. S. N.* i. p. 367; *Gosse, B. Jamaica*, p. 209; *Scl. Cat. Am. B.* p. 146; *Gray, Hand-l. B.* ii. p. 13; *Scl. & Salv. Nomencl. Av. Neotr.* p. 40; *Baird, Brewer, & Ridgw. N. Amer. B.* p. 234.
Corvus nasicus, *Gray, Hand-l. B.* ii. p. 13 (*nec Temm.*).

Adult male. General colour above cinereous brown, the bases of all the feathers ashy; wings and tail brown, slightly glossed with violet; head glossy blackish brown, the throat also blackish; rest of under surface of body cinereous brown; bill and legs black. Total length 16·5 inches, culmen 1·9, wing 9, tail 6·1, tarsus 1·85.

Hab. Jamaica.

a. Ad. sk.	Jamaica.	Purchased.
b. Ad. st.	Jamaica.	P. H. Gosse, Esq. [C.].
c. ♂ ad. sk.	Moneague, Jamaica (*W. T. March*).	Smithsonian Institution [P.].
d. Sternum.	Jamaica.	P. H. Gosse, Esq. [C.].

* *Cf.* also the "Key" in 'North-American Birds' (ii. p. 233), whence some of the above characters are derived.

2. Microcorax solitarius.

Corvus solitarius, *Würt. Naumannia*, ii. p. 55; *Bp. C. R.* xxxvii. p. 829; *Schl. Bijdr. Dierk. Amsterd.* fol. *art.* Corvus, pl. 1. fig. 17; *id. Mus. P.-B.* Coraces, p. 24; *Gray, Hand-l. B.* ii. p. 13.
Corvus palmarum, *Würt. Reis.* p. 73.

Adult. General colour black, with a strong violet or purplish gloss; bastard wing, primary-coverts, and primaries steel-black; tail black, externally glossed with violet; head all round and under surface of body glossy black, with slight purplish reflections; bill and legs black. Total length 15 inches, culmen 1·95, wing 9·35, tail 5·6, tarsus 1·95.

Hab. S. Domingo.

a.	Ad. sk.	S. Domingo.	M. A. Sallé [C.].
b.	Ad. sk.	S. Domingo.	M. A. Sallé [C.].

3. Microcorax leucognaphalus.

Corvus leucognaphalus, *Daud. Traité*, ii. p. 231; *Gray, Gen. B.* ii. p. 315; *Schl. Bijdr. Dierk. Amsterd.* fol. *art.* Corvus, pl. 1. fig. 28; *id. Mus. P.-B.* Coraces, p. 32; *Gray, Hand-l. B.* ii. p. 13; *Scl. & Salv. Nomencl. Av.* p. 40.
Corvus erythrophthalmus, *Bp. C. R.* xxxvii. p. 829 (*ex Würt. MS.*); *Gray, Hand-l. B.* ii. p. 13.

Adult. Above glossy black, with a faint violet reflection; the bastard wing, primary-coverts, and primaries steel-black, with somewhat of a purplish gloss; tail black, glossed with violet, the outer tail-feathers purplish black; head and neck all round, as well as the under surface of the body, glossy black, the feathers of the hinder neck rather silky in texture; the throat-feathers long, not pointed, but having the ends split into hair-like filaments; bases of the feathers over the whole body white; bill and legs black. Total length 18·7 inches, culmen 2·3, wing 12·1, tail 8·3, tarsus 2·05.

Hab. Porto Rico and S. Domingo.

a.	Ad. sk.	Porto Rico, 1869 (*G. Latimer*).	Smithsonian Institution [P.].
b.	Ad. sk.	S. Domingo.	M. A. Sallé [C.].

4. Microcorax nasicus.

Corvus nasicus, *Temm. Pl. Col.* ii. pl. 413; *Gray, Gen. B.* ii. p. 315; *Baird, Brewer, and Ridgw. N. Amer. B.* ii. p. 234.
Corvus americanus, *Lembeye, Aves de Cuba*, p. 65.

Adult. Nostrils scarcely concealed by the short nasal bristles. Entirely violaceous black, the feathers smoky grey beneath the surface. Wing 11·00, tail 7·75, culmen 2·45, depth of bill ·80, tarsus 1·95, middle toe 1·50, graduation of tail about 1·00; wing-formula 4, 3, 5, 6, 2, first quill shortest. (*Baird, Brewer, and Ridgway, N. Amer. B.* p. 234.)

Hab. Cuba.

10. PHYSOCORAX.

Type.
Physocorax, *Bp. C. R.* xxxviii. p. 829 .... P. moneduloides.

Range. Confined to New Caledonia.

1. Physocorax moneduloides. (Plate I.)

Corvus moneduloides, *Less. Traité*, p. 329; *Gray, P. Z. S.* 1859, p. 163; *id. B. Trop. Isl. Pacific Ocean*, p. 24; *id. Hand-l. B.* ii. p. 14.
Physocorax moneduloides, *Bp. C. R.* xxxviii. p. 829; *Verr. et Des Murs, Rev. et Mag. de Zool.* 1860, p. 432.

Adult. General colour purplish black, of a silky texture, including the wings and tail, the outer primaries and tail-feathers being glossed with green; head and neck all round greenish black, gradually shading off on the latter to the silky purplish black of the upper surface; bill and legs black; iris grey. Total length 15 inches, culmen 1·95, wing 10·5, tail 7·9, tarsus 2·05.

Obs. The male is one third larger than the female. (*Verreaux and Des Murs.*)

Hab. New Caledonia.

a. Ad. st. New Caledonia. M. Verreaux.

11. GYMNOCORAX.

Type.
Gymnocorvus, *Less. Traité*, p. 327 (1831).............. G. senex.
Gymnocorax, *Sundev. Av. Meth. Tent.* p. 44............ G. senex.

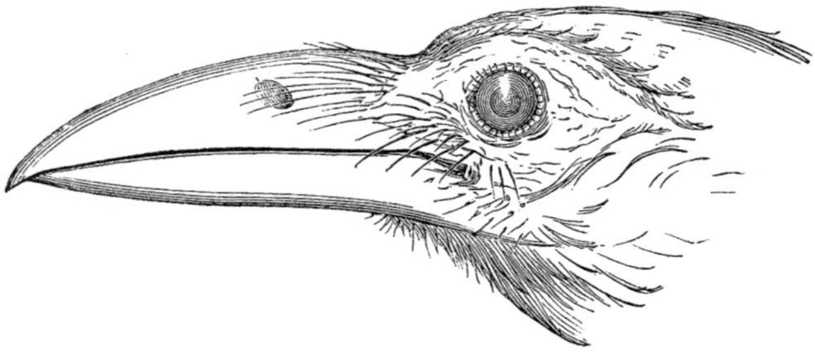

Head of adult, to show bare face.

Range. Confined to New Guinea.

1. Gymnocorax senex.

Corvus senex, *Less. Voy. Coquille, Ois.* p. 650, pl. 24; *Schl. Mus. P.-B.* Coraces, p. 36.
Gymnocorvus tristis, *Less. Traité*, p. 327.
Gymnocorvus senex, *Gray, Gen. B.* ii. p. 315; *Bp. Consp.* i. p. 387; *Gray, Cat. Mamm. &c. N. Guin.* p. 35; *id. Hand-l. B.* ii. p. 14.
Gymnocorax senex, *Sundev. Av. Meth. Tent.* p. 44.

Adult male. General colour above brownish black, rather greyer on the head and neck, and glossed with purple on the back, the wings and tail being purplish black, the latter ashy brown at the tip; primary-coverts also washed with grey; entire under surface of body ashy brown, the lower abdomen and under tail-coverts somewhat mixed with white; "bare skin of face whitish; bill whitish, washed with reddish; iris sky-blue" (*Bernstein*). Total length 21·5 inches, culmen 2·7, wing 13, tail 10·3, tarsus 2.

Adult female. Similar to the male, but rather larger. Total length 23 inches, culmen 2·9, wing 13·4, tail 10·5, tarsus 2·3.

Young female. General colour pale brown, inclining to yellowish brown on the head and neck; throat whity brown; rest of under surface drab-brown; quills and tail rather darker than the back; bill yellow; feet deep yellow, claws light horn-colour; eye bluish. Total length 23 inches, culmen 2·8, wing 13·2, tail 9·8, tarsus 2·3.

Hab. New Guinea and Salwati.

a. ♀ juv. sk.	Dorey.	A. R. Wallace, Esq. [C.].
b. Ad. sk.	Dorey.	A. R. Wallace, Esq. [C.].
c. ♀ ad. sk.	Dorey.	A. R. Wallace, Esq. [C.].
d. ♂ juv. sk.	Dorey.	A. R. Wallace, Esq. [C.].
e. ♂ ad. sk.	Salwati.	Mr. Hoedt [C.].

12. MACROCORAX.

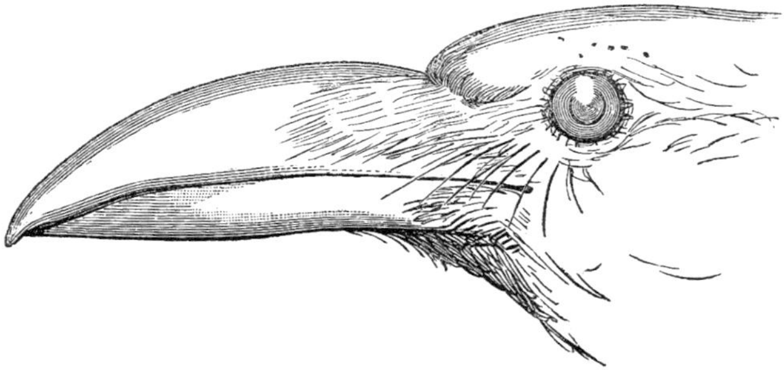

Head of *Macrocorax fuscicapillus.*

Range. Confined to the Aru Islands, Waigiou, and Ghemien.

1. Macrocorax fuscicapillus.

Corvus senex, *Schl. Bijdr. Dierk.* 4to, *Amsterd.* p. 10, pl. 3 (*nec Less.*).
Corvus orru, *Gray, Cat. Mamm. &c. N. Guin.* p. 35 (*nec Müll.*).
Corvus fuscicapillus, *Gray, P. Z. S.* 1859, p. 95; *Schl. Mus. P.-B.* Coraces, p. 22.
Gymnocorvus fuscicapillus, *Gray, Hand-l. B.* ii. p. 14.

Adult male (type of species). General colour above purplish black, inclining to finer and more bluish purple on the wings and tail, the gloss on the wings and on the rump having somewhat of a violet cast; head and neck all round of a deep umber-brown shade, blacker on the sides of the face and throat; the chest glossed with purple, and gradually shading off into the breast and abdomen, which are dull purplish black, as also the under wing- and tail-coverts; bill and feet black; iris azure-blue. Total length 23 inches, culmen 3·45, wing 13·5, tail 7·3, tarsus 2·4.

Young female. More dusky than the adult male, with less of a purple gloss and decidedly browner underneath; bill yellowish (in skin), inclining to black at the tip. Total length 22 inches, culmen 3·1, wing 13·1, tail 7·3, tarsus 2·4.

Obs. Professor Schlegel notes that the Waigiou birds are smaller than those from Aru. The Museum has a male from Waigiou which measures as follows:—total length 21 inches, culmen 3·2, wing 13, tail 7·5, tarsus 2·15. With the exception of the tarsus the dimensions are not much less.

Hab. Aru Islands, Waigiou, and Ghemien.

a. ♂ ad. sk.	Aru Islands, 1857.	A. R. Wallace, Esq. [C.].	(Type of species.)
b, c. ♀ juv. sk.	Aru Islands, 1857.	A. R. Wallace, Esq. [C.].	
d. ♂ juv. sk.	Waigiou, 1860.	A. R. Wallace, Esq. [C.].	

13. NUCIFRAGA.

	Type.
Nucifraga, *Briss. Orn.* ii. p. 58 (1760)	N. caryocatactes.
Caryocatactes, *Cuv. Règn. Anim.* i. p. 399 (1817) ..	N. caryocatactes.
Picicorvus, *Bp. Consp.* i. p. 384 (1850)	N. columbianus.

Range. Northern and Central Europe, Northern Asia, Himalaya Mountains, Western North America.

Key to the Species.

a. Outer tail-feather white, with black basal half; head brown or black; back brown, spotted with white.
- *a'.* Rump and upper tail-coverts uniform, with no mesial streaks of white; central primaries with a large patch of white near the base of the inner web.
  - *a''.* Outer tail-feathers tipped with white, forming a terminal band; sides of body brown, largely spotted with white, like the rest of the breast .... *caryocatactes*, p. 53.
  - *b''.* Outer tail-feathers white for their apical half; sides of body brown, with few or no spots of white, as also the lower back. *hemispila*, p. 54.
- *b'.* Rump and upper tail-coverts centred with triangular spots of white, like the rest of the back; primaries with no white patch on the inner web .... *multipunctata*, p. 55.

b. Outer tail-feather entirely white; entire head and upper parts uniform grey .... *columbiana*, p. 56.

1. Nucifraga caryocatactes.

Le Casse-noix, *Briss. Orn.* ii. p. 59; *Montb. Pl. Enl.* 60.
Corvus caryocatactes, *Linn. S. N.* i. p. 157; *Yarr. Brit. B.* ii. p. 122; *Nordm. in Démid. Voy. Russ. Mérid.* iii. p. 122.
Nucifraga guttata, *Vieill. N. Dict.* v. p. 354.
Caryocatactes maculatus, *Koch, Syst. baier. Zool.* p. 93.
Nucifraga caryocatactes, *Leach, Syst. Cat. Mamm. &c. B. M.* p. 18; *Temm. Man. d'Orn.* i. p. 117; *Naum. Vög. Deutschl.* ii. pl. 58. fig. 2; *Werner, Atlas*, Omnivores, pl. 10; *Gould, B. Eur.* iii. pl. 213; *Macg. Brit. B.* i. p. 583; *Gray, Gen. B.* ii. p. 313; *Blyth, Cat. B. Mus. A. S. B.* p. 90; *Bp. Consp.* i. p. 383; *Selys-Longchamps, Bull. Acad. R. Brux.* xi. p. 298, pl. iv.; *Schl. Vog. Nederl.* pl. 142; *Sundev. Sv. Fogl.* pl. xix. fig. 4; *Fritsch, Vög. Eur.* tab. 13. fig. 10; *Schl. Dier. Nederl.* iii. pl. 13. fig. 9; *id. Mus. P.-B.* Coraces, p. 37; *Degl. et Gerbe, Orn. Eur.* i. p. 207; *Doderl. Avif. Sicil.* p. 65; *Borggr. Vogelf. Norddeutschl.* p. 80; *Gould, B. Gt. Br.* iii. pls. 65, 66; *Gray, Hand-l. B.* ii. p. 9; *Von Tschusi, Verh. z.-b. Wien*, xxi. p. 83; *Vogel, Ber. St. Gall. Gesellsch.* 1871, ii. p. 156; *David, N. Arch. Mus.* vii. *Bull.* p. 9; *Swinh. P. Z. S.* 1871, p. 382; *Harting, Handb. Brit. B.* p. 118; *Salvad. Faun. Ital. Ucc.* p. 174; *Dresser, B. Eur.* part. xxviii.; *Irby, B. Gibr.* p. 129.
Caryocatactes nucifraga, *Nilss. Orn. Suec.* 1817, p. 90.
Nucifraga macrorhynchos, *Brehm, Lehrb. eur. Vög.* p. 104, *et Vög. Deutschl.* p. 181.
Nucifraga brachyrhynchos, *Brehm, t. c.* p. 103, *et V. D.* p. 182.
Nucifraga minor, *Brehm, Isis*, 1833, p. 970.
Nucifraga platyrhynchus, *Brehm, t. c.* p. 973.
Nucifraga hamata, *Brehm, t. c.* p. 975.
Caryocatactes guttatus, *Nilss. Skand. Faun.* i. p. 149; *Kjærb. Orn. Dan.* pl. xvi. fig. 1.
Caryocatactes caryocatactes, *Schl. Rev. Crit.* p. lv.
Nucifraga alpestris, *Brehm, Naum.* 1855, p. 274.
Nucifraga arquata, *Brehm, t. c.* p. 274.

Adult. Head uniform dark brown, the bristles over the nostrils and a few feathers just behind the eye streaked with white; hind neck and back brown, each feather spotted near the tip with a triangular white mark, small and less distinct on the lower back, disappearing on the rump, but again visible on the upper tail-coverts, which are glossed with green; wings blackish, the coverts glossed with green, the least ones with a large triangular spot of white, the greater series and the primary-coverts indistinctly tipped with white; a patch on the outer edge of the wing also white; the greater wing-coverts and secondaries glossed with green, the latter slightly tipped with white; tail greenish black, tipped with white, the latter much increasing in extent towards the outermost feathers; lores white; sides of face and of neck, as well as the under surface of the body, streaked with white, dilating into a broad triangular spot on the lower parts; throat brown, narrowly streaked with white; thighs brown, barred with white; under tail-coverts pure white; under wing-coverts brown, very broadly tipped with white in about equal proportions; inner lining of quills brown, with a large white patch near the base of the inner web of the median primaries; bill and feet black; iris dark brown. Total length 13 inches, culmen 1·95, wing 7·3, tail 5·2, tarsus 1·1.

Nestling. Head, neck, back, and underparts generally dull dirty brown, each feather having the shaft dirty white, which colour in many widens at the tip into a drop-shaped spot, all these feathers being very loose in texture; wings and tail as in the adult bird, but duller in colour; wing-coverts tipped with dull white; under tail-coverts dirty white. (*Dresser, B. Eur. l. c.*)

Hab. Northern and central portions of Europe and Northern Asia, ranging into Northern China.

a. Ad. st.	Kent.	William George, Esq. [P.].
b. Ad. sk.	Sweden, Oct. 19, 1859.	Prof. A. W. Malm [P.].
c. Ad. st.	Jura, France.	R. B. Sharpe, Esq. [P.]
d. Ad. st.	Germany.	Dr. Kaup [P.: *N. macrorhynchos*].
e. Ad. sk.	Germany.	Dr. Kaup [P.].
f. Jun. sk.	Germany.	Dr. Kaup [P.: *N. brachyrhynchus*].
g. Ad. st.	Russia.	Moscow Museum [P.].
h. Ad. sk.	Russia.	Moscow Museum [P.].
i. Ad. sk.	Kamtschatka.	Purchased.
k. Skeleton.	Norfolk.	Purchased.
l. Skull.	S. Germany.	Dr. Günther.

2. **Nucifraga hemispila.**

Nucifraga hemispila, *Vigors, P. Z. S.* 1830, p. 8; *Gould, Cent. Himal. B.* pl. 36; *Hodgs. in Gray's Zool. Misc.* p. 84; *Gray, Cat. Mamm. &c. Nepal Hodgs.* p. 102; *id. Gen. B.* ii. p. 313; *Blyth, J. A. S. B.* xv. p. 25; *id. Cat. B. Mus. A. S. B.* p. 90; *Bp. Consp.* i. p. 383; *Cab. Mus. Hein.* Th. i. p. 229; *Horsf. & Moore, Cat. B. Mus. E.I. Co.* ii. p. 563; *Jerd. B. Ind.* ii. p. 304; *Schl. Mus. P.-B.* Coraces, p. 38; *Gray, Hand-l. B.* ii. p. 9; *David, N. Arch. Mus. Bull.* p. 9; *Swinh. P. Z. S.* 1871, p. 382; *Hume, Nests & Eggs Ind. B.* p. 415.

Nucifraga immaculata, *Blyth, Ibis,* 1867, p. 36; *Gray, Hand-l. B.* ii. p. 9*.

Adult. Crown of head uniform black; back and scapulars brown, distinctly streaked on the mantle with longitudinal spots of white; the lower back, rump, and upper tail-coverts uniform brown, the latter glossed with greenish; wings glossy greenish black; tail greenish black, the two centre feathers not tipped with white, the rest with a broad white tip, gradually increasing in size towards the outermost, which are white for the terminal half; bristles over the nostrils blackish, streaked with white; lores white; feathers behind the upper part of the eye and sides of face and of neck

* *Nucifraga immaculata* was founded by the late Mr. Blyth on one of the Hodgsonian drawings (not *two*, as he says; at least I cannot find two, and Mr. Hodgson only refers to one). This drawing contains two figures (Nos. 519, 520, Hodgs. Cat.), one being a dark *N. hemispila*, and the other being a pale-coloured bird, the type of *N. immaculata*. No unspotted specimen exists in the collection; and I believe that it is an unfinished drawing of *N. hemispila*, the white spots not having been added.

blackish brown, longitudinally streaked with white; throat and under surface generally chocolate-brown, the chin narrowly streaked with white, the breast-feathers tipped with small triangular spots of white, the abdomen uniform brown; thighs blackish; under tail-coverts pure white; under wing-coverts dark brown, the outer ones tipped with white; inner lining of the wing dark brown, one or two of the middle primaries marked with white near the base of the inner web; bill horny black; feet black; iris reddish brown. Total length 16 inches, culmen 1·8, wing 8·4, tail 6·6, tarsus 1·75.

Young. Beak horn-coloured at tip; feathers of the upper surface with large triangular spots of pale ochraceous, as also the breast; chin and throat dull whitish, with distinct shaft-lines of white; wing-coverts narrowly tipped with ochraceous, otherwise the wings and tail much as in the adult bird; the abdomen more distinctly spotted with white, this part being uniform in the adult; white spot on the inner web of the primaries larger than in the old examples.

Hab. Himalaya Mountains.

a. Ad. sk.	Nepal.	B. H. Hodgson, Esq. [P.].
b, c, d. Ad. sk.	Nepal.	B. H. Hodgson, Esq. [P.].
e, f. Juv. sk.	Nepal.	B. H. Hodgson, Esq. [P.].
g. Ad. st.	Nepal.	B. H. Hodgson, Esq. [P.].
h. Imm. sk.	Nepal.	B. H. Hodgson, Esq. [P.].
i. Juv. sk.	Nepal.	B. H. Hodgson, Esq. [P.].
k. ♀ ad. sk.	Nepal.	Purchased.
l. Ad. sk.	Himalaya Mountains (*Capt. Tweedie*).	J. Gould, Esq.
m. Ad. sk.	Darjiling.	J. Gould, Esq.
n, o. ♂ ad. sk.	N.W. Himalayas.	Capt. Stackhouse Pinwill [P.].
p. ♀ ad. sk.	N.W. Himalayas.	Capt. Stackhouse Pinwill [P.].
q. Bones, various.	Nepal.	B. H. Hodgson, Esq.

3. Nucifraga multipunctata.

Nucifraga multipunctata, *Gould, P. Z. S.* 1849, p. 23; *id. B. Asia*, part i.; *Bp. Consp.* i. p. 383; *Gray, Hand-l. B.* ii. p. 9; *Henders. & Hume, Lahore to Yark.* p. 239.

Nucifraga multimaculata, *Jerd. B. Ind.* ii. p. 304.

Adult. Head dark chocolate-brown, with a few white streaks over the eye and on the hinder part of the nape; back and scapulars everywhere broadly streaked with white, the white occupying the greater part of the feather, the rump and upper tail-coverts as broadly streaked as the back; wings greenish black, the primaries rather browner, the lesser and median coverts with a broad triangular streak of white near the tip, the greater series as well as the primary-coverts and quills tipped with white; tail greenish black, tipped with white, gradually extending towards the outer feathers, which have very broad white tips, the outermost being edged with white along the exterior web; bristles hiding the nostrils black, narrowly streaked with white; lores white; ear-coverts, cheeks, and sides of neck dark brown, narrowly streaked with white, as also is the entire under surface of the body, the streaks being narrow on the throat but very broad on the breast, so that, as on the upper surface, the prevailing colour of the feathers is white, with a narrow edging of blackish brown; under tail-coverts pure white; under wing-coverts coloured like the breast, the inner lining of the quills ashy brown, with no white patch on the inner web of any of the primaries; bill horny brown; legs black. Total length 14 inches, culmen 1·8, wing 7·65, tail 6·2, tarsus 1·55.

Obs. As with the other species of Nutcracker, great difference is to be observed in the size of the bill. In the only two specimens the sex of which is known, the male has the longer and the thinner bill.

Hab. Cashmere.

a. Ad. sk.	[Cashmere.]	Dr. Jerdon [P.].
b. Ad. sk.	Cashmere.	R. B. Sharpe, Esq. [P.].
c, d. ♂ ♀ ad. sk.	Cashmere.	Purchased.
e. Ad. sk.	[Cashmere.]	J. Gould, Esq. (Type of species.)
f. ♀ ad. sk.	Cashmere, 12,000 feet, Oct. 1875.	Purchased.

4. Nucifraga columbiana.

Corvus columbianus, *Wils. Amer. Orn.* iii. p. 29, pl. 20. fig. 3; *Nutt. Man. Orn.* i. p. 218.
Corvus megonyx, *Wagler, Syst. Av.* Corvus, sp. 20.
Nucifraga columbiana, *Audub. Orn. Biogr.* iv. p. 459; *id. B. Amer.* iv. p. 127, pl. 235; *Gray, Hand-l. B.* ii. p. 9.
Picicorvus columbianus, *Bp. Consp.* i. p. 384; *Baird, B. N. Amer.* p. 573; *Lord, Natural. in Brit. Columb.* p. 126; *Dall & Bann. Tr. Chicago Acad.* i. p. 286; *Cooper, Orn. Calif.* p. 289; *Coues, Ibis,* 1872, p. 52; *id. Key N. Amer. B.* p. 162, fig. 104; *Baird, Brewer, & Ridg. N. Amer. B.* ii. p. 255, pl. 38. fig. 4; *Coues, B. North-west,* p. 207.

Adult female. General colour above ashy grey, paler on the head and neck, and becoming more dingy on the lower back and rump, the upper tail-coverts being slaty black; wings greenish black, the least coverts ashy grey, like the back, the outer secondaries broadly tipped with white; two centre tail-feathers greenish black, the next one on each side greenish black on the inner and white on the outer web, the four outer feathers pure white; bristles over the nostrils and lores ashy white, the sides of the face also washed with silvery white; chin white, rest of under surface ashy grey, darker on the thighs; under tail-coverts pure white; under wing-coverts blackish, as also the inner lining of the quills; bill and feet black. Total length 11 inches, culmen 1·6, wing 7·2, tail 4·7, tarsus 1·4.

Adult male. Similar to the female. Total length 10·5 inches, culmen 1·55, wing 7·15, tail 4·6, tarsus 1·4.

Young. Rather browner than the adult, and with a horny brown bill, only 1·2 inch in length; lower back with obsolete whitish edgings to the feathers; some of the small plumes on the upper and under wing-coverts also edged with whitish; on the breast-feathers a few remains of obsolete whitish edgings.

Hab. Rocky-Mountain region of Western North America.

a. Ad. st.	Black Hills, U. S. A.	Purchased.
b. Ad. sk.	Vancouver Island.	Dr. Lyall [P.].
c. Jun. sk.	Western side of the Rocky Mts., British Columbia.	J. K. Lord, Esq. [P.].
d, e. ♂ ♀ ad. sk.	British Columbia.	J. K. Lord, Esq. [P.].
f. Ad. sk.	49th parallel N. lat., Aug. 18, 1874 (*Dawson*).	N.-American Boundary Commission [P.].
g. ♀ ad. sk.	Buffalo Peak, Park Co. Colorado.	Purchased.

14. STREPERA.

		Type.
Strepera, *Less. Traité d'Orn.* p. 330 (1831)	....	S. anaphonensis.
Coronica, *Gould, Syn. B. Austr.* pt. i. (1837)	......	S. graculina.

Range. Confined to Australia and Van Diemen's Land.

Key to the Species.

a. Quills not broadly tipped with white; the white tip, if present, not exceeding 0·25 inch in breadth.
  a'. Base of tail white or pale rufous; primaries externally white at base, forming an alar speculum.
    a''. Base of tail white, as also the base of the primary-coverts .... *graculina*, p. 57.
    b''. Base of tail pale rufous; no white base to the primary-coverts .... *crissalis*, p. 58.
  b'. Base of tail blackish; no white speculum at the external base of the primaries.
    c''. Base of inner web of primaries white.
      a'''. Size large: tail more than 10 inches in length; colour black................ *arguta*, p. 59.
      b'''. Size smaller: wing not exceeding 8·5 inches in length.
        a[4]. Colour blackish brown .... *intermedia*, p. 59.
        b[4]. Colour grey .... { *cuneicaudata*, p. 60. / *plumbea*, p. 60. }
    d''. Primaries entirely black below; no white at all at base of inner web .... *melanoptera*, p. 61.
b. Quills with broad white tips nearly 1·5 inch in breadth; under tail-coverts blackish .... *fuliginosa*, p. 61.

1. Strepera graculina.

Corvus graculinus, *White, Voy. New S. Wales, App.* pl. to p. 251.
Coracias strepera, *Lath. Ind. Orn.* i. p. 173; *Daud. Traité*, ii. p. 267.
Noisy Roller, *Lath. Gen. Syn. Suppl.* ii. p. 121.
Le Grand Calibé ou Calibé bruyant, *Levaill. Ois. de Paradis*, pl. 24.
Gracula strepera, *Shaw, Gen. Zool.* vii. p. 462.
Corvus strepera, *Leach, Zool. Misc.* i. p. 78, pl. 86.
Barita strepera, *Temm. Man. d'Orn.* i. p. 41.

Cracticus streperus, *Vieill. Gal. Ois.* i. pl. 109; *Vig. & Horsf. Tr. Linn. Soc.* xv. p. 291.
Coronica strepera, *Gould, Syn. B. Austr.* part 1.
Strepera graculina, *Gray, List Gen. B.* 1841, p. 50; *Gould, B. Austr.* ii. pl. 42; *Gray, Gen. B.* ii. p. 302; *Blyth, Cat. B. Mus. A. S. B.* p. 105; *Cab. Mus. Hein.* Th. i. p. 227; *Gould, Handb. B. Austr.* i. p. 168; *Gray, Hand-l. B.* ii. p. 1.
Cracticus graculinus, *Schl. Mus. P.-B.* Coraces, p. 124.

Adult. General colour glossy black; tail black, the basal third of the feathers white, forming a broad band across the base of the tail, the feathers also tipped with white, which extends rather further up the inner web than the outer one; wing black, with a very large white alar speculum, formed by the bases of the primaries being white; under surface of body black, the under tail-coverts white; bill and feet black; iris beautiful yellow. Total length 18·5 inches, culmen 2·5, wing 11·2, tail 8·9, tarsus 2·15.

Obs. A Queensland specimen is decidedly smaller in every respect, and has a clearer shade of grey on the underparts. All the markings, however, are similar to those of a male specimen from New South Wales; and the probability is that the bird in question is a female. Mr. Gould informs us that the sexes are "similar in plumage, but that the female is always a trifle less in all her admeasurements, and the young birds have the corners of the mouth more fleshy and of a brighter yellow than the adults." The following are the dimensions of the two specimens in question:—

		Total length. in.	Wing. in.	Tail. in.	Tarsus. in.
a. ♂ ad.	New South Wales	18	10·5	8·0	2·15
b. [♀] ad.	Queensland	16·5	9·3	7·4	1·95

Hab. Australia.

a. Ad. st.	Australia.	
b. ♂ ad. sk.	Brisbane Water, New South Wales, Aug. 1847 (*J. Macgillivray*).	Voyage of the 'Rattlesnake.'
c. Ad. sk.	Queensland.	Purchased.
d, e. Skeletons.		Zoological Society.

2. Strepera crissalis. (Plate II.)

Strepera crissalis, *Gould, MS.*

Adult male. General colour black, with a slight gloss; under surface of body blackish, slightly glossed with grey, especially on the abdomen; upper and under wing-coverts black; quills also black, with broad whitish bases, especially to the primaries, where they form a large alar speculum; under tail-coverts buffy white, the longer ones pale rufous; tail black, with a broad basal band of pale rufous, all the feathers tipped with white, broadly on the inner web, and more narrowly on the outer. Total length 16·5 inches, culmen 2·45, wing 9·4, tail 7·6, tarsus 2.

Hab. Lord Howe's Island.

a. ♂ ad. sk.	Lord Howe's Island, Sept. 12, 1853 (*J. Macgillivray*).	J. Gould, Esq. [P.]. (Type of species.)

3. Strepera arguta.

Strepera arguta, *Gould, P. Z. S.* 1846, p. 19; *id. B. Austr.* ii. pl. 44; *Bp. Consp.* i. p. 367; *Gould, Handb. B. Austr.* i. p. 171; *Gray, Hand-l. B.* ii. p. 1.
Cracticus argutus, *Schl. Mus. P.-B.* Coraces, p. 123.

Young. Ashy brown, the feathers which are developing black on the upper parts, greyer on the lower; quills narrowly tipped with white; under tail-coverts white; tail as in adult.

Adult male. General colour dull blackish, with very slight gloss and rather browner on the sides of the neck; under surface of body blackish, the narrow shaft-streaks on the breast-feathers slightly metallic; wings black, the secondaries narrowly tipped with white, the primaries white for a considerable extent of the inner web; tail blackish, washed with grey externally, the two centre feathers with no perceptible white tip, but all the others broadly tipped with white on the inner web, extending on to the outer one at the tip; under tail-coverts white; bill and feet black; corner of the mouth yellow; iris orange-yellow. Total length 21 inches, culmen 2·75, wing 11 9, tail 9·8, tarsus 2·8.

Adult female. Much greyer than the male. Total length 21 inches, culmen 2·65, wing 12, tail 10·5, tarsus 2·85.

Obs. If we consider the big *Strepera arguta* of Van Diemen's Land to be the typical species of this group of the genus *Strepera*, we find three very closely allied species, whose exact relations time and a larger series of specimens are necessary to determine. I have separated the smaller form from Port Lincoln as *Strepera intermedia*, as well as the South-Australian *S. melanoptera*; but whether these both grade into one another or into *S. arguta* must be proved by the comparison of larger series. They seem to me at least to have distinct habitats.

Hab. Van Diemen's Land.

a. Juv. sk.	Spring Hill, interior of Van Diemen's Land, Jan. 2, 1838.	J. Gould, Esq. [C.].
b. ♀ ad. sk.	Norfolk, Van Diemen's Land, December 4, 1838.	J. Gould, Esq. [C.].
c. ♂ ad. st.	George Town, Van Diemen's Land, May 1839.	J. Gould, Esq. [C.].
d. Ad. sk.	Van Diemen's Land.	J. Gould, Esq. [C.].
e. Ad. sk.	Van Diemen's Land.	J. Gould, Esq. [C.].

4. Strepera intermedia.

Adult. General colour above and below blackish brown, blacker on the lores and region of the eye, and slightly shaded with grey on the breast and abdomen; vent and under tail-coverts white; thighs blackish brown; tail blackish, all the feathers tipped with white, the two centre ones very slightly, all the others very broadly, the white apical spot wider on the inner web than on the outer; wings blackish brown, the primary-coverts narrowly tipped with white, the quills blackish, also narrowly margined with white at the tip, the primaries broadly white at the base; bill and legs black. Total length 20 inches, culmen 2·6, wing 10·1, tail 8·4, tarsus 2·6.

Hab. South Australia: Port Lincoln, &c.

a. Ad. sk.	Port Lincoln.	J. Gould, Esq. [C.].
b. Ad. sk.	Port Lincoln.	J. Gould, Esq. [C.].

5. Strepera cuneicaudata.

Cracticus cuneicaudatus, *Vieill. N. Dict. d'Hist. Nat.* v. p. 356 (1816).
Barita anaphonensis, *Temm. Pl. Col. texte to art.* Barita (1824).
Strepera strepera, *Less. Traité,* p. 330.
Barita anaphonensis, *Less. Traité,* p. 345, pl. 47. fig. 1.
Strepera versicolor, *Gray, Gen. B.* ii. p. 302.
Gymnorhina anaphonensis, *Gray, Gen. B.* ii. p. 302.
Strepera anaphonensis, *Gould, B. Austr.* ii. pl. 45; *Bp. Consp.* i. p. 368; *Cab. Mus. Hein.* Th. i. p. 227; *Reich. Vög. Neuholl.* p. 237; *Gould, Handb. B. Austr.* i. p. 173; *Gray, Hand-l. B.* ii. p. 1.
Cracticus anaphonensis, *Schl. Mus. P.-B.* Coraces, p. 123.

Adult. Above brownish grey, clearer grey on the lower back and rump; the wing-coverts with indistinct tips of white; quills blackish brown, externally edged with ashy white, inclining to pure white on the secondaries, which have the tips also white, the primaries having the inner web white for about half their length; tail-feathers blackish brown, broadly edged with grey, and tipped with white at the extremity of each of the inner webs; lores, feathers round the eye, cheeks, ear-coverts, and chin much darker brown; rest of under surface ashy brown, with obsolete lighter streaks on the centre of the feathers; the thighs much deeper brown; under tail-coverts white; under wing-coverts ashy grey; lower surface of quills dark brown, the basal half of the primaries white. Total length 19 inches, culmen 2·5, wing 11, tail 8·7, tarsus 2·55.

Hab. South Australia.

a. Ad. st.	Australia.	Zoological Society.
b. Ad. sk.	New South Wales.	J. Gould, Esq.

Subsp. *α.* Strepera plumbea.

Strepera plumbea, *Gould, P. Z. S.* 1846, p. 20.

Adult. Of a deep leaden grey, very little paler below, the feathers round the eye, lores, sides of face, chin, and thighs blackish brown; lower back, rump, and upper tail-coverts rather clearer leaden grey, the tail-feathers being edged with this colour; quills as in *S. cuneicaudata,* but much blacker, as also the primary-coverts; iris brown. Total length 19 inches, culmen 2·7, wing 11·6, tail 9·5, tarsus 2·75.

Hab. Western Australia.

a. Ad. sk.	Western Australia.	J. Gould, Esq. [C.].
b. Ad. sk.	York, West Australia.	J. Gould, Esq. [C.].

6. Strepera melanoptera.

Strepera melanoptera, *Gould, P. Z. S.* 1846, p. 20; *Bp. Consp.* i. p. 368.

Adult male. General colour brownish black, rather lighter below, and darker on the crown and face; under tail-coverts pure white; all the tail-feathers with a broad white tip to the inner web, extending slightly onto the tip of the outer one; under surface of wings ashy black, with no white bases or tips to the feathers. Total length 19 inches, culmen 1·85, wing 11·2, tail 9·5, tarsus 2·6.

Obs. This species has been united by Mr. G. R. Gray, and even by Mr. Gould himself, to *S. arguta.* I have examined most carefully the series mentioned by the latter gentleman in his 'Handbook,' and I cannot see any variation in the amount of white in the wing of *S. arguta*; and I believe that the last-named species is the Tasmanian Hill-Crow, and that its place on the continent is occupied by *S. melanoptera.*

Hab. South Australia.

a. ♂ ad. sk.	South Australia, June 18, 1839.	J. Gould, Esq. [C.]. (Type of species.)
b. Ad. sk.	South Australia.	Purchased.
c. Ad. sk.	South Australia.	Purchased.

7. Strepera fuliginosa.

Cracticus fuliginosus, *Gould, P. Z. S.* 1836, p. 106; *Schl. Mus. P.-B.* Coraces, p. 125.
Coronica fuliginosa, *Gould, Syn. B. Austr.* part 1.
Strepera fuliginosa, *Gray, Gen. B.* ii. p. 302; *Gould, B. Austr.* ii. pl. 43; *Bp. Consp.* i. p. 367; *Gould, Handb. B. Austr.* i. p. 171; *Gray, Hand-l. B.* ii. p. 1.

Adult. General colour above and below glossy black, the under surface very slightly greyer; primaries with a faint white mark at the base of the inner web, the outer ones with a broad band of white at the tip; tail-feathers blackish, externally shaded with grey, and having a large spot of white at the tip of the inner web, extending very slightly onto the outer one; under tail-coverts blackish, like the rest of the under surface; bill and feet black; iris bright yellow. Total length 19 inches, culmen 2·75, wing 11·4, tail 8·5, tarsus 2·5.

Hab. Van Diemen's Land.

a. ♀ ad. sk.	Van Diemen's Land.	Antarctic Expedition [C.].
b. Ad. st.	Van Diemen's Land.	
c. Ad. sk.	Van Diemen's Land.	J. Gould, Esq. [C.].

15. PICA.

	Type.
Pica, *Briss. Orn.* ii. p. 35 (1760)	P. pica.
Cleptes, *Gambel, Journ. Acad. N. Sci. Philad.* 1847, p. 47	P. pica.

Range. Europe and Northern Asia, including the Himalayas; China and Formosa; North America.

Key to the Species.

a. Beak black.
  a'. Larger: rump white or greyish white; no bare spot behind the eye .... { *pica*, p. 62. / *leucoptera*, p. 66.
  b'. Smaller: rump black like the rest of the back; bare spot behind the eye blue .... *mauritanica*, p. 66.
b. Beak yellow; bare spot behind the eye yellow .. *nuttalli*, p. 66.

1. Pica pica.

La Pie, *Briss. Orn.* ii. p. 35; *Montb. Pl. Enl.* iii. pl. 488.

Corvus pica, *Linn. S. N.* i. p. 157; *Wils. Am. Orn.* iv. p. 75, pl. xxxv. fig. 2; *Temm. Man. d'Orn.* i. p. 113; *Naum. Vög. Deutschl.* ii. Taf. 56. fig. 2; *Werner, Atlas*, Omnivores, pl. 7; *Audub. Orn. Biogr.* iv. p. 408, pl. ccclvii.; *Nordm. in Démid. Voy. Russ. Mérid.* iii. p. 116; *Schl. u. Susem. Vög. Eur.* ii. Taf. 4. fig. 2; *Schl. Vog. Nederl.* pl. 141.

Corvus rusticus, *Scop. Ann.* i. p. 38.

Pica melanoleuca, *Vieill. N. Dict. d'Hist. Nat.* xxvi. p. 121; *Wagler, Syst. Av.* Pica, sp. 1; *Audub. B. Amer.* roy. 8vo, iv. p. 99, pl. ccxxvii.; *Macgill. Br. B.* i. p. 562.

Corvus hudsonius, *Sabine, App. Narr. Frankl. Journ.* p. 671.

Pica albiventris, *Vieill. Faun. Franç.* p. 119.

Pica europæa, *Boie, Isis*, 1826, p. 551; *Sundev. Sv. Fogl.* pl. xix. fig. 3.

Pica germanica, *Brehm, Vög. Deutschl.* p. 177.

Pica septentrionalis, *Brehm, t. c.* p. 178.

Pica hiemalis, *Brehm, t. c.* p. 178.

Garrulus picus, *Temm. Man. d'Orn.* iii. p. 63; *Drummond, Ann. N. H.* 1843, vol. xii. p. 414.

Pica hudsonica, *Bp. Comp. List B. Eur. & N. Amer.* p. 27; *Gray, Gen. B.* ii. p. 314; *Bp. Consp.* i. p. 383; *Baird, B. N. Amer.* p. 576, pl. xxv.; *Dall & Bann. Tr. Chic. Acad.* i. p. 286; *Gray, Hand-l. B.* ii. p. 10.

Pica caudata, *Keys. u. Blas. Wirb. Eur.* p. 45; *Gould, B. Eur.* iii. pl. 216; *Yarr. Br. B.* ii. p. 107; *Gray, Gen. B.* ii. p. 314; *Blyth, J. A. S. B.* xv. p. 26; *id. Cat. B. Mus. As. Soc. Beng.* p. 9; *Hutton, J. A. S. B.* xvi. p. 778; *Bp. Consp.* i. p. 382; *Midd. Sibir. Reis. Zool.* p. 158; *Jaub. et Barth. Lapomm. Rich. Orn.* p. 101; *Fritsch, Vög. Eur.* tab. 27. fig. 6; *Schrenck, Reis. Amurl. Vög.* p. 322; *Linderm. Vög. Griechenl.* p. 69; *Filippi, Viagg. Pers.* p. 350; *Radde, Reis. Sibir. Vög.* p. 206; *Gould, B. Gt. Br.* iii. pl. 216; *Bettoni, Ucc. Lomb.* pl. 101; *Doderl. Avif. Sicil.* p. 66; *Gray, Hand-l. B.* ii. p. 10; *Salvad. Faun. Ital. Ucc.* p. 175; *Shelley, B. Egypt*, p. 160.

Pica bottanensis, *Deless. Rev. Zool.* ii. p. 100; *Gray, Gen. B.* ii. p. 314; *Blyth, Cat. B. Mus. A. S. B.* p. 91; *Horsf. & Moore, Cat. B. Mus. E.I. Co.* ii. p. 551; *Jerd. B. Ind.* ii. p. 305; *Gould, B. Asia*, part xv.; *Gray, Hand-l. B.* ii. p. 10.

Pica megaloptera, *Blyth, J. A. S. B.* xi. p. 193.
Pica media, *Blyth, J. A. S. B.* xiii. p. 393; *id. Cat. B. Mus. A. S. B.* p. 91, *App.* p. 19; *Horsf. & Moore, Cat. B. Mus. E.I. Co.* ii. p. 554; *Gray, Hand-l. B.* ii. p. 10; *David, N. Arch. Mus.* vii. *Bull.* p. 9; *Swinh. P. Z. S.* 1871, p. 382.
Pica varia, *Schl. Rev. Crit.* p. 54; *id. Dier. Nederl.* pl. xiii. figs. 7, 7*a*; *id. Mus. P.-B.* Coraces, p. 39; *Kjærb. Orn. Dan.* pl. xii. fig. 2.
Pica sericea, *Gould, P. Z. S.* 1845, p. 2; *Gray, Gen. B.* ii. p. 314; *Bp. Consp.* i. p. 383.
Cleptes hudsonicus, *Gambel, Journ. Acad. N. Sci. Philad.* 1847, p. 47.
Pica tibetana, *Hodgs. Ann. N. H.* (2) iii. p. 203.
Pica varia japonica, *Schl. Faun. Jap.* Aves, p. 81.
Pica japonica, *Bp. Consp.* i. p. 383; *Gray, Hand-l. B.* ii. p. 10.
Pica butanensis, *Bp. Consp.* i. p. 383.
Pica bactriana, *Bp. Consp.* i. p. 383; *Horsf. & Moore, Cat. B. Mus. E.I. Co.* ii. p. 550; *Gould, B. Asia*, pt. xv.; *Gray, Hand-l. B.* ii. p. 10; *Henders. & Hume, Lahore to Yark.* p. 240; *Hume, Nests and Eggs Ind. B.* p. 416; *Scully, Str. F.* 1876, p. 158.
Pica chinensis, *Bp. Consp.* i. p. 383.
Cleptes pica, *Cab. Mus. Hein.* Th. i. p. 229.
Pica vulgaris, *Brehm, J. f. O.* 1858, p. 173.
Pica leucoptera, *Gould, B. Asia*, pt. xiv.; *Gray, Hand-l. B.* ii. p. 10.
Pica rustica, *Dresser, B. Eur.* part xxii.; *Irby, B. Gibr.* p. 129; *Dresser, Ibis*, 1875, p. 238; *Blanf. E. Persia*, p. 264.
Pica melanoleuca, var. hudsonica, *Coues, Key N. Amer. B.* p. 164.
Pica caudata, var. bactriana, *Severtz. Turkest. Jevotn.* p. 64.
Pica caudata, var. hudsonica, *Baird, Brewer, & Ridgw. N.-Amer. B.* p. 266.

Adult male (Sweden). Above glossy black, with a distinct shade of shining deep green on the crown and back, rather coppery on the former; across the lower back a slight indication of an ashy-white bar, some of the feathers being mottled with that colour; rump and upper tail-coverts purplish black; shoulders white, some of the lower feathers shaded with ashy brown; wings blue, the outer edge of the wing, the primary-coverts, and external web of primaries green; the secondary quills with an internal shade of bright green along the outer web near the base, the inner secondaries bright blue, shading into green on the inner web; the inner webs of the primaries white for the greater part of their extent, the tips only blackish for about an inch, the white gradually decreasing in extent on the secondaries, on which the dusky bases to the feathers encroach and confine the white, which on the last one marked with white is reduced to an ovate patch; tail dark green, tipped with dusky black, before which the green colour of the tail shades into coppery colour, and thence through purplish red to purplish blue, causing a strong subterminal band on all excepting the outermost feather, which is dusky blackish, externally deep blue, inclining to purple near the base; head and neck all round and chest glossy purplish black, the throat streaked with numerous stripes of grey; rest of under surface pure white, the lower abdomen and under tail-coverts purplish black, the thighs and under wing-coverts black, shaded with green; bill and legs black; iris brown. Total length 16 inches, culmen 1·5, wing 7·9, tail 9·8, tarsus 1·95.

Obs. Having examined with care the large series of Magpies possessed by the Museum, I can only recognize three species; and this is the conclusion arrived at by Mr. Dresser, who also worked with the same series. The following is a synopsis of the measurements, not only of the birds in the National collection, but also of those lent to me by Mr. Swinhoe and Dr. Anderson:—

	Culmen. in.	Wing. in.	Tail. in.	Tarsus. in.
1. Great Britain	1·4 –1·5	7·6–7·8	10·0–10 2	2
2. Sweden	1·5	7·4–7·8	9·3– 9·5	1·85–1·95
3. Belgium	1·5	7·7	10·3	1·9
4. France	1·5	8·1	11·3	2
5. Silesia	1·5	7·7	10·1	2
6. Sicily	1·55	8·2	10·8	2
7. Turkey..........	1·55	7·6	9·5	1·9
8. Erzeroum........	1·4	7·65	10·2	1·9
9. Persia	1·5	8·4	11·5	2
10. Cashmere........	1·4 –1·6	7·8–8·4	9·7–10·5	1·7 –1·9
11. Central Asia	1·35–1·5	8·2–9·3	9·7–12	1·85–2
12. Japan	1·4 –1·45	7·5–8	9·2–10·2	1·9 –2
13. N. China........	1·35–1·55	7·9–8·6	9·3–10	1·85–2·2
14. Formosa	1·45–1·55	7·9–8·6	9·0–9·5	1·95–2·1
15. Hainan..........	1·5 –1·6	7·4–7·9	8·5–9·4	1·85–2
16. Szechuen	1·4	7·8	10	1·9
17. Burmah, Bhamau.	1·45–1·5	7·9–8·3	9·0–10·5	1·95–2·05
18. Kakhyen hills....	1·5 –1·6	8·2–8·7	10·4–10·5	2·15–2·2
19. North America ..	1·35–1·5	7·6–8·5	9·5–12·3	1·7 –2

I believe that the above table of measurements effectually shows that there is no dependence to be placed on the size of the birds to warrant their being subdivided into seven species, as is done again by Mr. G. R. Gray in the 'Hand-list.' The distinctions of colour pointed out by Mr. Gould and others are equally worthless when a sufficient number of specimens have been got together. The white on the wing varies also to such an extent that *P. leucoptera* of Central Asia can only be regarded as a barely recognizable race; and Dr. Severtzoff tells me that some Magpies killed by himself in Central Russia exhibited a tendency to a greater amount of white on the primaries.

A curious intermediate race occurs in Southern Spain, whence Mr. Howard Saunders possesses an interesting series, which show a tendency to a wholly black rump and a bare spot behind the eye. Two Portuguese specimens also exhibit these peculiarities; but even here the characters are not constant; for in some specimens an ashy grey band is apparent on the rump, though this is never so conspicuous as in true *Pica pica*, the nearest approach to any appearance of white being in a Seville male, which has a slight shade near the base of some of the rump-feathers; a bird from Granada also has a certain appearance of white on these plumes.

a. Ad. sk.	England.	J. Gould, Esq. [C.].
b. ♂ ad. st.	Devonshire.	Col. Montagu [P.].
c. ♂ ad. sk.	Hampshire, March 1862.	J. Gould, Esq. [C.].

d. ♀ ad. sk.	Avington, Hants, May 19, 1876.	E. Shelley, Esq. [P.].
e, f. Pull. st.	Cambridgeshire.	Mr. J. Baker [C.].
g. Ad. st.	Cambridgeshire.	Mr. J. Baker [C.].
h. Ad. sk.	Sweden (*H. Wheelwright*).	J. Gould, Esq.
i. ♂ ad. sk.	Wermland, April 10, 1874.	Purchased.
k. ♀ ad. sk.	Wermland, April 1874.	R. B. Sharpe, Esq. [P.].
l. Ad. sk.	Belgium, Jan. 3, 1845.	Brussels Museum [P.].
m. Pull. sk.	Belgium.	Brussels Museum [P.].
n. Ad. st.	Montgeron, Seine-et-Oise, France.	R. B. Sharpe, Esq. [P.].
o. ♂ ad. sk.	Near Seville, Spain, Dec. 1871.	Lieut.-Colonel Irby [P.].
p, q. Ad. sk.	Oporto, Portugal.	
r. ♂ ad. sk.	Turkey.	T. Robson, Esq. [P.].
s. Ad. sk.	Erzeroum.	
t. ♂ ad. sk.	Razin, S.E. of Karman, S.E. Persia, April 30, 1872.	W. T. Blanford, Esq. [C.].
u. ♂ ad. sk.	Shiraz, June 1869.	W. T. Blanford, Esq. [C.].
v, w. ♂ ♀ ad. sk.	Valley of Aster, Cashmere, Sept. 1875, "12,000 feet."	Purchased.
x. Ad. sk.	India.	Hon. E. India Co. [P.].
y. Ad. sk.	Bhamau, Upper Burmah, Feb. 11, 1868.	Dr. J. Anderson [C.].
z. Ad. sk.	Bhamau, Feb. 6, 1868.	Dr. J. Anderson [C.].
a'. ♂ ad. sk.	Ponsee, Kakhyen hills, March 29, 1868.	Dr. J. Anderson [C.].
b'. ♀ ad. sk.	Onon river, Jan. 12, 1873.	Dr. Dybowski [C.].
c'. ♂ ad. sk.	Kultuk, S. Lake Baikal, Dec. 25, 1871.	Dr. Dybowski [C.].
d', e'. Pull. sk.	China.	Capt. Blakiston [C.].
f'. Ad. sk.	China.	Horticultural Soc. [P.].
g'. Ad. st.	China.	Sir A. Smith [P.].
h'. Ad. sk.	China.	J. Gould, Esq. (Type of *P. sericea*.)
i'. Ad. sk.	Fokien, China.	J. Gould, Esq.
k'. ♂ ad. sk.	Shanghai.	J. Gould, Esq.
l'. Ad. sk.	Hainan.	R. Swinhoe, Esq. [P.].
m'. Ad. sk.	Japan.	
n'. Ad. st.	Japan.	Purchased.
o'. Ad. st.	North America.	Sir J. Richardson [P.].
p'. Ad. sk.	North America (*Blakiston*).	J. Gould, Esq.
q'. Ad. sk.	Fort Dufferin, Manitoba (*Dawson*).	N. A. Boundary Commission [P.].
r'. Ad. sk.	49th parallel (*Dawson*).	N. A. Boundary Commission [P.].
s', t', u'. Ad. sk.	British Columbia.	J. K. Lord, Esq. [C.].
v', w'. ♂ ♀ ad. sk.	West side of Rocky Mountains.	J. K. Lord, Esq. [P.].
x', y'. Ad. sk.	N. California.	
z'. Skeleton.	England.	
a''. Juv. skeleton.	England.	
b''. Sternum.	England.	G. Ingall, Esq. [P.].
c'', d''. Skulls.	S. Germany.	Dr. Günther.

Subsp. α. Pica leucoptera.

Similar to *P. pica*, but having the white on the quills extended sometimes to the very tip of the inner web, never reaching less than to 0·3 inch of the tip; on the throat the base of the feathers white; tail in adult bird coppery green. (For measurements, see *résumé* of specimens, *anteà*, p. 64.)

Hab. Turkestan, Thibet, and Ladák.

a. Ad. sk.	[E. Siberia] *.	J. Gould, Esq.	(Type of species.)
b. ♂ ad. sk.	Turkestan (*Severtzoff*).	J. Gould, Esq.	
c. Ad. sk.	Thibet (*Hay*).	J. Gould, Esq.	
d. Ad. sk.	Ladák.	Purchased.	

2. Pica mauritanica.

Pica mauritanica, *Malh. Mém. Soc. H. N. Moselle, Metz*, 1845, p. 7; *Levaill. jun. Exp. Sci. Algér. Ois.* pl. 8; *Bp. Consp.* i. p. 383; *Schl. Mus. P.-B.* Coraces, p. 41; *Loche, Expl. Sci. Algér. Ois.* p. 41; *Gray, Hand-l. B.* ii. p. 13; *Dresser, B. Eur.* part xxii.; *Irby, B. Gibr.* p. 129.

Adult. Above glossy black, with a very distinct shade of oil-green on the back and scapulars; lower back, rump, and upper tail-coverts purplish black; scapulars pure white; entire crown and nape dull metallic oil-green, glossed with purple shades; wing-coverts dull oil-green, the greater series glossed with purplish blue near the base; quills purplish blue, internally more or less golden-green, the inner webs blackish, excepting the innermost secondaries, which are entirely purplish blue, the primaries externally dull green, the inner webs for the most part white, the bases and tips blackish, the latter glossed with green; tail dull green, purplish at tip and inclining to purple or to blue on the edges of the feathers, with a faint subterminal shade of copper, very indistinct; entire sides of head and neck, throat, and fore neck dull greenish black, the throat with hair-like whitish shafts; breast white; lower abdomen, vent, under wing- and under tail-coverts and thighs black; behind the eye a bare space of bright cobalt-blue; bill and legs black; iris brown. Total length 17 inches, culmen 1·5, wing 6·2, tail 9·5, tarsus 1·9.

Hab. Algeria and Morocco.

a. Ad. sk.	Algeria.	J. Gould, Esq.
b. Ad. sk.	Algeria.	G. Munby, Esq. [P.].
c. Ad. st.	Algeria.	Purchased.
d. Ad. sk.	Algeria.	Purchased.

3. Pica nuttalli.

Corvus nuttalli, *Audub. Orn. Biogr.* iv. p. 450; *id. B. Amer.* pl. 362.

Pica nuttalli, *Audub. B. Amer.* roy. 8vo, iv. p. 104, pl. ccxxviii.; *Gray, Gen. B.* ii. p. 314; *Bp. Consp.* i. p. 303; *Newb. Pacif. Rail. Rep.* vi. p. 84, pl. 26; *Baird, B. N. Amer.* p. 578, pl. 26; *Gray, Hand-l. B.* ii. p. 10; *Cooper, Orn. Calif.* p. 295.

* Locality doubtless erroneous.

Cleptes nuttalli, *Gambel, Journ. Acad. N. Sci. Philad.* 1847, p. 46.
Pica melanoleuca, var. nuttalli, *Coues, Key N. Amer. B.* p. 164; *id. B. North-west*, p. 213.
Pica caudata, var. nuttalli, *Baird, Brewer, & Ridgw. N. Amer. B.* ii. p. 270, pl. 38. fig. 2.

Adult. Above velvety black, the crown glossed with bronzy green, the back with dark green, the hind neck, rump, and upper tail-coverts with purple; across the lower back a band of dusky grey; scapulars pure white; wings glossy black, with dark green reflections on the least wing-coverts, primary-coverts, and primaries; the greater coverts intense purple, varying with a shade of blue, the secondaries purplish blue, varying to deep green with the light, and shading into dull bronzy purple on the outer webs of the innermost, which are distinctly green interiorly; all the primaries white for the greater part of the inner web; centre tail-feathers dull bronzy purple, shaded with blue on the inner web and with green at the base, all the other feathers dull greenish black, the outer webs bronzy purple on the outer edge, shading through indigo to bright green, and thence to copper and bright purple, these varying shades intensified towards the upper portion of the feather, the extreme end of the web being green; sides of head, neck all round, throat, and chest velvety black, glossed slightly with purple, rest of under surface white; thighs greenish black; under wing- and tail-coverts purplish black, the edge of the wing greenish; bill bright yellow; legs black. Total length 13·5 inches, culmen 1·35, wing 7·65, tail 9·5, tarsus 1·65.

Obs. None of the Museum specimens having their sexes accurately determined, I have described the brightest-looking with pure yellow bill as being probably the adult male. The bronze-coloured tail-feathers look like a sign of immaturity, as all but one of the other specimens have the centre tail-feathers green, and the white on the quills reaches within half an inch of the extremities of the feathers; one specimen has one of the long median tail-feathers green and the other bronzy.

Hab. California.

a.	Ad. sk.	Santa Clara, Co. California (*Dr. J. G. Cooper*).	J. Gould, Esq.
b, c.	Ad. sk.	California.	Purchased.
d–i.	Ad. sk.	California.	Admiral Sir E. Belcher P.].

16. CYANOPOLIUS.

	Type.
Cyanopolius, *Bp. Report Br. Assoc.* 1849, p. 75	C. cooki.
Cyanopica, *Bp. P. Z. S.* 1850, p. 86	C. cyanus.
Dolometis, *Cabanis, Mus. Hein.* Th. i. p. 222 (1850)	C. cyanus.

Range. Europe: Spain and Portugal. Asia: Eastern Siberia, North China, and Japan.

Key to the Species.

a. With distinct white tips to the two centre tail-feathers *cyanus*, p. 68.
b. With no white tips to the centre tail-feathers .... *cooki*, p. 69.

1. Cyanopolius cyanus.

Corvus cyanus, *Pall. Reis. Russ. Reichs.* iii. *App.* p. 694; *id. Zoogr. Rosso-Asiat.* p. 391, pl. xvi.; *Temm. & Schl. Faun. Jap.* Aves, pl. xlii.

Blue Crow, *Lath. Gen. Syn.* i. part 1, p. 394.

La Pie bleue à tête noire, *Levaill. Ois. d'Afr.* ii. pl. 58.

Pica melanocephalos, *Wagler, Syst. Av.* Pica, sp. 7; *Gray, Hand-l. B.* ii. p. 10.

Pica cyana, *Gray, Gen. B.* ii. p. 314; *Schl. Faun. Japon.* p. 81; *Schrenck, Reis. Amurl.* p. 318; *Radde, Sibir. Reis.* p. 205; *Schl. Mus. P.-B.* Coraces, p. 41; *Gray, Hand-l. B.* ii. p. 10; *David, N. Arch. Mus.* vii. *Bull.* p. 9.

Cyanopica vaillanti, *Bp. P. Z. S.* 1850, p. 85.

Cyanopica pallasii, *Bp. t. c.* p. 85.

Cyanopica cyaneus, *Bp. Consp.* i. p. 382.

Cyanopica melanocephalus, *Bp. t. c.* p. 382.

Dolometis cyana, *Cab. Mus. Hein.* Th. i. p. 222.

Dolometis melanocephala, *Cab. Mus. Hein.* Th. i. p. 222.

Cyanopica cyana, *Swinh. P. Z. S.* 1861, p. 336.

Cyanopolius cyanus, *Swinh. P. Z. S.* 1871, p. 352.

Adult male. General colour above clear ashy, with a greyish lustre especially on the upper tail-coverts; head and nape, as well as the sides of the face and ear-coverts, glossy blue-black; cheeks and throat white; the rest of the under surface delicate ash-colour, inclining to white on the centre of the body; wings pale cobalt, the least coverts ashy, like the back; primaries black, edged with blue near the base and with white towards the tip; the lower surface of the wing pale greyish brown, white at the base of the inner web; tail pale cobalt-blue, with a broad white terminal band to the two centre feathers, the others imperceptibly tipped with white; bill and feet black: "iris red" (*Reeves*). Total length 15 inches, culmen 1, wing 5, tail 9, tarsus 1·35.

Young. General colour above dusky drab-brown, with pale sandy buff margins to the feathers, broader on the scapulars; head and sides of face black, obscured by sandy buff margins to the feathers; round the hind neck the feathers are sandy buff, forming a partial collar; wing-coverts drab-brown, with the same sandy margins as on the back, the outer ones slightly washed with dull azure-blue; the primary-coverts dull azure, with narrow white tips; quills black; secondaries externally azure-blue, the primaries more or less broadly margined with white towards the tip; tail dull azure-blue, tipped with white; cheeks and throat sandy white; breast and sides of body greyish drab-brown, mottled with dull white margins to the feathers, the abdomen and under tail-coverts dull white; bill and legs dull horn-brown.

Obs. The tint of colour on the back and breast of *C. cyanus* is much greyer and less ruddy than in *C. cooki*; but the species are extremely close. The white tips to the tail-feathers are, as a rule, a clearly distinguishing character; but I have seen some Chinese specimens which wanted them, while Mr. Howard Saunders possesses a Spanish bird with the centre tail-feathers tipped with white. This, however, is the only one he has ever seen out of hundreds of specimens examined by him.

Hab. Eastern Siberia, Japan, North China.

a. ♂ ad. sk.	River Onon, Feb. 18, 1872.	Dr. Dybowski [C.].
b. Ad. sk.	Yokohama, Japan.	H. Pryer, Esq. [C.].
c, d. Ad. st.	North China.	Mr. Fortune [C.].
e. Ad. sk.	Shanghai.	Shanghai Museum [P.].
f. Skeleton.		

2. Cyanopolius cooki.

Pica cyanea, *Wagler, Syst. Av.* Pica, sp. 6 (*nec Pall.*); *Cook, P. Z. S.* 1831, p. 96; *Keys. & Blasius, Wirb. Eur.* p. xlv; *Gould, B. Eur.* iii. pl. 217; *Bree, B. Eur.* i. p. 140, cum tab.
Corvus cyanus, *Schl. u. Susem. Vög. Eur.* ii. Taf. 5.
Cyanopolius cooki, *Bp. Report Brit. Assoc.* 1849, p. 75.
Cyanopica cooki, *Bp. P. Z. S.* 1850, p. 86; *id. Consp.* i. 352; *Fritsch, Vög. Eur.* tab. xxvii. fig. 9; *Dresser, B. Eur.* part xxvii.; *Irby, B. Gibr.* p. 129; *Bree, B. Eur.* 2nd ed. ii. p. 4, cum tab.
Dolometis cooki, *Cab. Mus. Hein.* Th. i. p. 222.
Pica cooki, *Gray, Hand-l. B.* ii. p. 10.

Adult male. Similar to *C. cyanea*, but decidedly browner and not so grey, particularly on the under surface; head more purplish brown; white tips to secondaries and centre tail-feathers almost imperceptible or entirely absent; bill and legs black. Total length 13·5 inches, culmen 1, wing 5·25, tail 8, tarsus 1·35.

Hab. Southern Spain.

a. Ad. sk.	South Spain.	Capt. Cook, R.N. [P.]. (Type of species.)
b. Ad. sk.	Spain.	Purchased.
c. ♂ ad. sk.	Madrid, June 1863 (*H. Saunders*).	R. B. Sharpe, Esq. [P.].

17. UROCISSA.

	Type.
Urocissa, *Cab. Mus. Hein.* Th. i. p. 87 (1850)....	U. erythrorhyncha.
Calocitta, *Bp. Consp.* i. 381 (1850, *nec Gray*)....	U. erythrorhyncha.

Range. Himalaya Mountains, Burmese and Indo-Chinese countries to China and Formosa.

Key to the Species.

a. Underparts white or yellowish white, sometimes greyish.
  a'. With coral-red bill.
    a''. Back purplish, with a strong shade of azure-blue.
      a'''. Bill small; subterminal bar on upper tail-coverts very narrow, not equal to half the width of the subterminal band; primaries tipped white on the outer web .... *occipitalis*, p. 70.
      b'''. Bill larger; subterminal bar on tail-

coverts very broad and equal to the black band in width; primaries not tipped with white .... *magnirostris*, p. 71.

b''. Back lavender-brown; subterminal bar on upper tail-coverts broad and equal to terminal black bar; primaries tipped with white on outer web .... *erythrorhyncha*, p. 71.

b'. With bright yellow bill .... *flavirostris* and allies, pp. 72–74.

b. Underparts bright deep azure-blue, like the upper parts .... *cærulea*, p. 74.

1. Urocissa occipitalis.

Pica erythrorhyncha, *Vigors, P. Z. S.* 1831, p. 173 (*nec Gm.*); *Gould, Cent. Himal. B.* pl. 41.
Cyanurus erythrorhynchus, *Hodgs. in Gray's Zool. Misc.* p. 84.
Psilorhinus sinensis, *Gray, Cat. Mamm. &c. Nepal Hodgs.* 1846, p. 101; *Blyth, J. A. S. B.* xv. p. 27.
Psilorhinus occipitalis, *Blyth, J. A. S. B.* xv. p. 26; *id. Cat. B. Mus. A. S. B.* p. 93; *Hutton, J. A. S. B.* xvii. p. 681.
Psilorhinus albicapillus, *Blyth, J. A. S. B.* xv. p. 28.
Urocissa occipitalis, *Cab. Mus. Hein.* Th. i. p. 87; *Gould, B. Asia*, pt. xiii.; *Jerd. B. Ind.* ii. p. 309; *Hume, Nests and Eggs Ind. B.* p. 419.
Urocissa albicapilla, *Cab. Mus. Hein.* Th. i. p. 87.
Urocissa sinensis, *Horsf. & Moore, Cat. B. Mus. E.I. Co.* ii. p. 577 (pt.).
Cissa occipitalis, *Gray, Hand-l. B.* ii. p. 7.

Adult. General colour above pale purplish brown, the wings brighter and more azure-blue; quills blackish on the inner web, externally azure-blue, the secondaries terminally banded with white, the primaries outwardly margined with whitish ash-grey, shading off into azure-blue near the tip of the outer web, which has also a terminal white spot; upper tail-coverts azure-blue, tipped with black, before which is a subterminal shade of pale bluish grey; tail-feathers azure-blue very broadly tipped with white, all but the two centre feathers having a broad subterminal band of black, before which a white spot is apparent on the inner web of most of them; head all round, sides of neck, throat, and fore neck black, the latter feathers slightly shaded with blue on the lower margin; plumes of the crown tipped with white; from the nape extending onto the hinder neck a large occipital patch of white; rest of under surface of body silky white, including the under wing- and tail-coverts; flanks inclining to light bluish grey, as also on the thighs; bill-coral red; legs orange; iris fine red. Total length 24 inches, culmen 1·4, wing 7·5, tail 16, tarsus 1·95.

Hab. Himalaya Mountains.

a. Ad. sk.	Himalaya Mountains.	J. Gould, Esq.
b. Ad. sk.	Himalaya Mountains.	Secretaryof State foIndia [P.].
c. Juv. sk.	Himalaya Mountains.	J. R. Reeves, Esq. [P.].
d. Ad. st.	Nepal.	B. H. Hodgson, Esq. [P.].
e, f, g. Ad. sk.	Nepal.	B. H. Hodgson, Esq. [P.].
h. Juv. sk.	Nepal.	B. H. Hodgson, Esq. [P.].

i. ♂ ad. sk.	N.W. Himalaya.	Capt. Stackhouse Pinwill [P.].
k, l, m. ♀ ad. sk.	N.W. Himalaya.	Capt. Stackhouse Pinwill [P.].

2. Urocissa magnirostris.

Psilorhinus magnirostris, *Blyth, J. A. S. B.* xv. p. 27; *id. Cat. B. Mus. A. S. B.* p. 93.
Urocissa magnirostris, *Cab. Mus. Hein.* Th. i. p. 87, note; *Gould, B. Asia*, part xiii.; *Jerd. B. Ind.* ii. p. 311; *Godwin-Aust. J. A. S. B.* 1874, p. 170; *Hume, Str. F.* 1874, p. 480, 1875, p. 145; *Blyth & Walden, B. Burmah*, p. 88.
Calocitta magnirostris, *Bp. Consp.* i. p. 381.
Cissa magnirostris, *Gray, Hand-l. B.* ii. p. 7.

Adult. Very similar to *U. occipitalis*, but with a very much stouter bill and tarsus. The colour is much darker, the wings less blue, and the white spots at the tips of the primaries are either nearly obsolete or absent altogether. Total length about 20 inches, culmen 1·6, wing 7·4, tarsus 1·95.

Hab. Burmah and Siam.

a. Ad. sk.	Siam (*Schomburgk*).	J. Gould, Esq.
b. Skeleton.		Zool. Society.

3. Urocissa erythrorhyncha*.

Le Geai de la Chine à bec rouge, *Montb. H. N. Ois.* iii. p. 115; *id. Pl. Enl.* 622.
Red-billed Jay, *Lath.*; *Gen. Syn.* i. p. 390.
Black-headed Roller, *Lath. Gen. Syn. Suppl.* p. 86.
Corvus erythrorhynchus, *Gm. S. N.* i. p. 372 (*ex Lath.*); *Daud. Traité*, p. 240, pl. xv.
Coracias melanocephalus, *Lath. Ind. Orn.* i. p. 170.
La Pie bleue, *Levaill. Ois. d'Afr.* ii. pl. 57.
Psilorhinus sinensis, *Gray, Gen. B.* ii. p. 308.
Calocitta sinensis, *Bp. Consp.* i. p. 381.
Urocissa sinensis, *Cab. Mus. Hein.* Th. i. p. 87; *Gould, B. Asia*, part xiii.; *Swinh. P. Z. S.* 1871, p. 382.
Calocitta erythrorhyncha, *Pelz. Reise Novara, Vög.* p. 86.
Cissa erythrorhyncha, *Schl. Mus. P.-B.* Coraces, p. 70.
Cissa sinensis, *Gray, Hand-l. B.* ii. p. 7.
Urocissa brevivexilla, *Swinh. P. Z. S.* 1873, p. 688.

Adult male. Above lavender-brown, slightly shaded with bluish purple; wings dull azure, brighter on the quills, which are dark brown internally, and all of the primaries tipped with white on the outer web, the secondaries having a terminal bar of the same; upper tail-coverts lavender-brown, shading off into pale grey before the tip, which has a broad black band; tail azure-blue, broadly tipped with white, all but the two centre feathers having a wide subterminal

* This is the bird generally called *Urocissa sinensis* by authors; but I think this name ought to be rejected. The *Cuculus sinensis* of Linnæus was founded on Brisson's "*Coucou bleu de la Chine*" (Orn. iv. p. 137, pl. xiv. A. fig. 2). This bird was described and figured from a native Chinese drawing, and is a fancy picture of the Blue Pie of China, and should be discarded on account of its thorough inaccuracy.

band of black, which is again preceded on the inner web by a more or less distinct bar of white; head, sides of face, and entire throat black, all the feathers of the fore part of the crown tipped with lavender-grey, the hinder part of the crown, occiput, and nape being entirely of the latter colour; rest of under surface light greyish, with somewhat of a bluish lustre, the flanks more decidedly grey, the under wing- and tail-coverts white, the former slightly shaded with grey; bill and feet coral-red; iris deep burnt sienna, tinged with crimson. Total length 21 inches, culmen 1·5, wing 7·4, tail 16, tarsus 1·8.

Nestling. Showing much of the colours of the adult bird, but browner, and having not only the whole of the crown, with the exception of the frontal plumes, white, but also the throat.

Obs. Mr. Swinhoe described, in 1873, a new species of *Urocissa*, from the "western hills of Pekin;" but on examining the type of his *U. brevivexilla* I find that there is a perfect gradation between the Amoy and Pekin birds: those from Amoy are darker on the breast, and the Pekin ones are very pale; but the Ningpo examples are exactly intermediate.

Hab. China.

a. Ad. sk.	China.	J. Gould, Esq.
b. Ad. sk.	China.	Capt. Blakiston [C.].
c. ♂ ad. sk.	Ningpo, China.	J. R. Reeves, Esq. [P.].
d. Ad. sk.	Ningpo, China.	J. R. Reeves, Esq. [P.].
e. Ad. sk.	North China.	Mr. Fortune [C.].
f. Pull. sk.	Amoy.	Purchased.

4. Urocissa flavirostris.

Psilorhinus flavirostris, *Blyth, J. A. S. B.* xv. p. 28; *id. Cat. B. Mus. A. S. B.* p. 93.

Urocissa flavirostris, *Cab. Mus. Hein.* Th. i. p. 87; *Horsf. & Moore, Cat. B. Mus. E.I. Co.* ii. p. 578; *Gould, B. Asia*, part xiii.; *Jerd. B. Ind.* ii. p. 310; *Cock & Marsh. Str. F.* 1873, p. 357; *Henders. & Hume, Lahore to Yark.* p. 242; *Hume, Nests and Eggs Ind. B.* p. 419.

Calocitta flavirostris, *Bp. Consp.* i. p. 381.

Urocissa cucullata, *Gould, B. Asia,* part xiii.

Cissa flavirostris, *Schl. Mus. P.-B.* Coraces, p. 70; *Gray, Hand-l. B.* ii. p. 7.

Adult. Upper surface lavender-coloured, with a slight tinge of azure-blue, the wings resembling the back for the most part, the primaries externally greyish white for their apical two thirds, the secondaries tipped with white, less distinctly so the primaries, where it is confined to the outer web; upper tail-coverts azure-blue tipped with black, before which is a faint subterminal bar of bluish white, very narrow; tail azure-blue, broadly tipped with white, all but the two centre feathers subterminally barred with black, before which a faint appearance of a white bar is seen on the inner web; head and neck all round, including the mantle above and the fore neck below, jet-black, with a large nuchal patch of white slightly glossed with a shade of blue; rest of the under surface ashy grey, paler on the abdomen and under tail-coverts; under wing-coverts yellowish, the edge of the wing washed with lavender, like the upper wing-coverts; bill yellow; legs orange-yellew; iris brownish red. Total length 20 inches, culmen 1·45, wing 7·3, tail 16, tarsus 1·9.

Obs. Mr. Hume writes concerning the present bird:—"I do not consider that Gould's species *U. cucullata* can stand, as specimens are to be obtained throughout the Himalayas, from Darjiling to Murree, presenting the characteristics of both supposed species, as well as all possible intermediate forms" (Lahore to Yark. p. 242). I am fortunate in having the type of Mr. Gould's *Urocissa cucullata* before me, as well as a very fair series of *U. flavirostris* and *U. occipitalis*; and I must say that it is very difficult to distinguish *U. cucullata* specifically from *U. flavirostris*, while the former bird would seem to grade into *U. occipitalis*.

The differences between *U. occipitalis* and *U. flavirostris* have been summarized by Dr. Jerdon as consisting in the "generally duller hue, yellow bill, and small occipital mark" of the latter bird, whose dimensions also are much smaller. To these characters may be added, that the crown and upper part of the mantle in *U. flavirostris* are both of them jet-black, the nuchal patch being somewhat crescentic in shape, well defined and separating the black crown from the interscapulary region. In *U. occipitalis* the black does not occupy so much of the mantle, the white occipital patch descends on the hinder neck longitudinally and divides the black; the feathers of the crown also show some bluish white apical spots. In this species, which has usually the bill distinctly coral-red, there is also a shade of lighter blue preceding the black tip to the upper tail-coverts; but in true *U. flavirostris* these are generally uniform bluish with a black tip, though I admit that this is a character which varies considerably.

Urocissa cucullata resembles *U. flavirostris* in its yellow bill and in having a blackish collar round the hind neck, though this black does not extend so low down on the mantle as it does in *U. flavirostris*. The way in which the white of the occipital patch descends nearly to the edge of the black on the hinder neck is almost similar to the form of this same patch in *U. occipitalis*; and *U. cucullata* further resembles the latter bird in its light under surface: true *U. flavirostris* is very dark and grey in shade below. Altogether *U. cucullata* seems to me to be an intermediate form of doubtful specific value; and I therefore prefer to keep it united for the present to *U. flavirostris*, merely separating in the list of specimens those examples which ought probably to be referred to it.

The Museum contains from Cashmere three specimens, probably representing the young of *U. cucullata*, which differ from the adult in being of a fine yellow underneath, and in having the tail-feathers tipped with the same; the tail is shaded with bright cobalt instead of purplish blue; and the subterminal line on the upper tail-coverts is also yellow. The upper surface of the body has a slight yellowish cast; but it appears, from admixture of black in the yellow colour of the bill, that the birds, though full-grown, are not quite adult; at the same time they differ from the young plumage of the allied species, which are not distinctly yellow when immature, as far as the specimens in the Museum show.

Hab. Himalaya Mountains.

a. U. flavirostris.

a. Ad. st.	Himalaya Mountains.	Purchased.
b, c. Ad. sk.	Nepal.	B. H. Hodgson, Esq. [P.].
d. Ad. sk.	Sikhim.	J. Gould, Esq.
e. Ad. sk.	Darjiling.	J. Gould, Esq.

β. U. cucullata.

f. Ad. sk.	Himalaya Mountains.	Purchased.
g. Ad. sk.	Kumaon (*Strachey*).	Secretary of State for India [P.].
h. Ad. sk.	Kooloo (*Tweedie*).	J. Gould, Esq. (Type of *U. cucullata.*)
i, k. Ad. sk.	Ladakh.	Purchased.
l. Ad. sk.	Cashmere.	E. M. Langworthy, Esq. [P.].
m, n. ♂ ♀ juv. sk.	Cashmere.	
o. Juv. sk.	Cashmere.	R. B. Sharpe, Esq. [P.].

5. Urocissa cærulea.

Urocissa cærulea, *Gould, P. Z. S.* 1862, p. 282; *Swinh. Ibis,* 1863, p. 384; *Gould, B. Asia,* part xvi.; *Swinh. P. Z. S.* 1871, p. 382.
Cissa cærulea, *Schl. Mus. P.-B.* Coraces, p. 71; *Gray, Hand-l. B.* ii. p. 7.

Adult. General colour above and below deep azure-blue, the under tail-coverts paler and shading off into whitish at the tips; head and neck all round, including the fore neck, black; quills black, externally blue like the back, tipped with white, which forms a tolerably broad terminal bar on the secondaries; upper tail-coverts azure-blue like the back, broadly tipped with black, preceded by a subterminal bar of greyish blue; tail blue, the two centre tail-feathers broadly tipped with white, all the other ones having the white tip preceded by a broad subterminal bar of black, much broader on the outer feathers; on some of the tail-feathers a faintly indicated white spot is seen on the inner web before the black bar; "bill and legs bright red-lead, the former tipped paler; sole-pads light and dingy; claws light reddish brown" (*Swinhoe*). Total length 22 inches, culmen 1·7, wing 7·9, tail 15·5, tarsus 1·85.

Hab. Formosa.

a, b. Ad. sk.	N. Formosa.	R. Swinhoe, Esq. [C.].
c. Ad. sk.	Formosa (*Swinhoe*).	J. Gould, Esq.

18. CRYPTORHINA.

	Type.
Cryptorhina, *Wagl. Syst. Av. art.* Cryptorhina (1827)	C. afra.
Ptilostomus, *Swains. Classif. B.* ii. p. 266 (1837)	C. afra.

Range. North-eastern Africa and Senegambia.

1. Cryptorhina afra.

La Pie du Sénégal, *Briss. Orn.* ii. p. 40, pl. 3. fig. 2; *Montb. Pl. Enl.* 538.
Corvus afer, *Linn. S. N.* i. p. 157 (*ex Briss.*).
Corvus senegalensis, *Linnæus, Syst. Av.* i. p. 158 (*ex Briss.*).
Coracias nigra, *Lath. Ind. Orn.* i. p. 172.
Le Piacpiac, *Levaill. Ois. d'Afr.* pl. 54; *Sundev. Crit. om Levaill.* p. 30.
Pica nigra, *Vieill. N. Dict.* xxxi. p. 130.
Lanius acuticaudatus, *Vieill. N. Dict. t. c.* p. 134.
Cryptorhina piacpiac, *Wagler, Syst. Av.* Cryptorhina, sp. 1.
Cryptorhina pœcilorhynchus, *Wagler, t. c.* sp. 2; *Gray, Hand-l. B.* ii. p. 9.
Ptilostomus senegalensis, *Swains. B. W. Afr.* i. p. 135; *Rüpp. Syst. Uebers.* p. 74; *Gray, Gen. B.* ii. p. 311; *Bp. Consp.* i. p. 369; *Hartl. Orn. W.-Afr.* p. 113; *Antin. Cat. descr. Ucc.* p. 58; *Layard, B. S. Afr.* p. 167; *Heugl. Orn. N.O.-Afr.* p. 491; *Sharpe, Cat. Afr. B.* p. 37.
Ptilostomus pœcilorhynchus, *Gray, Gen. B.* ii. p. 311; *Bp. Consp.* i. p. 369; *Cab. Mus. Hein.* Th. i. p. 217; *Heugl. Syst. Uebers.* p 35.
Glaucopis senegalensis, *Schl. Mus. P.-B.* Coraces, p. 77.

Adult male. General colour silky black, with a slight purplish gloss; wing-coverts faintly glossed with green; quills brown with blackish tips, the secondaries glossy purplish black with indications of bars under certain lights; upper and under tail-coverts brown; tail brown, with indications of bars according to light; under surface not quite so glossy as the upper, the under wing-coverts brown; inner lining of quills ashy brown, paler near the bases; bill black; feet leaden black; iris crimson, with a slight shade of lilac. Total length 17·5 inches, culmen 1·35, wing 6·35, tail 11·2, tarsus 1·9.

Female. Altogether smaller than the male, and distinguished by the bill being yellow in life, tipped with black. Total length 12·5 inches, culmen 1·25, wing 5·9, tail 7·5, tarsus 1·7.

Obs. The specimen from the White Nile is a much larger and finer bird than any of the West-African ones. Total length 19 inches, culmen 1·4, wing 6·6, tail 12, tarsus 1·9.

Hab. North-eastern Africa and Senegambia.

a. [♂] ad. st.	Africa.	J. Gould, Esq. [P.].
b. [♀] ad. sk.	River Gambia.	
c. [♂] ad. sk.	River Gambia.	R. B. Sharpe, Esq.
d. [♀] ad. sk.	River Gambia.	R. B. Sharpe, Esq.
e. [♂] ad. sk.	White Nile.	R. B. Sharpe, Esq.
f. Skeleton.		Zoological Society.

19. DENDROCITTA.

	Type.
Dendrocitta, *Gould, P. Z. S.* 1833, p. 565	D. leucogastra.
Crypsirhina, *Swains. Class. B.* ii. p. 266 (1837)	D. rufa.
Vagabunda, *Kaup, J. f. O.* 1854, p. lv.	D. rufa.

Range. India and the Indo-Chinese countries, China, Hainan, Formosa, Sumatra.

Key to the Species.

a. With no white alar speculum.
a'. Secondaries externally whitish grey like the wing-coverts; breast entirely orange; central tail-feathers grey, with black tip .... *rufa*, p. 76.
b'. Secondaries black like the rest of the quills; breast grey; abdomen and under tail-coverts orange; tail-feathers all black .... *frontalis*, p. 78.
b. With a distinct white alar speculum.
c'. Under tail-coverts orange, in strong contrast to the belly, which is ashy grey or white.
a''. Back orange; hind head and neck pure white like breast .... *leucogastra*, p. 79.
b''. Back brown; hind head and neck grey; breast ashy or brown.
a'''. Tail black at tip, grey for at least half or nearly the whole of its length.
a[4]. Larger: tail 8·5–9·4 inches; belly ashy, very little paler than the breast; upper tail-coverts grey .... *himalayensis*, p. 79.
b[4]. Smaller: tail 6·7–7·7 inches; belly white in contrast to the breast; upper tail-coverts whitish also .... *formosæ*, p. 80.
b'''. Tail entirely black, or at most with a faint shade of grey near the base of the centre tail-feathers .... *sinensis*, p. 81.
d'. Under tail-coverts orange like the abdomen.
c''. Head brown, with a white occipital patch; centre tail-feathers grey, with a black tip.. *occipitalis*, p. 81.
d''. Head black, the neck uniform; tail-feathers all uniform .... *bayleyi*, p. 82.

1. Dendrocitta rufa.

La Pie rousse de la Chine, *Sonn. Voy. Ind.* ii. p. 186, pl. 106.
Rufous Crow, *Lath. Gen. Syn Suppl.* p. 84.
Grey-tailed Roller, *Lath. Gen. Syn. Suppl.* p. 86.
Lanius rufus, *Scop. Del. Faun. et Flor. Insubr.* ii. p. 86.
Corvus rufus, *Lath. Ind. Orn.* i. p. 161; *Daud. Traité*, p. 245.
Coracias vagabunda, *Lath. Ind. Orn.* i. p. 171.
La Pie rousse, *Levaill. Ois. d'Afr.* ii. p. 31, pl. 59.
Pica rufa, *Vieill. N. Dict. d'Hist. Nat.* xxvi. p. 130; *Steph. Gen. Zool.* xiv. p. 64 (1826); *Wagl. Isis*, 1829, p. 750; *Sundev. Ann. N. H.* xviii. p. 168.
Pica vagabunda, *Vieill. N. Dict.* xxvi. p. 121; *Wagl. Syst. Av.* Pica, sp. 5; *Gould, Cent. Himal. B.* pl. 42; *Gray in Hardw. Ill. Ind. Zool.* i. pl. 25.
Dendrocitta vagabunda, *Gould, P. Z. S.* 1833, p. 57, *et Tr. Z. S.* i. p. 89; *Gray, Cat. Hodgs. Coll. Nepal*, p. 101; *Bp. Consp.* i. p. 369.
Crypsirhina vagabunda, *Swains. Classif. B.* ii. p. 266; *Jerd. Madr. Journ.* xi. p. 19; *Blyth, J. A. S. B.* xii. p. 932, xiii. p. 389; *Hodgs. in Gray's Zool. Misc.* p. 84.
Crypsirhina rufa, *Swains. Classif. B.* ii. p. 266; *Blyth, J. A. S. B.* xv. p. 30.
Dendrocitta rufa, *Hartl. Syst. Verz.* p. 63 (1844); *Blyth, Cat. B. Mus. A. S. B.* p. 92; *Cab. Mus. Hein.* Th. i. p. 216; *Horsf. & Moore, Cat. B. Mus. E.I. Co.* ii. p. 565; *Jerd. B. Ind.* ii. p. 314; *Gray,*

Hand-l. B. ii. p. 8; *Godwin-Aust. J. A. S. B.* 1870, p. 110; *Hume, Str. F.* 1873, p. 206; *Adam, t. c.* p. 386; *Hume, Str. F.* 1874, p. 480; *id. Nests &c. Ind. B.* p. 421; *Blyth & Wald. B. Burm.* p. 88.
Temnurus rufus, *Gray, Gen. B.* ii. p. 310.
Temnurus vagabundus, *Gray, t. c.* p. 310.
Crypsirhina pallida, *Blyth, J. A. S. B.* xv. p. 30.
Dendrocitta pallida, *Blyth, Cat. B. Mus. A. S. B.* p. 336; *Horsf. & Moore, Cat. B. Mus. E.I. Co.* ii. p. 568; *Jerd. B. Ind.* ii. p. 315; *Gray, Hand-l. B.* ii. p. 8.
Vagabunda rufa, *Kaup, J. f. O.* 1854, p. lv; *Ball, Str. F.* 1874, p. 418.
Glaucopis rufa, *Schl. Mus. P.-B.* Coraces, p. 77.

Adult. Above orange-brown, shading off into brighter orange-buff on the lower back, rump, and upper tail-coverts; head and neck all round sooty brown, this colour gradually shading off into the orange-brown of the back; the lores and throat blackish; rest of under surface of body orange-buff, including the thighs and under tail-coverts; under wing-coverts and edge of the wing grey, slightly washed with orange; wings black, the least coverts dull grey, the median and greater series clear whitish grey, as well as the outer webs of the inner secondaries, forming a broad bar down the wing; tail-feathers grey, broadly tipped with black; bill black; legs dark slaty; iris blood-red. Total length 16 inches, culmen 1·25, wing 5·9, tail 10, tarsus 1·45.

Young. Differs from the adult in being very much paler and more yellow than the adult, in having the grey on the wings more or less obscured with a wash of brown, and in having the tail-feathers tipped with greyish buff. The head and neck are dull brown, instead of blackish.

Obs. I do not think Blyth's species, *D. pallida*, can be upheld for a moment. It is, to judge from the specimens now before me, the young bird of *D. rufa*, although no mention is made of the white tips to the tail-feathers. The only variation in the latter species seems to take place in the South-Indian specimens, which are altogether smaller, darker, have the brown plumage of the head extended onto the mantle above and the breast below, and further show a white subterminal bar before the black tip on the centre tail-feathers. In Himalayan specimens the grey fades off lighter just before the black tip, but is not so decidedly white as in Madras examples. The latter are also rather smaller; but I cannot bring myself to believe that the southern birds are specifically distinct from the Himalayan ones, as the specimens from Central India appear to be intermediate.

The pale coloration, which must have suggested to Mr. Blyth the name of *pallida*, appears to be equally unworthy of consideration as a specific character; for pale and dark-coloured birds appear from every locality, and must be either females or birds in their second plumage. Unfortunately the specimens in the Museum are not determined as regards sex and date of capture in such a way as to enable me to decide this question; but as the young birds with the light tips to the tail-feathers are pale-coloured, we may reasonadly suppose that the colour of this Magpie intensifies with age.

Hab. The whole of India, ranging through Assam and the Burmese countries to Tenasserim.

a. Ad. st.	India.	
b. Juv. sk.	India.	
c, d. Ad. sk.	India.	J. Gould, Esq.
e. Pull. sk.	Himalayas (*Tytler*).	Secretary of State for India [P.].
f. ♀ juv. sk.	Shikapore (*Griffith*).	Secretary of State for India [P.].
g, h. Ad. sk.	N.W. Himalayas.	Capt. Stackhouse Pinwill [P.].
i. ♀ ad. sk.	N.W. Himalayas.	Capt. Stackhouse Pinwill [P.].
k. Juv. sk.	Seetapore.	Purchased.
l, m. Juv. sk.	Kamptee, Central India.	Dr. B. Hinde [P.].
n. Juv. sk.	Malabar.	
o, p, q. Ad. sk.	Madras.	Rev. H. H. Baker [P.].
r. Juv. sk.	Madras.	J. Gould, Esq.
s, t. Ad. sk.	Travancore.	Capt. Biddulph [C.].
u. Ad. sk.	Behar.	B. H. Hodgson, Esq. [P.].
v. Juv. sk.	Nepal.	B. H. Hodgson, Esq. [P.].
w, x. Ad. sk.	Nazeerah, Upper Assam (*Dr. Reid*).	R. B. Sharpe, Esq. [P.].
y, z. Juv. sk.	Nazeerah, Upper Assam (*Dr. Reid*).	R. B. Sharpe, Esq. [P.].
a'. Ad. sk.	Katha, Upper Burmah, Jan. 19, 1868.	Dr. J. Anderson [C.].
b'. Ad. sk.	Pegu.	W. T. Blanford, Esq. [P.].
c'. Skeleton.		Zoological Society.

2. Dendrocitta frontalis.

Dendrocitta frontalis, *McClell. P. Z. S.* 1839, p. 163; *Bp. Consp.* i. p. 369; *Horsf. & Moore, Cat. B. Mus. E.I. Co.* ii. p. 569; *Jerd. B. Ind.* ii. p. 316; *Gray, Hand-l. B.* ii. p. 8.
Crypsirhina frontalis, *Blyth, J. A. S. B.* xii. p. 933, xv. p. 30.
Crypsirhina altirostris, *Blyth, J. A. S. B.* xii. p. 932, xv. p. 30.
Temnurus frontalis, *Gray, Gen. B.* ii. p. 310, pl. 75.
Temnurus altirostris, *Gray, Gen. B.* ii. p. 310.
Dendrocitta altirostris, *Blyth, Cat. B. Mus. A. S. B.* p. 92; *Bp. Consp.* i. p. 369.
Glaucopis frontalis, *Schl. Mus. P.-B.* Coraces, p. 75.

Adult. Upper surface orange-chestnut, clearer on the rump, and inclining to deeper chestnut on the upper tail-coverts; hinder part of the head and the entire hind neck, mantle, and sides of neck whitish grey; all the rest of the head, sides of face, and entire throat black, forming a distinct hood; fore neck and breast whitish grey; rest of the lower surface of the body bright orange-chestnut, much deeper on the lower tail-coverts; thighs and under wing-coverts deep grey, slightly washed with orange; upper wing-coverts clear blue-grey, the bastard wing and the primary-coverts black; quills and tail black; bill and feet black; iris brownish red. Total length 14 inches, culmen 1·05, wing 5·15, tail 9·3, tarsus 1·2.

Hab. Eastern Himalayas and Assam.

a. Ad. sk.	Himalayas (*A. Grote*).	J. Gould, Esq.
b, c. Ad. sk.	Nepal.	B. H. Hodgson, Esq. [P.].
d. Ad. st.	Darjiling.	B. H. Hodgson, Esq. [P.].
e. Ad. sk.	Darjiling.	J. Gould, Esq.

3. Dendrocitta leucogastra.

Dendrocitta leucogastra, *Gould, P. Z. S.* 1833, p. 57, et *Tr. Z. S.* i. p. 89, pl. 12; *Blyth, Cat. B. Mus. A. S. B.* p. 91; *Bp. Consp.* i. p. 369; *Cab. Mus. Hein.* Th. i. p. 217; *Horsf. & Moore, Cat. B. Mus. E.I. Co.* ii. p. 570; *Jerd. B. Ind.* ii. p. 317; *Gray, Hand-l. B.* ii. p. 8; *Hume, Nests and Eggs Ind. B.* p. 424.

Crypsirhina leucogastra, *Sw. Classif. B.* ii. p. 266; *Jerd. Madr. Journ.* xi. p. 19.

Temnurus leucogaster, *Gray, Gen. B.* ii. p. 310.

Glaucopis leucogaster, *Schl. Mus. P.-B.* Coraces, p. 74.

Adult. Back and scapulars orange-brown; crown of head, sides of face, cheeks, throat, and fore neck black; hinder part of crown and the whole of the hinder part and sides of the neck white; rump and upper tail-coverts white, the latter slightly shaded with grey; tail black, inclining to grey near the base of the inner web, the two long central feathers grey for two thirds of their length, the terminal third black; wings entirely black, a large speculum caused by the white bases to the primaries; the rest of the under surface of the body pure white; thighs black; under tail-coverts orange-rufous; under wing-coverts black like the inner lining of the quills, which are black below with a white band across the base of the primaries; bill black; legs and feet dark plumbeous; iris blood-red. Total length 17 inches, culmen 1·3, wing 5·7, tail 11·2, tarsus 1·25.

Hab. Southern India.

a. Ad. st.	India.	Major-Gen. Hardwicke [P.].
b. Ad. sk.	India.	J. R. Reeves, Esq. [P.]
c. Ad. sk.	Malabar.	J. Gould, Esq. (Type of species.)

4. Dendrocitta himalayensis.

Dendrocitta sinensis, *Gould, P. Z. S.* 1833, p. 57; *id. Tr. Z. S.* i. p. 89; *Gray, Cat. Mamm. &c. Nepal Hodgs.* p. 102; *Hutton, J. A. S. B.* xvii. p. 682 (1848); *Blyth, Cat. B. Mus. A. S. B.* p. 92; *Bp. Consp.* i. p. 369; *Cab. Mus. Hein.* i. p. 217; *Horsf. & Moore, Cat. B. Mus. E.I. Co.* i. p. 568; *Godw.-Austen, J. A. S. B.* 1870, p. 110.

Pica sinensis, *Gould, Cent. Himal. B.* pl. 43; *Gray, in Hardw. Ill. Ind. Zool.* i. pl. 26.

Crypsirhina sinensis, *Swains. Classif. B.* ii. p. 266; *Hodgs. in Gray's Zool. Misc.* p. 84; *Blyth, J. A. S. B.* xii. p. 932, xv. p. 30.

Temnurus sinensis, *Gray, Gen. B.* ii. p. 310.

Dendrocitta himalayensis, *Blyth, Ibis,* 1865, p. 45; *Gray, Hand-l. B.* ii. p. 8; *Hume, Str. F.* 1874, p. 480; *Blyth & Walden, B. Burm.* p. 38.

Glaucopis sinensis, *Schl. Mus. P.-B.* Coraces, p. 76.

Adult. Back and scapulars pale reddish brown, gradually shading off into grey on the rump and upper tail-coverts; crown of head and hind neck bluish grey, a little duller on the mantle; the forehead black, as also the sides of the face and throat, the ear-coverts and lower throat a little browner; rest of under surface of body ashy grey, shading off paler on the abdomen; thighs blackish washed with bluish grey; under tail-coverts orange; under wing-coverts glossy black; wings above glossy black, with a distinct white speculum, all the primaries, except the two outermost, being white at the base of the external web; tail glossy black, the two centre tail-feathers bluish grey for two thirds of their length, the next pair also grey at base; bill horny black; legs dusky black; iris reddish brown. Total length 13·5 inches, culmen 1·3, wing 5·45, tail 8·7, tarsus 1·15.

Young. In general similar to the adult, but paler, and having the feathers of the upper surface tipped with ochraceous, the brown plumes as well as the grey being thus marked; the under tail-coverts ochraceous buff instead of orange.

Hab. Himalaya Mountains and Assam.

a. Ad. st.	Himalaya Mountains.	J. R. Reeves, Esq. [P.].
b. Juv. sk.	Near Purneah.	Capt. Stackhouse Pinwill [P.].
c. Ad. sk.	N.W. Himalayas.	Capt. Stackhouse Pinwill [P.].
d. ♂ ad. sk.	N.W. Himalayas.	Capt. Stackhouse Pinwill [P.].
e–k. Ad. sk.	Nepal.	B. H. Hodgson, Esq. [P.].
l. Juv. sk.	Nepal.	B. H. Hodgson, Esq. [P.].
m, n. Ad. sk.	Darjiling.	J. Gould, Esq.
o. Ad. sk.	Nazeerah, Upper Assam (*Dr. Reid*).	R. B. Sharpe, Esq. [P.].
p. Bones, various.	Nepal.	B. H. Hodgson, Esq. [P.].

5. Dendrocitta formosæ.

Dendrocitta sinensis, var. formosæ, *Swinh. Ibis,* 1863, p. 387; 1865, p. 234; 1866, pp. 296, 394.
Glaucopis formosæ, *Schl. Mus. P.-B.* Coraces, p. 76 (1867).
Dendrocitta formosæ, *Swinh. P. Z. S.* 1871, p. 382.

Adult. General colour above rather rufescent brown, deeper on the lower back; the rump whitish, the upper tail-coverts shading into grey; forehead and feathers round the eye blackish; crown of head and hind neck dull grey, shading off from the mantle into the brown of the back; sides of face and neck, throat, and breast dull smoky brown, paler on the latter; abdomen white; flanks greyish brown; thighs blackish brown, slightly shaded with grey; under tail-coverts deep orange-buff; under wing-coverts blackish, with a distinct shade of grey; wings above black, with a distinct purplish gloss; a distinct alar speculum, caused by the white bases to the primaries, all but the two outermost of which have a large spot on the outer as well as the inner web; tail black, the centre feathers with a distinct shade of bluish grey for more than half their length, the same shade being also slightly developed near the base of the next pair. Total length 12 inches, culmen 1·3, tail 7·3, tarsus 1·15.

Hab. Formosa.

a. Ad. sk.	Formosa.	R. Swinhoe, Esq. [C.].

6. Dendrocitta sinensis.

La Pie de Macao, *Sonn. Voy. Ind.* ii. p. 187.
Macao Crow, *Lath. Gen. Syn. Suppl.* p. 84.
Corvus sinensis, *Lath. Ind. Orn.* i. p. 161; *Daud. Traité,* ii. p. 244; *Shaw, Gen. Zool.* vii. p. 369.
Dendrocitta sinensis, *Gray, Hand-l. B.* ii. p. 8; *Swinh. P. Z. S.* 1871, p. 382.

Adult. Back and scapulars brown, shading off into greyish on the lower back; the rump whiter grey; the upper tail-coverts also clear grey; crown of head and hind neck dull grey, shading off a little deeper on the mantle; a broad frontal band and feathers below the eye and near the base of the bill black, all the rest of the sides of the face and of the throat dull ashy brown, extending also onto the chest, and shading off into grey on the lower breast and abdomen; thighs blackish brown, shaded with grey; under tail-coverts orange; under wing-coverts blackish: wings above black, with a greenish steel gloss, the least coverts slightly shaded with grey; on the wing a very distinct speculum, caused by the white bases to the primaries extending onto the outer as well as the inner web; tail black, slightly glossed like the wings. Total length 13·5 inches, culmen 1·25, wing 5·4, tail 7·4, tarsus 1·15. (*Mus. R. Swinhoe*).

Obs. The true *D. sinensis* is a very much smaller bird than *D. himalayensis*; and the following are the measurements of some specimens collected by Mr. Swinhoe:—

		Total length. in.	Wing. in.	Tail. in.	Tarsus. in.
a. Ad.	Ningpo, Dec. 23, 1872	13·5	5·4	7·4	1·15
b. ♀ ad.	„ Sept. 30, 1872	12·5	5·55	7·4	1·2
c. Ad.	Hainan, Feb. 1868	10·6	4·9	6·7	1·0
d. Ad.	„ „ „	10·0	4·8	6·0	1·05
e. Ad.	„ „ „	11·0	5·15	7·0	1·05

Hab. China.

a. Ad. sk.	Hainan, February 1868.	R. Swinhoe, Esq. [P.].
b. Ad. sk.	China.	J. Gould, Esq.

7. Dendrocitta occipitalis. (Plate III.)

Glaucopis occipitalis, *Müll. Tijdschr.* 1835, p. 343, pl. ix. fig. 1; *Schl. Mus. P.-B.* Coraces, p. 75.
Dendrocitta rufigaster, *Gould, P. Z. S.* 1837, p. 80; *Blyth, Cat. B. Mus. A. S. B.* p. 336; *Gray, Hand-l. B.* ii. p. 8.
Crypsirhina rufigastra, *Blyth, J. A. S. B.* xii. p. 933.
Dendrocitta occipitalis, *Bp. Consp.* i. p. 369; *Gray, Hand-l. B.* ii. p. 8.

Adult. Back and scapulars warm brown, inclining to grey on the lower back; the rump and upper tail-coverts whitish grey, clearer on the former; head all round smoky brown, blackish on the crown and paler on the forehead, nasal bristles, and sides of face; occiput and nape greyish white, shading off into clearer grey on the mantle;

wings glossy black, with a very distinct alar speculum formed by the white bases to the primaries, all but the last two of which have a white base on the outer as well as the inner web; outer tail-feathers entirely black, the two centre ones grey, with somewhat of a sea-green lustre under certain lights, and having a broad black tip, the intermediate ones grey, also tipped with black, this colour increasing obliquely towards the outer feathers; under surface of body pale orange-buff, increasing to deeper orange on the under tail-coverts; under wing-coverts blackish brown. Total length 17·5 inches, culmen 1·2, wing 5·55, tail 11·5, tarsus 1·2.

Young. Similar to the adult, but the colours much duller, especially on the head, where the brown colour is paler.

Hab. Sumatra.

a. Ad. sk.	Sumatra.	Purchased.	
b. Juv. sk.	Sumatra.	Purchased.	
c. Ad. sk.	Sumatra.	J. Gould, Esq.	(Type of *D. rufigaster.*)

8. **Dendrocitta bayleyi.**

Dendrocitta baylei, *Tytler, J. A. S. B.* 1863, p. 88; *Beav. Ibis,* 1867, p. 329; *Gray, Hand-l. B.* ii. p. 8; *Ball, Str. F.* 1873, p. 75; *Hume, Str. F.* 1874, p. 245; *Walden, Ibis,* 1874, p. 145, pl. vi.

Dendrocitta bazlei, *Blyth, Ibis,* 1863, p. 88 (*lapsu*).

Adult male. Back and scapulars dull tawny brown, becoming slightly more orange on the rump; head black, shading off into leaden grey on the hind neck and on the upper breast, the rest of the under surface bright orange-chestnut; thighs blackish; under wing-coverts blackish, the inner ones shaded with grey; wings above black, the lesser and median coverts slightly shaded with leaden grey; a very distinct alar speculum, caused by the outer webs of all the secondaries except the innermost being white at the base; upper tail-coverts leaden black; tail entirely black, with a slight leaden-grey shade near the base; "bill, legs, and feet black, the soles of the latter plumbeous grey; iris bright yellow, sometimes golden yellow" (*Hume*). Total length 12·5 inches, culmen 1, wing 4·55, tail 7·7, tarsus 1.

Adult female. Only differing from the male in being slightly duller-coloured.

Young. Coloured like the adults, but not so bright; the head and throat browner, the chest reddish-brown instead of grey.

Hab. Andaman Islands.

a. ♂ ad. sk.	South Andamans, April 18, 1873 (*Ramsay*).	R. B. Sharpe, Esq. [P.].
b. ♀ juv. sk.	Port Blair, S. Andamans, June 18th, 1873 (*Wimberley*).	R. B. Sharpe, Esq. [P.].

20. CRYPSIRHINA.

	Type.
Crypsirhina, *Vieill. Analyse*, p. 36 (1816)	C. varians.
Phrenotrix, *Horsf. Tr. Linn. Soc.* xiii. p. 162 (1820).....	C. varians.

Range. Java; Tenasserim Provinces and Burmah, eastwards to Cochin China.

Key to the Species.

a. Greenish black above and below .... *varians*, p. 83.
b. Clear grey above and below; head black .... *cucullata*, p. 84.

1. Crypsirhina varians.

Le Temia, *Levaill. Ois. d'Afr.* ii. pl. 56.
Variable Crow, *Lath. Gen. Syn. Suppl.* ii. p. 119.
Corvus varians, *Lath. Ind. Orn. Suppl.* p. xxvi.
Corvus temia, *Daud. Traité*, p. 244.
Corvus caudatus, *Shaw, Nat. Miscell.* pl. 561.
Crypsirhina varians, *Vieill. Analyse*, p. 36; *id. et Oud. Gal. Ois.* pl. 106; *Vigors, Mem. Raffles*, p. 665; *Gray, Gen. B.* ii. p. 310; *Blyth, J. A. S. B.* xii. p. 933, xvi. p. 464, xvii. p. 249; *id. Cat. B. Mus. A. S. B.* p. 92; *Bp. Consp.* i. p. 369; *Cab. Mus. Hein.* Th. i. p. 216; *Horsf. & Moore, Cat. B. Mus. E.I. Co.* ii. p. 564; *Blyth, Ibis*, 1862, p. 194; *Gray, Hand-l. B.* ii. p. 9; *Blanf. Ibis*, 1870, p. 469; *Hume, Str. F.* 1874, p. 480, 1875, p. 146.
Phrenotrix temia, *Horsf. Tr. Linn. Soc.* xiii. p. 163; *id. Zool. Research in Java*, pl. 50.
Glaucopis varians, *Temm. Pl. Col. art.* Glaucopis; *Schl. Mus. P.-B.* Coraces, p. 73.
Cryptorhina varians, *Wagler, Syst. Av.* Cryptorhina, sp. 6.
Temia levaillantii, *Lesson, Traité*, p. 342.

Adult male. General colour, both above and below, of a bottle-green, with olive reflections, the head and nape glossed with deep green; forehead, lores, a circlet of feathers round the eye, and a spot at the base of the lower mandible velvety black; primaries black, the secondaries olive-green, like the back; tail black, glossed with olive-green externally; lower abdomen and under tail-coverts rather browner than the rest of the under surface, the lower surface of the wings and tail brownish black; bill and feet black; iris bright blue. Total length 13 inches, culmen 1, wing 4·7, tail 7·8, tarsus 1·1.

Obs. Some specimens are decidedly browner than others; but whether these are females or immature I have not been able to decide. The brown shade shows itself on the lower parts and on the tail.

Hab. Java; Burmese countries from Tenasserim northwards, eastwards also to Cochin China.

a. ♂ ad. sk.	East Java.	A. R. Wallace, Esq. [C.].
b. Ad. sk.	West Java.	A. R. Wallace, Esq. [C.].
c. Juv. sk.	Java.	Dr. Horsfield [C].
d. Ad. sk.	Java.	J. Gould, Esq.
e. Ad. st.	Tenasserim.	J. C. D. Packman, Esq. [P.].
f. Ad. st.	Lower Pegu.	W. T. Blanford, Esq. [P.].
g. Ad. sk.	Tonghoo, Burmah.	Purchased.
h. Ad. sk.	Tonghoo, Burmah.	J. Gould, Esq.
i. Skeleton.		Purchased.

2. Crypsirhina cucullata.

Crypsirhina cucullata, *Jerdon, Ibis,* 1862, p. 20; *Gould, B. Asia,* pt. xv.; *Blanford, Ibis,* 1870, p. 468; *Gray, Hand-l. B.* ii. p. 9; *Hume, Str. F.* iii. p. 147; *Blyth & Wald. J. A. S. B.* 1875, *Suppl.* p. 88.
Glaucopis cucullata, *Schl. Mus. P.-B.* Coraces, p. 74.

Adult. General colour above clear grey, shading off browner on the scapulars, lower back and rump, wing-coverts and innermost secondaries, the two latter shading off into white on their terminal margin; primary-coverts and quills black, the secondaries rather broadly margined with white; head all round, including the throat, dark greenish black, the nape and hind neck white, less distinct on the sides of the latter, and forming only an ill-defined collar across the fore neck; rest of under surface brownish grey, the under wing-coverts and inner bases of the quills whitish; two centre tail-feathers black, washed with brown near the base of the outer web, the remainder of the tail-feathers pale ashy brown, darker towards the base, and inclining to white at the tips; bill black; legs brownish black; iris very pale glaucous blue. Total length 10·5 inches, culmen 0·85, wing 4, tail 6·8, tarsus 0·95.

Young. Differs from the adult in having the hood dusky ashy instead of black; lores, ear-coverts, and chin blackish brown, and the top of the head darker brown; no white line round the neck; the grey of the plumage duller and dingier, and the quills and central tail-feathers altogether duller-coloured (*Hume*); basal portion of bill orange, as well as the edges of the eyelids and inside of the mouth (*Oates*).

Hab. Burmah and Upper Pegu.

a, b. Ad. sk.	Upper Burmah.	T. C. Jerdon, Esq. [P.]. (Types of species.)
c. Ad. st.	Upper Burmah.	W. T. Blanford, Esq. [P.].
d. Ad. sk.	Burmah.	Capt. Stackhouse Pinwill [P.].
e. Ad. sk.	Burmah.	J. Gould, Esq.

21. CISSA.

	Type.
Cissa, *Boie, Isis,* 1826, p. 975, note	C. sinensis.
Corapica, *Less. Traité,* p. 352 (1831)	C. sinensis.
Chlorisoma, *Swains. Classif. B.* ii. p. 230 (1837)	C. thalassina.

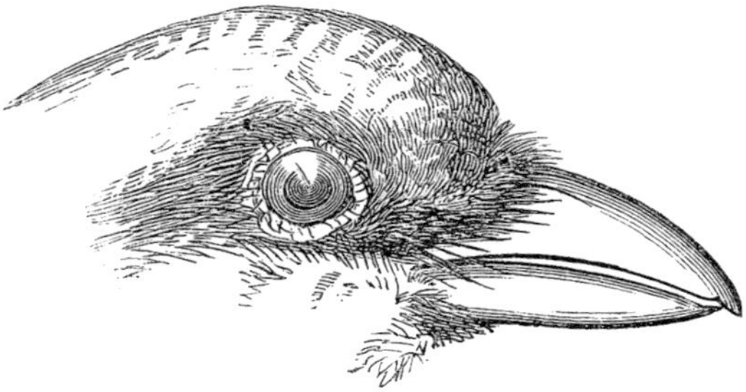

Head of *Cissa chinensis.*

Range. Ceylon, Himalaya Mountains, Burmese countries, Sumatra.

Key to the Species.

a. Head and neck blue or green; through the eye a black band, encircling the nape.
  a'. Inner secondaries ashy grey (in skin), tipped with a broad white spot; tail with black spots. *chinensis*, p. 85.
  b'. Inner secondaries entirely bluish white, the base only greyish black; no black spots on tail .... *thalassina*, p. 86.
b. Head and neck chestnut; no black nape-band; back and breast bright blue........................ *ornata*, p. 87.

1. Cissa chinensis.

Le Rollier de la Chine, *Briss. Orn.* ii. p. 77, pl. 6. fig. 2; *Pl. Enl.* 620.
Chinese Roller, *Lath. Gen. Syn.* i. p. 414.
Coracias chinensis, *Bodd. Tabl. Pl. Enl.* p. 38.
Corvus speciosus, *Shaw, Gen. Zool.* vii. p. 364.
Pica speciosa, *Wagler, Syst. Av.* Pica, sp. 30.
Corapica bengalensis, *Less. Traité*, p. 353.
Kitta venatoria, *Gray, Ill. Ind. Zool.* i. pl. 2; *McClell. P. Z. S.* 1839, p. 164; *Blyth, J. A. S. B.* xii. p. 938; *Bp. Consp.* i. p. 166.
Chlorisoma sinensis, *Swains. Classif. B.* ii. p. 231.
Cissa sinensis, *Blyth, J. A. S. B.* xiii. p. 934; *Gray, Gen. B.* ii. p. 308; *id. Cat. Mamm. &c. Nepal, Hodgs.* p. 101; *Horsf. & Moore, Cat. B. Mus. E.I. Co.* ii. p. 575; *Jerd. B. Ind.* ii. p. 312; *Schl. Mus. P.-B.* Coraces, p. 68; *Blyth & Wald. B. Burm.* p. 89.
Cissa venatoria, *Blyth, J. A. S. B.* xv. p. 29; *id. Cat. B. Mus. A. S. B.* p. 92; *Gould, B. Asia*, part ix.
Kitta speciosa, *Bp. Consp.* i. p. 166.
Citta sinensis, *Licht. Nomencl. Av.* p. 9.
Cissa speciosa, *Gray, Hand-l. B.* ii. p. 6; *Hume, Nests & Eggs Ind. B.* p. 421; *id. Str. F.* 1875, p. 146.

Adult. General colour light green (in life), pale blue (in skin), the head very much crested; lores, feathers round the eye extending in a band round the nape and forming a nuchal band, black; cheeks, sides of neck, and under surface of body light green or blue; wings reddish brown, the least wing-coverts green, like the scapulars, all the quills dull brown on the inner web, the innermost secondaries broadly tipped with bluish white, before which is a very distinct subterminal bar of black; tail pale blue or green, the two middle feathers white at tip, the rest broadly tipped with white, before which is a very broad band of black; bill and legs bright coral-red; bare skin round the eye vermilion (*Jerdon*); edge of eyelid coral-red, rest of it yellowish brown (*Oates*); iris dark reddish brown (*Jerdon*), blood-red (*Oates*). Total length 15·5 inches, culmen 1·45, wing 5·8, tail 7–8·5, tarsus 1·9.

Obs. The curious way in which this bird changes its colour from green to blue has often been remarked. Mr. Blyth, who has kept it alive, thus writes:—"When nearly moulted it is of a lovely green, with the wings bright sanguine red, and the bill and legs deep coral; but whether alive (wild or in confinement) or mounted as a stuffed specimen and exposed to the light, the green soon changes to verdigris-blue."

Hab. South-eastern Himalayas, ranging through the Burmese provinces to Tenasserim.

a. Ad. st.	Nepal.	B. H. Hodgson, Esq. [P.].
b–g. Ad. sk.	Nepal.	B. H. Hodgson, Esq. [P.].
h. Ad. sk.	Nepal.	B. H. Hodgson, Esq. [P.].
i. Ad. sk.	Himalayas.	Indian Museum.
k. Ad. sk.	"Moalmaza."	
l. Bones, various.	Nepal.	B. H. Hodgson, Esq.

Subsp. *α.* **Cissa minor.**

Cissa minor, *Cab. Mus. Hein.* i. p. 86; *Gray, Hand-l. B.* ii. p. 7.

Dr. Cabanis has separated the Sumatran *Cissa* as a distinct species. The birds in the Museum measure as follows:—

	Total length.	Culmen.	Wing.	Tail.	Tarsus.
	in.	in.	in.	in.	in.
a. Ad. Sumatra......	13·0	1·5	5·35	7·8	1·75
b. Juv. Sumatra......	11·5	1·2	4·95	6·0	1·65

Both these Sumatran specimens are smaller than the ordinary Himalayan skins; but beyond that fact there is not much ground for specific separation, especially as the Sumatran examples differ so much *inter se.* The character given by Dr. Cabanis, viz. that the white on the inner secondaries extends up the outer web, as in the Javan *Cissa thalassina*, does not hold good; for these quills are as distinctly spotted as in *C. chinensis.*

Hab. Sumatra.

a, b. Ad. sk.	Sumatra.	Leiden Museum [P.].
c. Ad. sk.	[Sumatra.]	Purchased.

2. **Cissa thalassina.**

Kitta thalassina, *Temm. Pl. Col.* 401; *Bp. Consp.* i. p. 166.
Corapica thallasina, *Less. Traité*, p. 353.
Chlorisoma thalassina, *Swains. Classif. B.* ii. p. 230.
Cissa thalassina, *Gray, Gen. B.* ii. p. 308; *Cab. Mus. Hein.* Th. i. p. 86; *Schl. Mus. P.-B.* Coraces, p. 67; *Gray, Hand-l. B.* ii. p. 7.

Adult male. Beautiful light green above and below, more or less inclining to verdigris-blue; lores and feathers round the eye, extending in a band round the nape, black; least wing-coverts green like the back, the rest of the wing reddish, shaded with green on the greater coverts, the innermost secondaries bluish white, the bases of the outer webs reddish brown, gradually extending towards the tips of the feathers as they approach the secondaries proper; tail green or bluish green, the outermost feather slightly whitish at tip; bill and feet red. Total length 12 inches, culmen 1·4, wing 5·5, tail 5, tarsus 1·75.

Adult female. Similar to the male. Total length 11·5 inches, culmen 1·5, wing 5·4, tail 4·7, tarsus 1·75.

Hab. Java.

a. Ad. st.	Java.	
b. ♂ ad. sk.	W. Java.	A. R. Wallace, Esq. [C.].
c. ♀ ad. sk.	W. Java.	A. R. Wallace, Esq. [C.].

3. Cissa ornata.

Pica ornata, *Wagl. Isis*, 1829, p. 749.
Cissa puella, *Blyth, Cat. B. Mus. A. S. B.* p. 93 (*ex Layard, MS.*); *Layard, Ann. & Mag. N. H.* (2) xiii. p. 213.
Cissa pyrrhocyanea, *Gould, B. Asia*, part 1 (*ex Licht. MS.*).
Kitta ornata, *Bp. Consp.* i. p. 166.
Kitta speciosa, *Bp. t. c.* p. 370 (*lapsu*).
Citta ornata, *Licht. Nomencl. Av.* p. 9.
Cissa ornata, *Blyth, Ibis*, 1867, p. 299; *Schl.* Coraces, p. 69; *Gray, Hand-l. B.* ii. p. 7; *Holdsw. P. Z. S.* 1872, p. 461.

Adult. General colour deep blue, shading off into bright cobalt on the rump and upper tail-coverts; head and neck all round, as well as the chest, chestnut, the rest of the under surface deep blue, shading off into brilliant cobalt on the abdomen and flanks; thighs ultramarine; wing-coverts rich ultramarine, the rest of the wing chestnut on the outer web, blackish on the inner one, the innermost secondaries shaded with blue on the inner web, the first primary black, washed with blue externally; tail deep blue, more or less shaded with cobalt, broadly tipped with white, before which is a tolerably broad bar of black; under wing-coverts blue, shading off into grey on the innermost, the inner lining of the quills blackish, externally edged with chestnut, and inclining to rufous near the base of the inner web; bill red; bare skin round the eye crimson; feet coral-red; iris light brown. Total length 17 inches, culmen 1·6, wing 6·5, tail 10·5, tarsus 1·8.

Young. All the colours duller than in the adult; the wing-coverts greyish black, with a slight wash of blue; lower back, rump, and under surface of body grey, with only a slight tinge of blue; bill blackish.

Hab. Ceylon.

a. Ad. sk.	Ceylon.	E. L. Layard, Esq. [C.].
b, c. Ad. sk.	Ceylon.	Hugh Cuming, Esq.
d. Juv. sk.	Ceylon.	Hugh Cuming, Esq.

22. CALOCITTA.

	Type.
Calocitta, *Gray, List Gen. B.* 1841, p. 50	C. formosa.
Cyanurus, *Bp. Consp.* i. p. 380 (1850, *nec Swains.*)	C. formosa.

Head of *Calocitta formosa.*

Range. Confined to Central America.

Key to the Species.

a. Under surface of body white, with a crescentic band of dull blue across the chest; throat white; cheeks and ear-coverts white .... *formosa,* p. 88.
b. Throat and fore neck black, with a malar patch of white washed with blue; rest of under surface of body white; cheeks and ear-coverts black ...... *colliei,* p. 89.

1. Calocitta formosa.

Pica formosa, *Swains. Phil. Mag.* 1827, p. 437.
Pica bullockii, *Wagler, Syst. Av.* Pica, sp. 4.
Garrula gubernatrix, *Temm. Pl. Col.* ii. pl. 436.
Cyanurus formosus, *Swains. Faun. Bor.-Amer. Birds,* p. 495.
Calocitta bullockii, *Gray, List Gen. B.* 1841, p. 50; *Cab. Mus. Hein.* Th. i. p. 225.
Psilorhinus gubernatrix, *Gray, Gen. B.* ii. p. 308.
Psilorhinus bullockii, *Gray, t. c.* p. 308.
Cyanurus bullockii, *Bp. P. Z. S.* 1837, p. 115; *id. Consp.* i. p. 380; *Gray, Hand-l. B.* ii. p. 6.
Calocitta formosa, *Scl. & Salv. Ibis,* 1859, p. 22; *Scl. Cat. Amer. B.* p. 145; *Scl. & Salv. Nomencl. Av. Neotr.* p. 40; *Lawr. Mem. Bost. Soc. N. H.* ii. p. 285.

Adult. General colour above bluish grey, brighter round the hinder neck and on the sides of the latter; wings more decidedly blue than the back, with a slight shade of greyish purple, especially on the innermost secondaries, which are also decidedly shaded with cobalt on the inner webs, the rest of the quills brownish on their inner webs; upper tail-coverts dull cobalt; tail-feathers rich purplish cobalt, the four outermost rectrices white for the terminal third; crown of head black washed with blue, with a long crest of slightly recurved feathers springing from the forehead and reaching beyond the nape; lores and feathers above the eye bluish white; feathers round the eye dusky; above the hinder part of the latter a small black spot; cheeks, ear-coverts, and sides of face creamy white, with a broad black band running down the sides of the neck behind the ear-coverts, and trending below the latter on their hinder margin; from below the eye a narrow perpendicular streak of bluish black crosses the cheek; entire under surface of body creamy white, the chin slightly washed with blue, and the chest crossed with a crescentic band of dull blue inclining to black on the sides of the chest; flanks purplish grey; under wing-coverts creamy white, the edge of the wing greyish blue. Total length 18 inches, culmen 1·5, wing 6·95, tail 12, tarsus 1·7.

Nestling. Upper surface dusky grey, the downy feathers edged with whitish at the tip, the new feathers dull verditer-green; wing-coverts and quills greenish; tail dull cobalt, the outer feathers tipped with white; head and neck all round and entire under surface of body white.

Hab. Central America: Mexico, Guatemala, Costa Rica, Nicaragua, Honduras.

a. ♂ ad. sk.	Oaxaca, Mexico.	A. Boucard, Esq. [C.].
b. ♀ jun. sk.	San Juan del Rio, Pacific side of Mexico (*Rébouch*).	A. Boucard, Esq. [C.].
c. Ad. sk.	Honduras.	Purchased.
d. Pull. sk.	Honduras.	Mr. Dyson [C.].
e. Ad. st.	Nicaragua.	A. Sallé, Esq. [C.].
f. Ad. st.	Central America.	G. Barclay, Esq. [P.].

2. Calocitta colliei.

Pica colliei, *Vigors in Zoology of Beechey's Voyage*, p. 22, pl. 7.
Pica bullockii, *Audub. B. Amer.* iv. p. 105, pl. 227.
Cyanurus colliei, *Bp. Consp.* i. p. 381; *Gray, Hand-l. B.* ii. p. 6.
Calocitta colliei, *Finsch, Abhandl. Ver. Brem.* 1870, p. 334; *Scl. & Salv. Nomencl. Av. Neotr.* p. 40; *Lawr. Mem. Bost. Soc. N. H.* ii. p. 285.
Calocitta elegans, *Finsch, Abhandl. Ver. Brem.* 1870, p. 334.

Adult. General colour above dull blue; the wings like the back but a little brighter on the edges, the primaries externally washed with cobalt; tail-feathers purplish cobalt, the four outermost broadly tipped with white; crown of head deep blue, with a very long crest of black feathers slightly washed with blue, and extending from the forehead beyond the nape; lores, region of the eye, ear-coverts, sides of neck, and entire throat and fore neck black, the latter here and there washed with blue; above and below the eye a streak of bluish-white plumes; a broad malar stripe of white, blue on its upper margin; down the sides of the neck a narrow band of bluish white feathers running behind the ear-coverts; rest of under surface of body creamy white, the flanks purplish grey; under wing-coverts white, the edge of the wing streaked with blue; bill and legs black. Total length 24 inches, culmen 1·5, wing 7·75, tail 17·7, tarsus 1·75.

Young. Similar to the adult, but having the long crest-feathers very broadly tipped with white, and all the black feathers of the throat obscured by very broad bluish white edgings.

Hab. Western Mexico.

a. Ad. sk.	[Central America.]	
b. Ad. sk.	[Central America.]	Dr. J. E. Gray [P.].
c. Jun. sk.	San Blas.	Capt. Kellett and Lieut. Wood [P.].
d. Ad. sk.	San Blas.	A. Collie, Esq. [P.]. (Type of species).

23. PLATYSMURUS.

	Type.
Platysmurus, *Reich. Syst. Av.* tab. lxxx. fig. 6 (1850)	P. leucopterus.
Glenargus, *Cab. Mus. Hein.* Th. i. p. 216 (1850)	P. leucopterus.

Range. From Burmah, through Tenasserim, down the Malayan peninsula; Sumatra, Borneo.

Key to the Species.

a. Black, with a white bar down the wing.......... *leucopterus*, p. 90.
b. Entirely black; head crested .... *aterrimus*, p. 91.

1. Platysmurus leucopterus.

Glaucopis leucopterus, *Temm. Pl. Col.* 265; *Schl. Mus. P.-B.* Coraces, p. 72.
Cryptorhina leucoptera, *Wagl. Syst. Av.* Cryptorhina, sp. 4.
Temnurus leucopterus, *Less. Traité*, p. 341; *Gray, Gen. B.* ii. p. 310; *Blyth, Cat. B. Mus. A. S. B.* p. 92; *Gray, Hand-l. B.* ii. p. 8; *Blyth & Wald. Cat. B. Burma*, p. 88; *Hume, Str. F.* 1874, p. 480.
Crypsirhina leucoptera, *Blyth, J. A. S. B.* xii. p. 933; *Bp. Consp.* i. p. 369.
Platysmurus leucopterus, *Reich. Syst. Av.* tab. lxxx. fig. 6; *Horsf. & Moore, Cat. B. Mus. E.I. Co.* ii. p. 564; *Salvad. Ucc. Born.* p. 279.
Glenargus leucopterus, *Cab. Mus. Hein.* Th. i. p. 216.

Adult. Everywhere black, excepting a white bar down the wing formed by the tips of the greater coverts and the basal two thirds of two of the central secondaries being white, the amount on the latter quills varying in extent in different individuals; forehead much crested, the feathers being harsh in texture and somewhat recurved; round the eye a circlet of scaly black feathers. Total length 16 inches, culmen 1·25, wing 7·6, tail 7·8, tarsus 1·55.

Hab. Malayan peninsula and Sumatra, ranging northwards into Tenasserim.

a. Ad. st.	Malacca.	Major-Gen. Hardwicke [P.].
b. Ad. sk.	Malacca.	Capt. Stackhouse Pinwill [P.].
c. Ad. sk.	Malacca.	W. Harvey, Esq. [P.].
d. Jun. sk.	Malacca.	W. Harvey, Esq. [P.].
e. ♀ ad. sk.	Malacca.	A. R. Wallace, Esq. [C.].
f. ♀ ad. sk.	Sumatra.	A. R. Wallace, Esq. [C.].

2. **Platysmurus aterrimus.**

Glaucopis aterrimus, *Temm. Pl. Col.* livr. 57*a*.
Temnurus aterrimus, *Less. Traité*, p. 341; *Gray, Gen. B.* ii. p. 310; *id. Hand-l. B.* ii. p. 8.
Crypsirhina aterrima, *Bp. Consp.* i. p. 369; *Scl. P. Z. S.* 1863, p. 218.
Glenargus aterrimus, *Cab. Mus. Hein.* Th. i. p. 216, note.
Glaucopis aterrima, *Schl. Handl. Dierk.* i. p. 323; *id. Mus. P.-B.* Coraces, p. 72.
Platysmurus aterrimus, *Salvad. Ucc. Born.* p. 279; *Sharpe, Ibis,* 1877, p. 20.

Adult. Entirely black, with no white on the wing at all; frontal feathers developed into a very fine crest; "bill and feet black; iris crimson" (*Everett*). Total length 12·5 inches, culmen 1·35, wings 6·85, tail 7·5, tarsus 1·45.

Hab. Borneo.

a. Ad. sk.	Borneo.	Sir J. Brooke [P.].
b. Ad. sk.	South Borneo.	J. Gould, Esq.
c. Ad. sk.	Banjermassing.	J. Motley, Esq. [C.].
d. Ad. sk.	Belidah, Sarawak.	A. Everett, Esq. [C.].
e, f. ♂ ♀ ad. sk.	Bintulu, Oct. 1875.	A. Everett, Esq. [C.].

24. TEMNURUS.

Type.
Temnurus, *Less. Traité*, p. 341 (1831) .... T. truncatus.

Range. Confined to Cochin China.

1. **Temnurus truncatus.**

Glaucopis temnura, *Temm. Pl. Col.* 357; *Schl. Mus. P.-B.* Coraces, p. 72, note.
Temnurus truncatus, *Less. Traité*, p. 341; *Gray, Gen. B.* ii. p. 310; *Bp. Consp.* i. p. 369; *Gray, Hand-l. B.* ii. p. 8.
Crypsirhina temnura, *Blyth, J. A. S. B.* xii. p. 933.

Adult. Plumage bronzy black; tail graduated, each feather cut square at the tip (*Lesson*).

Hab. Cochin China (*Diard*).

25. GARRULUS.

Type.
Garrulus, *Briss. Orn.* ii. p. 46 (1760).............. G. glandarius.
Glandarius, *Koch, Syst. baier. Zool.* p. 99 (1816) .... G. glandarius.

Range. Europe and Asia, including all the countries of the Mediterranean basin, Central Asia, and the Himalaya Mountains; Burmese countries; China, Formosa, and Japan.

Key to the Species.

a. Throat white, vinaceous, or isabelline, not black.
- a'. Crown of head streaked with black.
  - a''. Back vinaceous: forehead whitish, with black streaks.
    - a'''. Space in front of the eye whitish, or vinaceous.
      - a^4. Larger: wing 7·3 inches; throat white; feathers of forehead and crown edged with white .... *glandarius*, p. 93.
      - b^4. Smaller: wing 6·5 inches; throat vinaceous; feathers of forehead and crown edged with vinaceous ...... *hyrcanus*, p. 94.
    - b'''. Space in front of eye black like the cheek-stripe .... *japonicus*, p. 95.
  - b''. Back grey, contrasting with rufous neck.
    - c'''. Nasal bristles and feathers round the eye white, the former tipped with black; throat white; head and neck vinaceous, broadly streaked with black, especially on the forehead .... *minor*, p. 96.
    - d'''. Nasal bristles and feathers over the eye foxy red, the former tipped with black; throat white, washed with rufous; head and neck bright foxy red, streaked with black .... *brandti*, p. 96.
- b'. Crown of head uniform black.
  - c''. Forehead pure white, without any spots; cheeks and throat pure white, the former with a slight shade of vinaceous here and there .... *atricapillus*, p. 97.
  - d''. Forehead whitish, spotted or streaked with black.
    - e'''. Secondaries externally white for basal half.
      - c^4. Back greyish; neck all round rich chestnut; lores, feathers round the eye, and ear-coverts pure white.... *cervicalis*, p. 98.
      - d^4. Back vinous, the sides of the neck uniform with the back; the ear-coverts and sides of face vinous like the sides of the neck .... *krynickii*, p. 99.
    - f'''. Secondaries externally black, barred with blue, exactly like the primary-coverts; eyebrow, ear-coverts, sides of neck, and throat white; rest of under surface dull tawny, the throat contrasting with the breast .... *leucotis*, p. 99.
- c'. Crown of head uniform vinous fawn-colour.
  - e''. Nasal bristles fawn-colour like the forehead, which has a few brown or rufous streaks; wing 6·35 inches .... *bispecularis*, p. 100.

f''. Nasal bristles very distinctly tipped with black; forehead slightly streaked with the latter; wing 6·95 inches .... *sinensis*, p. 101.
g''. Nasal bristles entirely black; wing 5·9 inches .... *taivanus*, p. 101.

b. Throat black, with white stripes.
d'. Head and ear-coverts black; back vinaceous; upper tail-coverts grey; primary-coverts white................................ *lanceolatus*, p. 101.
e'. Head and ear-coverts, back and upper tail-coverts rufous; bill yellow; primary-coverts barred with black and blue .... *lidthi*, p. 102.

1. Garrulus glandarius.

Le Geay, *Briss. Orn.* ii. p. 47; *Montb. Pl. Enl.* 481.

Corvus glandarius, *Linn. S. N.* i. p. 156; *Werner, Atlas*, Omnivores, pl. 8; *Yarr. Brit. B.* ii. p. 116; *Démid. Voy. Russie Mér.* iii. p. 115.

Glandarius pictus, *Koch, Syst. Baier. Zool.* p. 99.

Garrulus glandarius, *Vieill. N. Dict. d'Hist. Nat.* xii. p. 471; *Naum. Vög. Deutschl.* ii. pl. 58. fig. 1; *Gould, B. Eur.* iii. pl. 214; *Macgill. Hist. Brit. B.* i. p. 576; *Kjærb. Orn. Dan.* pl. xii. fig. 3; *Gray, Gen. B.* ii. p. 306; *Bp. Consp.* i. p. 375; *id. P. Z. S.* 1850, p. 81; *Cab. Mus. Hein.* i. p. 220; *Schl. Vog. Nederl.* pl. 143; *Horsf. & Moore, Cat. B. Mus. E.I. Co.* ii. p. 578; *Sundev. Sv. Fogl.* pl. xix. fig. 1; *Fritsch, Vög. Eur.* tab. 27. fig. 10; *Gould, B. Gt. Br.* iii. pl. 64; *Schl. Dier. Nederl.* iii. pl. 13. fig. 8; *Schl. Mus. P.-B.* Coraces, p. 57; *Degl. et Gerbe, Orn. Eur.* i. p. 215; *Borggr. Vogelf. Norddeutschl.* p. 80; *Doderl. Avif. Sicil.* p. 66; *Gray, Hand-l. B.* ii. p. 2; *Salvad. Faun. Ital. Ucc.* p. 176; *Dresser, B. Eur.* part xvi.; *Irby, B. Gibr.* p. 132.

Lanius glandarius, *Nilss. Orn. Suec.* p. 75.

Glandarius germanicus, *Brehm, Vög. Deutschl.* p. 180.

Glandarius septentrionalis, *id. t. c.* p. 180.

Glandarius robustus, tæniurus, *et* leucocephalus, *Brehm, Naum.* 1855, p. 273.

Adult male. Fore part of head white, all the feathers with a broad central streak of black; nasal bristles and lores rusty white; round the eye a circlet of whitish plumes, those behind the latter slightly mottled with black bases; cheeks black, forming a broad moustache; rest of the sides of the face and of the neck, as well as the hind head and upper portion of mantle, vinous red, the occipital plumes slightly shaded with a greyish lustre; the rest of the back pale grey, with a slight vinous tinge on the scapulars and lower back; rump and upper tail-coverts pure white; tail black, shaded with grey near the base, and having slight indications of bluish grey bars; least upper wing-coverts light grey, tinged with vinous and inclining to chestnut on the median series, the bastard wing and primary-coverts barred with black and bright cobalt-blue, the greater series black, the outermost barred externally with cobalt like the primary-coverts; quills black, the primaries externally washed with ashy grey; the secondaries white for more than half of the outer web and shaded with cobalt near the shaft, the inner ones showing slight indications of blue barring; innermost secondaries black, the extreme inner feather deep chestnut, tipped with black; throat white; rest of under surface of body pale vinous red, shaded slightly with grey; thighs whitish; vent and under tail-coverts pure white; under wing-coverts very deep vinous; inner lining of quills greyish black, much paler along the inner web. Total length 13 inches, culmen 1·15, wing 7·3, tail 6·6, tarsus 1·7.

Young. Similar to the adults, but the colours not so vivid.

Hab. Europe.

a. Var. st.	England.	Mr. Bullock.
b. Ad. sk.	England.	J. Gould, Esq. [C.].
c. ♀ ad. sk.	Preston Hall.	J. Gould, Esq. [C.].
d. ♂ ad. sk.	Hawkstone, Salop, Jan. 1869.	J. Gould, Esq. [C.].
e. Pull. st.	Cambridgeshire.	Mr. J. Baker [C.].
f, g. Ad. st.	Oakley Eye, Sussex.	Rev. J. Kirby [P.].
h, i. Ad. st.	Cookham, Berks.	Mr. J. Briggs [P.].
k, l. ♂ ad. sk.	Avington, Hants, Nov. 25, 1872.	Edward Shelley, Esq. [P.]
m. Ad. st.	Devonshire.	Col. Montagu [P.].
n. Albino.	Germany.	Dr. Kaup [P.].
o. ♂ ad. sk.	Wermland, Feb. 28, 1874.	Purchased.
p, q. Ad. sk.	Sweden (*Wheelwright*).	J. Gould, Esq. [P.].
r. Ad. sk.	Belgium, Jan. 14, 1849.	Brussels Museum [P.].
s. ♂ ad. sk.	Luz, Hautes-Pyrénées, May 20, 1872.	A. B. Brooke, Esq. [C.].
t. ♂ ad. sk.	Turkey.	T. Robson, Esq. [C.].
u. ♂ ad. sk.	Petnahore, Turkey, Dec. 8, 1870 (*Robson*).	R. B. Sharpe, Esq. [P.].
v. ♂ ad. sk.	Ortakeuy, Turkey, March 3, 1871 (*Robson*).	R. B. Sharpe, Esq. [P.].
w. Skull.		
x. Skull.	England.	J. Ray, Esq.
y. Sternum.	England.	J. Ray, Esq.
z, a'. Skulls.	S. Germany.	Dr. Günther.

2. Garrulus hyrcanus.

Garrulus hyrcanus, *Blanford, Ibis*, 1873, p. 225; *Dresser, B. Eur.* pt. xx. (*Review of genus* Garrulus); *Blanford, E. Persia*, p. 265, pl. xviii.

Adult male. General colour vinous red, with a slight greyish lustre, the scapulars obscurely waved under certain lights; forehead and crown streaked with black, the margins pale fulvous or isabelline, the hinder streaks narrower and waved with grey, nasal bristles isabelline, tipped with black; feathers round the eye isabelline buff, tipped with brown; ear-coverts pale vinous red; cheeks black, forming a very broad moustache; wing-coverts vinous red like the back, the median series inclining to chestnut at tip; bastard wing, outer web of greater coverts and the primary-coverts cobalt-blue, barred with black, the interspaces here and there inclining to whitish, inner greater coverts velvety black; quills black, the primaries externally whity brown, the secondaries white, with faint indications of blue bars, the terminal third black, the innermost secondary deep chestnut, tipped with black, the next two feathers also distinctly dashed with chestnut on the outer web; hinder rump and upper tail-coverts pure white; tail black, with indications of bluish grey bars near the base; under surface of body deep vinous red, paler and inclining to isabelline on the throat; lower abdomen, vent, under tail-coverts, and thighs white; under wing-coverts deep vinous, a little deeper than the sides of the body, which are more richly coloured than the breast. Total length 11·5 inches, culmen 1·3, wing 6·4, tail 5·4, tarsus 1·7.

Young female. Similar to the male, but rather duller in colour, the streaks on the head not nearly so broad or distinct. Total length 11·8 inches, wing 6·5, tail 5·7, tarsus 1·65.

Hab. Northern Persia.

a. ♂ ad. sk.	Mazenderan.	Major St. John [C.]. (One of the typical specimens.)
b. ♀ jun. sk.	Anan, Mazenderan, August 12, 1872.	W. T. Blanford, Esq. [C.].

3. Garrulus japonicus.

Garrulus glandarius japonicus, *Schl. Faun. Jap.* p. 83, pl. 43 (1850).
Garrulus japonicus, *Bp. Consp.* i. p. 375; *id. P. Z. S.* 1850, p. 375; *Cab. Mus. Hein.* Th. i. p. 220; *Schl. Mus. P.-B.* Coraces, p. 60; *Dresser, B. Eur. art.* Garrulus, part xx.

Adult. Above deep vinaceous red, the forehead and fore part of crown white, with broad central markings of black; lores, feathers round the eye, and a broad malar patch black; ear-coverts and sides of neck deep vinaceous red, like the back; lower rump and upper tail-coverts white; tail brownish black, with scarcely any remains of greyish blue bars near the base; scapulars and wing-coverts vinaceous, the median series waved near the tips with narrow bars of grey and chestnut; greater series black, the exterior ones barred externally with cobalt-blue and black, the bastard wing and primary-coverts entirely cobalt, barred with black; quills black, the primaries externally ashy white, increasing in extent on the outer web towards the outer feathers, the outer secondaries pure white for the greater part of the exterior web, barred near the base with cobalt-blue and black; the inner secondaries black, like the adjoining coverts, the last one chestnut on the inner web; throat, centre of abdomen, thighs, and under tail-coverts pure white, the breast buff, tinged with vinaceous; the sides of the body and under wing-coverts deep vinaceous red. Total length 11·5 inches, culmen 1·15, wing 6·25, tail 5·6, tarsus 1·5.

Hab. Japan.

a. Ad. st.	Japan.	Purchased.
b. Ad. sk.	Nagasaki, Japan.	Capt. St. John [C.].
c. Ad. sk.	Nagasaki, Japan.	J. Gould, Esq.

4. Garrulus minor.

Garrulus minor, *J. Verr. Rev. et Mag. de Zool.* 1857, p. 439, pl. 14; *Loche, Cat. Mamm. &c. Alger.* p. 52; *id. Expl. Scient. Algér. Ois.* p. 122.

Adult. Head and neck vinous red, the forehead whitish, with broad black centres to the frontal plumes as well as to those of the crown, the hindermost of which have indistinct transverse rays; nasal bristles buffy white, tipped with black; a small eyebrow, feathers round the eye, and ear-coverts white, the latter vinous red on their hinder part; a broad malar streak of black; back and scapulars grey, the latter with a slight tinge of vinous; lower rump and upper tail-coverts white; tail brownish black, with a few greyish blue bars near the base; lesser wing-coverts grey like the back, the median ones inclining to chestnut at the tip, some of the latter with faint transverse rays of grey; bastard wing and primary-coverts bright cobalt, with narrow black cross bars, some of the interspaces inclining to silvery white, the rest of the wing-coverts velvety black, the outer ones barred with black and cobalt, like the primary-coverts; quills black, the primaries externally ashy white, the secondaries white for basal half of outer web, shaded with blue near the extreme base, the innermost secondaries black, with chestnut base; throat white; breast and sides of body pale vinous red, deeper on the flanks, the shaft-streaks white; centre of the abdomen, thighs, and under tail-coverts white, the latter pure white; under wing-coverts deep vinous; bill brownish. Total length 12 inches, culmen 1·1, wing 6·8, tail 6, tarsus 1·65.

Hab. Algeria.

a. Ad. st.	Algeria.	Purchased.

5. Garrulus brandti.

Garrulus brandti, *Eversm. Add. Pall. Zoogr.* iii. p. 8; *Hartl. Rev. Zool.* 1845, p. 52; *Bp. Consp.* i. p. 376; *id. P. Z. S.* 1850, p. 82; *Cab. Mus. Hein.* Th. i. p. 220; *Schl. Mus. P.-B.* Coraces, p. 59; *Whitely, Ibis,* 1867, p. 200, pl. iii.; *Gray, Hand-l. B.* ii. p. 2; *Swinh. P. Z. S.* 1871, p. 382; *David, N. Arch. Mus.* vii. *Bull.* p. 9; *Dresser, B. Eur.* pt. xvi.

Garrulus glandarius, var. brandti, *Midd. Sibir. Reis.* p. 157; *Radde, Reis. Ost-Sibir.* ii. p. 204; *Schrenck, Reis. Amurl.* p. 316.

Adult (Altai Mountains). Upper surface of body grey, somewhat tinged with vinous on the rump, the upper tail-coverts white; head and neck rufous, the nasal bristles tipped with black, the frontal plumes spotted and the crown streaked with black; feathers round the eye mottled with black markings; ear-coverts and sides of neck rufous; cheeks black, forming a very broad moustache; throat fulvescent, with white shaft-streaks; rest of under surface pale vinous red, the fore neck and chest washed with rufous like the neck, the flanks washed with grey; the lower abdomen and thighs isabelline colour; under tail-coverts white; under wing-coverts rather more rufous than the breast; least wing-coverts greyish like the scapulars, the median ones inclining externally to chestnut; bastard wing and primary-coverts cobalt-blue, barred with black, the interspaces more or less inclining to white, and shaded with purple along the black bars; the greater series velvety black, the outermost barred with black and cobalt, like the others; quills blackish, the primaries externally ashy white, the secondaries white for two thirds of outer web, faintly barred with black and cobalt near the base, the innermost secondaries chestnut-red, black at tip, the next two feathers with a dash of chestnut on the outer web; tail black, with a few indications of greyish blue bars near the base; bill black; "iris dusky brown" (*Dybowski, MS.*). Total length 12·5 inches, culmen 1·2, wing 6·75, tail 6·2, tarsus 1·5.

Obs. There seems to be no difference in specimens from Japan, Siberia, and the Altai Mountains. A pair collected in the Baikal district by Dr. Dybowski measure as follows:—

	Total length. in.	Wing. in.	Tail. in.	Tarsus. in.
a. ♂ ad. Kultuk............	13·0	6·85	6·3	1·6
b. ♀ ad. Kultuk	13·5	6·7	6·2	1·55

Hab. Ural Mountains, across the whole of Siberia to Japan and Northern China.

a. Ad. sk.	Altai Mountains.	Purchased.
b. ♂ ad sk.	Ussuri river (*Dybowski*).	Warsaw Museum [P.].
c, d. ♂ ♀ ad. sk.	Kultuk, southern part of Lake Baikal.	Dr. Dybowski [P.].
e. Ad. sk.	Hakodadi, Japan, Oct. 15, 1865 (*Whitely*).	R. B. Sharpe, Esq. [P.].
f. ♂ ad. sk.	Hakodadi, Japan, Oct. 23, 1865 (*Whitely*).	J. Gould, Esq.
g. ♀ ad. sk.	Hakodadi, Japan, Nov. 22, 1865 (*Whitely*).	J. Gould, Esq.

6. Garrulus atricapillus.

Garrulus atricapillus, *Geoffr. St.-Hilaire, Etud. Zool.* fasc. i.; *Dresser, B. Eur.* part xx.
Garrulus stridens, *Hempr. & Ehrenb. Symb. Phys.* fol. L.
Garrulus melanocephalus, *Géné, Mem. Accad. Torino*, xxxvii. p. 298; *Bp. P. Z. S.* 1850, p. 82; *Bree, B. Eur.* i. p. 144.
Garrulus iliceti, *Licht. Nomencl.* p. 9.

Adult female. General colour above clear vinaceous, washed with grey, the rump and upper tail-coverts pure white: wings and tail much as in *G. cervicalis*; nasal bristles white, washed with vinaceous at the tip; forehead, feathers round the eye, a broad eyebrow, and the sides of the face and neck, as well as the throat, white; a broad malar bar of black; rest of under surface pale vinaceous, deeper on the sides of the body and under wing-coverts, and inclining to buffy white on the centre of the abdomen, thighs, and under tail-coverts. Total length 12·7 inches, culmen 1·2, wing 6·7, tail 6, tarsus 1·6.

Adult male. Similar to the female, showing less grey on the back and rather more on the chest; the face and throat somewhat washed with a vinous tinge. Total length 14·5 inches, culmen 1·4, wing 7·6, tail 7, tarsus 1·75.

Obs. Garrulus atricapillus and *G. krynickii* are by no means so different as Mr. Dresser demonstrates in his 'Birds of Europe.' Some specimens of the latter have the forehead nearly as white as in Syrian or Palestine birds; and although certain examples of *G. krynickii* are very rich vinous on the sides of the face, others are whitish, with very little vinous tinge. Although I keep them distinct for the present, I should not be surprised to learn that a large series showed a gradual transition from one form to the other.

Hab. Syria and Palestine eastwards into Persia.

a. ♀ ad. sk.	Samaria.	Canon Tristram [C.].
b. ♂ ad. sk.	Plain of Esdraelon.	Canon Tristram [C.].
c. ♂ ad. sk.	Oak-forest, near Shiraz, June 1870.	Major St. John [C.].

7. Garrulus cervicalis.

Garrulus cervicalis, *Bp. C. R.* xxxvii. p. 828; *Loche, Expl. Sci. Algér.* i. p. 120 (1867); *Dresser, B. Eur.* part xix. (1873); *Bp. C. R.* xxxvii. p. 828 (1853).

Garrulus melanocephalus, *Bp. Consp.* i. p. 375 (*nec Géné*); *Malh. Bull. Soc. N. H. Moselle, Metz*, 7th cahier, p. 24 (1853); *Schl. Mus. P.-B.* Coraces, p. 59 (pt.).

Garrulus atricapillus, *Levaill. Jun., Expl. Sci. Algér. Atlas*, pl. 6 (1855, *nec Geoffr.*).

Garrulus krynicki, *Fritsch, Vög. Eur.* tab. 27. fig. 11 (*nec Kalecz.*).

Adult. Crown of head black, the forehead white, with black centres, the occiput with a few black streaks, and slightly shaded with grey; nape, hind neck, and sides of neck deep vinaceous red; the entire back pale ashy grey, slightly tinged with vinaceous on the rump, the lower part of which is white, like the upper tail-coverts; tail brownish black, bluish grey at base, and barred with the same colour for more than half the length of the centre feathers; wing-coverts grey, the least coverts barred with chestnut, the median series broadly chestnut at tip; the greater series velvety black, the outermost barred with blue and black, the bastard wing and the primary-coverts also cobalt-blue, narrowly barred with black; quills brownish black, externally greyish white; the secondaries white for two thirds of the outer web, with slight indications of blue bars, the innermost black, inclining to chestnut on all but the tip of the last secondary; lores, a narrow eyebrow, feathers round the eye, and ear-coverts pure white; cheeks black, forming a very broad moustache; throat white; the rest of the under surface fulvous, with indistinct white shaft-streaks, the sides of the body vinaceous, as also the under wing-coverts; abdomen, thighs, and under tail-coverts white. Total length 14 inches, culmen 1·15, wing 6·9, tail 6·2, tarsus 1·7.

Hab. Algeria.

a. Ad. sk.	Algeria.	Purchased.
b. Ad. st.	Algeria.	M. Malherbe [E.].
c. Ad. sk.	Algeria.	J. Gould, Esq.
d. Sternum.	Algeria.	Purchased.

8. Garrulus krynickii.

Garrulus glandarius, var. (pileo nigro), *Hohenacker, Bull. Soc. Imp. Mosc.* x. p. 141.
Garrulus krynicki, *Kalenicz. Bull. Mosc.* xii. p. 319, pl. 9; *Bp. Consp.* i. p. 375; *Schl. Mus. P.-B.* Coraces, p. 58; *Degl. & Gerbe, Orn. Eur.* i. p. 216; *Dresser, B. Eur.* part xvii.; *Bree, B. Eur.* ii. p. 8.
Garrulus melanocephalus, *Schl. & Susem. Vög. Eur.* ii. Taf. 6 (*nec Géné*).
Garrulus glandarius melanocephalus, *Sch. Faun. Jap.* p. 83.

Adult male. Above vinaceous, washed with grey on the scapulars; the sides of the face and neck also deep vinaceous, the feathers round the eye whitish, the lores and frontal plumes white, slightly washed with vinaceous and broadly streaked with black; the crown of the head and nape jet-black; a broad malar patch black; under surface of body deep vinaceous, especially rich on the sides of the body; the throat, centre of the abdomen, and thighs vinaceous white; upper and under tail-coverts pure white; wings and tail as in *G. glandarius*, but showing rather more white on the secondaries. Total length 13·5 inches, culmen 1·2, wing 7·05, tail 6·2, tarsus 1·7.

Hab. Turkey and Asia Minor.

a. Ad. sk.	Xanthus, Asia Minor.	Sir C. Fellows [P.].
b. Ad. sk.	Xanthus, Asia Minor.	Sir C. Fellows [P.].
c. ♂ ad. sk.	Asia Minor.	T. Robson, Esq. [C.].
d. ♂ ad. sk.	Belgrade, Turkey, March 14, 1870 (*Robson*).	J. Gould, Esq.
e. Ad. sk.	[Turkey*.]	J. Gould, Esq.

9. Garrulus leucotis. (Plate IV.)

Garrulus leucotis, *Hume, Str. F.* 1874, pp. 106, 443, 480; *Blyth & Wald. B. Burm.* p. 89.

Adult female. Above ashy fawn-colour on the back, inclining to purer fawn-colour on the scapulars and lower back; rump and upper tail-coverts white; nasal bristles and forehead white, streaked with black, the crown and nape entirely black; lores, feathers above the eye, ear-coverts, sides of face, and entire throat white; a broad malar streak of black; rest of under surface fawn-colour, the under tail-coverts white; under wing-coverts chestnut, the lower series ashy like the lining of the wings; least and median series of upper wing-coverts chestnut, the greater series black, the inner ones inclining to chestnut near the base, like the adjoining secondaries;

* Marked "Syria;" but apparently one of Mr. Robson's skins from Turkey or Asia Minor: it is labelled, in Prince Bonaparte's handwriting, "*G. melanocephalus.*"

quills black, the primaries margined with whity brown towards the tip, the secondaries barred with black and cobalt, the innermost deep chestnut at the base; all the feathers of the bastard wing, the primary-coverts, and the outermost greater coverts barred with black and cobalt, exactly like the secondaries, but much narrower; tail black; bill blackish horny, whitish at tip; feet whitish horny; iris wood-brown. Total length 12·5 inches, culmen 1·25, wing 6·7, tail 5·7, tarsus 1·6.

Hab. Burmese countries.

a. ♀ ad. sk.	Tonghoo, Burmah, Nov. 17, 1874 (*Wardlaw Ramsay*).	Marquess of Tweeddale [P.].

10. Garrulus bispecularis.

Garrulus bispecularis, *Vigors, P. Z. S.* 1830, p. 7; *Gould, Cent. Himal. B.* pl. 38; *Royle, Himal.* pl. —; *Hodgs. in Gray's Zool. Misc.* p. 84; *Bp. P. Z. S.* 1850, p. 82; *id. Consp.* i. p. 376; *Cab. Mus. Hein.* Th. i. p. 220; *Horsf. & Moore, Cat. B. Mus. E.I. Co.* ii. p. 579; *Jerd. B. Ind.* ii. p. 307; *Schl. Mus. P. B.* Coraces, p. 60; *Gray, Hand-l. B.* ii. p. 3; *Henders. & Hume, Lahore to Yark.* p. 242; *Hume, Nests and Eggs Ind. B.* p. 416.

Garrulus ornatus, *J. E. Gray, in Hardw. Ill. Ind. Zool.* i. pl. 23. fig. 2; *Gray, Gen. B.* ii. p. 306; *Blyth, J. A. S. B.* xv. p. 31; *id. Cat. B. Mus. A. S. B.* p. 94; *David, N. Arch. Mus.* vii. *Bull.* p. 9.

Adult. General colour vinaceous fawn-colour, much paler on the forehead and nasal bristles, which incline to isabelline; a broad malar streak of black; under surface of body pale vinaceous fawn-colour, the throat and abdomen inclining to isabelline; thighs whitish; upper and under tail-coverts white; least wing-coverts vinaceous chestnut, deeper on the median coverts, which are ashy black at base, the outermost having some transverse rays of grey; bastard wing barred with black and cobalt; primary-coverts externally barred with black and cobalt; rest of the coverts velvety black, the innermost of all chestnut, with a black tip; quills black, the primaries white along the outer edge, the secondaries also externally barred with cobalt for about two thirds of the outer web; tail black, with a few indications of grey bars near the base; "bill dusky horny; legs dull yellowish; iris red-brown" (*Jerdon*). Total length 12 inches, culmen 1·15, wing 6·35, tail 6·, tarsus 1·55.

Obs. Specimens from the North-western Himalayas are rather paler on the head and have longer bills than those from Nepal and Sikkim.

Hab. Himalaya Mountains.

a. Ad. st.	Himalaya Mountains.	Major-General Hardwicke [P.].
b. Ad. sk.	Himalaya Mountains (*E. V. Harcourt*).	J. Gould, Esq.
c, d, e. Ad. sk.	Nepal.	B. H. Hodgson, Esq. [P.].
f. Ad. sk.	Darjiling.	Purchased.
g, h. Ad. sk.	N.W. Himalayas.	Capt. Stackhouse Pinwill [P.].
i. Ad. sk.	Ladakh.	Purchased.

11. Garrulus sinensis*.

Garrulus ornatus, *Swinh. Ibis*, 1862, p. 263 (*nec Gray*).
Garrulus sinensis, *Swinh. P. Z. S.* 1863, p. 304 (*ex Gould MS.*); *id. P. Z. S.* 1871, p. 381; *David, N. Arch. Mus. Bull.* vii. p. 9.

Adult. Above deep vinaceous red, both above and below, slightly paler on the throat and centre of the abdomen, the latter shading off into the white under tail-coverts; upper tail-coverts white; wings much as in *G. bispecularis*, but the blue more brilliant and the black bars less broad; tail black, with a few greyish bars near the base; nasal plumes vinaceous, with distinct tips of black; forehead also minutely streaked with black; a broad black malar stripe; bill blackish; "iris pinkish" (*Reeves, MS.*). Total length 12·5 inches, culmen 1·25, wing 6·95, tail 6·1, tarsus 1·65.

Hab. China and Japan.

a. Ad. sk.	Nagasaki, Japan.	Capt. St. John [C.].
b. Ad. sk.	North China.	Mr. Fortune [C.].
c, d. Ad. sk.	China.	J. Gould, Esq.
e. Ad. sk.	Fokien, China.	J. Gould, Esq.
f. Ad. sk.	Near Ningpo (*Reeves*).	J. Gould, Esq.
g, h. Ad. sk.	Ningpo.	R. Swinhoe, Esq. [C.].
i, k. Ad. sk.	North China.	Capt. Blakiston [C.].
l. Bones, various.	Nepal.	B. H. Hodgson, Esq. [P.].

12. Garrulus taivanus.

Garrulus taivanus, *Gould, P. Z. S.* 1862, p. 282; *Swinh. Ibis*, 1863, p. 386; *Gould, B. Asia*, pt. xvi.; *Swinh. P. Z. S.* 1871, p. 381.
Garrulus insularis, *Swinh. P. Z. S.* 1863, p. 304.

Adult male. Above vinaceous red, the back and scapulars strongly washed with grey; wing-coverts black, with chestnut margins, the remainder of the wings and tail as in *G. sinensis*; upper and under tail-coverts pure white; nasal plumes jet-black; a few of the frontal feathers also streaked with black; entire under surface of body deep vinaceous, a little paler on the chin; thighs whitish; "bill bluish grey on rather more than the basal half, apical portion black; legs light ochreous brown, with brown claws; irides light clear blue" (*Swinhoe*). Total length 10·5 inches, culmen 1·1, wing 5·9, tail 5·3, tarsus 1·4.

Hab. Island of Formosa.

a. ♂ ad. sk.	Formosa.	R. Swinhoe, Esq. [C.].

13. Garrulus lanceolatus.

Garrulus lanceolatus, *Vigors, P. Z. S.* 1830, p. 7; *Gould, Cent. Himal. B.* pls. 39, 40; *Hodgs. in Gray's Zool. Misc.* p. 84; *Hutton, J. A. S. B.* xvii. pt. ii. p. 5; *Bp. P. Z. S.* 1850, p. 81; *id. Consp.* i.

* This name does not appear to have been published by Mr. Gould.

p. 376; *Cab. Mus. Hein.* Th. i. p. 221; *Horsf. & Moore, Cat. B. Mus. E.I. Co.* ii. p. 579; *Jerd. B. Ind.* ii. p. 308; *Schl. Mus. P.-B.* Coraces, p. 61; *Gray, Hand-l. B.* ii. p. 3; *Hume, Nests and Eggs Ind. B.* p. 417; *Brooks, Str. F.* 1875, p. 253.
Garrulus gularis, *J. E. Gray, in Hardw. Ill. Ind. Zool.* pl. 10; *Gray, Gen. B.* ii. p. 306; *Blyth, J. A. S. B.* xv. p. 31; *id. Cat. B. Mus. A. S. B.* p. 94.
Garrulus vigorsii, *J. E. Gray, in Hardw. Illustr.* pl. 9.

Adult. Above vinaceous grey, the entire head, nape, and sides of face jet-black; throat blackish, with broad lanceolate streaks of white, the fore neck shading off into bluish grey, with a few obsolete whitish streaks, the rest of the under surface clear vinaceous red; scapulars a little greyer than the back; wing-coverts velvety black, the least series somewhat shaded with grey; primary-coverts white; the bastard-wing black, barred with cobalt near the base, the tips white; quills black, tipped with white, the secondaries more broadly, the primaries edged with grey towards the tip, and all barred with deep cobalt and black on the outer web, this blue colour increasing on the outer secondaries and extending nearly to the tips of these feathers; the two innermost secondaries grey, somewhat shaded with cobalt along the upper margin of a broad subterminal black bar: upper tail-coverts grey; tail dull cobalt-blue, narrowly banded across with black, the blue colour disappearing on the outer tail-feathers and towards the tip of all the feathers, which have a broad terminal band of white; under wing-coverts greyish black. Total length 12 inches, culmen 1, wing 5·7, tail 5·7, tarsus 1·5.

Hab. Himalaya Mountains.

a, b. Ad. sk.	Himalaya Mountains.	J. Gould, Esq.
c, d. Ad. sk.	Nepal.	B. H. Hodgson, Esq. [P.].
e. Jun. sk.	Nepal.	B. H. Hodgson, Esq. [P.].
f, g. ♂ ad. sk.	N.W. Himalayas.	Capt. Stackhouse Pinwill [P.].
h. ♀ juv. sk.	N.W. Himalayas.	Capt. Stackhouse Pinwill [P.].
i, k. Ad. sk.	Cashmere.	E. M. Langworthy, Esq. [P.].
l. Ad. sk.	Ladakh.	Purchased.

14. Garrulus lidthi.

Garrulus lidthi, *Bp. Consp.* i. p. 376; *Schl. Bijdr. Zool. Genootsch. Amsterd.* 4to, p. 80, pl. 17; *Bp. P. Z. S.* 1850, p. 80, pl. xvii.; *Schl. Hand-l. Dierk. Vog.* pl. 4. fig. 47; *id. Dirent.* p. 116, fig.; *id. Mus. P.-B.* Coraces, p. 61; *Salvad. Atti R. Accad. Torin.* vii. p. 473.

Reddish vinaceous; entire head and neck, wings, and tail dark azure-blue; forehead and lores blackish; feathers of the throat lanceolate, the barbules lax, the shafts white; wing-coverts banded with black; quills and tail-feathers blackish towards the tip, the latter itself being white; bill whitish, high, much compressed; tail rather lengthened, the feathers equal. (*Bonaparte, l. c.*)

Hab. Japan.

26. PERISOREUS.

	Type.
Perisoreus, *Bp. Comp. List B. Eur. & N. Amer.* p. 27 (1838)	P. infaustus.
Dysornithia, *Swains. Faun. Bor.-Amer. Birds,* p. 296 (1831)	P. canadensis.

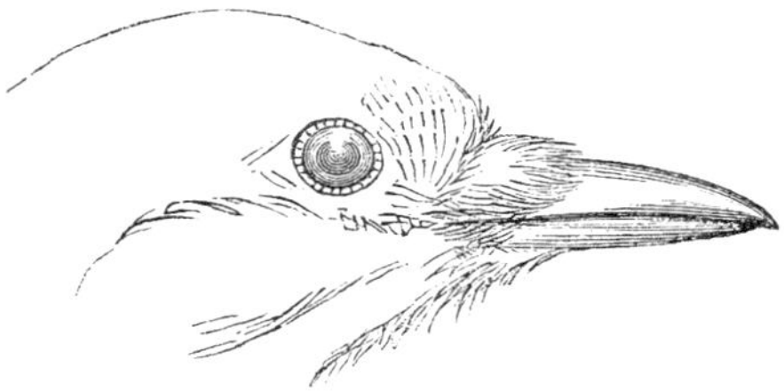

Head of *Perisoreus infaustus.*

Range. Northern Europe, across Northern Asia to Saghalien; North America.

Key to the Species *.

a. Tail rufous, the middle feathers only grey; forehead brown; sides of neck greyish .... *infaustus,* p. 103.
b. Tail grey, tipped with white; forehead and sides of neck whitish.
  a'. Dusky nuchal hood plumbeous black, and reaching forward to, or in front of, the eyes.
    a''. Dorsal feathers without any whitish shafts in old and young .... *canadensis,* p. 104.
    b''. Dorsal feathers with whitish shafts in both old and young .... *obscurus,* p. 105.
  b'. Dusky nuchal hood bluish plumbeous, not reaching to the eyes .... *capitalis,* p. 106.

1. Perisoreus infaustus.

Lanius infaustus, *Linn. S. N.* i. p. 138.
Geai de Sibérie, *Montb. Pl. Enl.* iii. pl. 608.
Siberian Jay, *Lath. Gen. Syn.* i. pt. i. p. 391.
Corvus sibiricus, *Bodd. Tabl. Pl. Enl.* p. 37; *Gm. S. N.* i. p. 373.
Corvus infaustus, *Sparrm. Mus. Carls.* iv. pl. 76; *Temm. Man. d'Orn.* i. p. 115; *Werner, Atlas,* Omnivores, pl. 9.
Corvus russicus, *Pall. Zoogr. Rosso-Asiat.* i. p. 395.
Garrulus infaustus, *Vieill. N. Dict. d'Hist. Nat.* xii. p. 478; *Naum. Vög. Deutschl.* ii. Taf. 350; *Gould, B. Eur.* iii. pl. 215; *Schl. u. Su-*

* The characters of the American species are derived from the great work on North-American Birds by Messrs. Baird, Brewer, and Ridgway. Specimens of all three are in the Museum; but the characters seemed to be so well expressed that I could not improve upon them.

sem. Vög. Eur. ii. Taf. 6 *a.* fig. 2; *Midd. Sibir. Reis. Zool.* p. 157; *Schrenck, Reis. Amurl.* p. 315; *Radde, Sibir. Reis.* p. 203; *Sundev. Sv. Fogl.* pl. xix. fig. 2; *Schl. Mus. P.-B.* Coraces, p. 65.
Pica infausta, *Wagler, Syst. Av.* Pica, sp. 20.
Perisoreus infaustus, *Bp. Comp. List B. Eur. & N. Amer.* p. 27; *Gray, Gen. B.* ii. p. 306; *Bp. Consp.* i. p. 375; *Fritsch, Vög. Eur.* tab. 13. fig. 9; *Degl. & Gerbe, Orn. Eur.* i. p. 218; *Gray, Hand-l. B.* ii. p. 3; *Dresser, B. Eur.* part xix.

Adult. Bristles over the nostrils ashy fulvous; head, including the sides of the face, dark brown; back dull ashy grey, slightly washed with olivaceous, especially on the rump; the upper tail-coverts clear olivaceous; tail chestnut, slightly fulvescent at tip, the two centre feathers ashy grey, the two next and the two outermost greyish towards the tip of the *outer* web; least wing-coverts ashy grey, like the back, as well as the bastard wing; the primary- and greater coverts orange-chestnut, tipped with ashy grey, the innermost entirely ashy grey; quills brown, externally margined with clear grey, some of the inner primaries rufous near the base of the outer web, the secondaries clear grey externally, the innermost entirely so; a spot on the anterior part of the cheeks and chin yellowish white; rest of under surface ashy grey, shading off gradually into orange-chestnut on the lower abdomen, flanks, and under wing- and tail-coverts. Total length 11 inches, culmen 0·95, wing 5·45, tail 5·4, tarsus 1·45.

Young. Similarly coloured to the adult, but the colours much more dingy; head and sides of face dusky brown, the ear-coverts streaked with fulvous.

Hab. Northern Europe, across the whole of Northern Asia.

a. Ad. st.	Northern Bothnia.	Prof. Wahlberg [C.].
b. Ad. sk.	Northern Bothnia.	Prof. Wahlberg [C.].
c. Juv. st.	Northern Bothnia.	Prof. Wahlberg [C.].
d. Ad. sk.	Sweden.	J. Gould, Esq.
e. Juv. sk.	Northern Russia.	Baron A. von Hügel [P.].
f. ♂ pull. sk.	Ust Zylma, Great Petchora river, May 30, 1875 (*H. Seebohm & J. A. Harvie-Brown*).	Capt. Feilden and J. A. Harvie-Brown, Esq. [P.].
g. ♂ skeleton.	Jemtland, Feb. 1847.	R. B. Sharpe, Esq.
h. Skeleton.		Purchased.
i. Skeleton.		Zoological Society.

2. Perisoreus canadensis.

Le Geay brun de Canada, *Briss. Orn.* ii. p. 54, pl. iv. fig. 2.
Corvus canadensis, *Linn. S. N.* i. p. 158; *Wils. Am. Orn.* iii. p. 33, pl. xxi. fig. 1; *Audub. Orn. Biogr.* ii. p. 53, v. p. 208, pl. 107.
Garrulus fuscus, *Vieill. N. Dict. d'Hist. Nat.* xii. p. 479.
Pica nuchalis, *Wagler, Syst. Av.* Pica, sp. 14.
Garrulus canadensis, *Swains. & Rich. Faun. Bor.-Amer. Birds,* p. 295; *Audub. B. Amer.* roy. 8vo, iv. p. 121, pl. 234.
Garrulus brachyrhynchus, *Swains. & Rich. Faun. Bor.-Amer. Birds,* p. 296, pl. 55.

Dysornithia canadensis, *Swains. t. c.* pp. 131, 495.
Perisoreus canadensis, *Bp. Comp. List B. Eur. and N. Amer.* p. 27; *Gray, Gen. B.* ii. p. 306; *Bp. Consp.* i. p. 375; *Cab. Mus. Hein. Th.* i. p. 219; *Baird, B. N. Amer.* p. 590; *Gray, Hand-l. B.* ii. p. 3; *Coues, Key N.-Amer. B.* p. 166; *Baird, Brewer, & Ridgw. N.-Amer. B.* ii. p. 299, pl. 41. fig. 3; *Coues, B. N. West,* p. 221.

Adult male. Upper surface dull grey, somewhat tinged with brown on the mantle; nasal bristles, forehead, and anterior part of crown whitish, slightly tinged with brown, and then shading off from grey into black, of which colour the hinder part of the crown and nape is composed; the plumes round the eye also streaked with blackish; cheeks, throat, sides of face and of neck white; rest of under surface ashy brown, inclining to white on the vent, the under wing-coverts greyish; wings above dull grey like the back, the primaries blackish, some of the wing-coverts and all the secondaries narrowly tipped with white; tail rather more slaty grey than the back and tipped with white; "bill, legs, and feet black; iris dark brown" (*B. R. Ross, MS.*). Total length 11 inches, culmen 0·95, wing 5·65, tail 5·7, tarsus 1·35.

Adult female. Not perceptibly different from the male. Total length 10·5 inches, wing 5·4, tail 5·5, tarsus 1·35.

Young. Different from the adult. Entirely dark plumbeous, blacker on the ear-coverts; the wing-coverts with indistinct brownish edgings; the abdomen lighter grey than the breast, with fulvescent tips to the feathers; quills and tail-feathers clearer grey than the back, with white tips.

Hab. Canada, Maine, Labrador, to the Yukon.

a, b. Ad. sk.	North America.	Dr. Ridgeway [P.].
c. Juv. sk.	North America.	Dr. Rae [P.].
d, e. ♂ ♀ ad. sk.	Fort Simpson, April 7, 1862.	B. R. Ross, Esq. [P.].
f. Ad. sk.	Fort Simpson, March 26, 1860.	B. R. Ross, Esq. [P.].
g. Juv. sk.	Fort Simpson, June 1862.	B. R. Ross, Esq. [P.].
h, i, k. Ad. sk.	49th parallel, August 1874 (*Dawson*).	N.-A. Boundary Commission [P.].
l. ♀ ad. sk.	West side of Rocky Mountains.	J. K. Lord, Esq. [P.].

3. **Perisoreus obscurus.** (Plate V. fig. 2.)

Perisoreus canadensis, *Newb. P. Rail. Rep.* vi. p. 85 (*nec L.*); *Dall & Bann. Tr. Chicago Acad.* i. p. 286; *Cooper, B. Calif.* i. p. 397; *Finsch, Abhandl. Bremen,* iii. p. 40.
Perisoreus canadensis, var. obscurus, *Ridgw. Bull. Essex Inst.* v. p. 199; *Baird, Brewer, & Ridgw. N.-Amer. B.* ii. p. 302; *Coues, B. N. West,* p. 221.

Adult. Above dusky brown, with whity-brown streaks to all the feathers of the back and scapulars; the wing-coverts and quills externally greyish, with pale whitish tips, very indistinct; tail-feathers ashy brown, with scarcely perceptible whitish tips; nasal bristles, forehead, and lores white; the crown and hind head dusky brown sides of face entirely white; sides of neck whitish; hind neck ashy brown, rather lighter than the back; throat white; rest of under surface pale brownish, lighter and more whity brown in the centre of the abdomen; under wing-coverts ashy brown. Total length 10·2 inches, culmen 0·85, wing 5·55, tail 5·3, tarsus 1·4.

Young. Entirely plumbeous brown, the feathers of the back also streaked with whitish along the shafts.

Hab. British Columbia, Oregon, Washington territory (*Ridgway*).

a, b. Ad. sk.	British Columbia.	J. K. Lord, Esq. [C.].
c. Juv. sk.	Straits.	Capt. Kellett and Lieut. Wood [P.].

4. **Perisoreus capitalis.** (Plate V. fig. 1.)

Perisoreus canadensis, *Hayden, Rep.* 1862, p. 171 (*nec L.*); *Allen, Bull. Harv. Coll.* iii. p. 179.

Perisoreus canadensis, var. capitalis, *Baird, MS.*; *Ridgw. Bull. Essex Inst.* v. p. 199; *Baird, Brewer, & Ridgw. N.-Amer. B.* ii. p. 302, pl. 41. fig. 4; *Coues, B. N. West,* p. 221.

Adult female. General colour above greyish, the wing-coverts coloured like the back, the greater and median series slightly tipped with whitish; primary-coverts and quills dark brown, externally bluish grey, shading off into whity brown towards the tips of the primaries; the secondaries more broadly tipped with white, the innermost secondaries almost entirely grey; tail grey, broadly tipped with white; nasal bristles buffy white; crown of head white, the hinder part of the crown and nape grey, forming an ill-defined horseshoe mark; sides of neck pure white, the hinder neck grey; lores, entire sides of face and throat white; rest of under surface pale ashy brown, greyish on the sides of the body, the under tail-coverts greyish white; under wing-coverts dark ashy grey. Total length 10·5 inches, culmen 0·85, wing 6·2, tail 5·8, tarsus 1·45.

Adult male. Similar to the female. Total length 10·3 inches, culmen 0·9, wing 6·1, tail 5·8, tarsus 1·45.

Hab. Rocky Mountains of the United States.

a, b. ♂ ♀ ad. sk.	Alimosa Creek, Colorado, August 31 (*C. E. Aiken*).	Smithsonian Institution [P.].

27. **CYANOCITTA.**

	Type.
Cyanocitta, *Strickl. Ann. N. H.* xv. p. 261 (1845)	C. cristata.
Cyanogarrulus, *Bp. Consp.* i. p. 376 (1850)............	C. cristata.
Lophocorax, *Kaup, J. f. O.* 1854, p. lv	C. diademata.

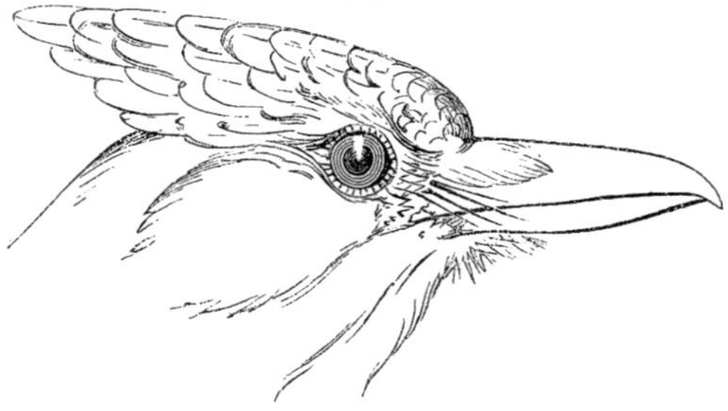

Head of *Cyanocitta cristata.*

Range. North America and Mexico.

Key to the Species.

a. Cheeks and abdomen whitish; tail-feathers broadly tipped with white, as also the greater wing-coverts and secondaries .... *cristata*, p. 107.
b. Cheeks blackish; abdomen bluish; no white tips to secondaries or tail-feathers.
  a'. With no whitish spot over the eye .... *stelleri*, p. 108.
  b'. With a patch of silky white over the eye.
    a''. Crest black, like the cheeks, throat, and fore neck .... *macrolopha*, p. 110.
    b''. Crest blue-black; cheeks black, breast shaded with blue .... *coronata*, p. 111.
    c''. Crest bright blue, lores black; sides of face black, shaded with blue .... *diademata*, p. 111.

Although admitting that extreme forms can be found to justify the separation of *C. macrolopha* into three races, yet there are in the series now before me some examples so thoroughly intermediate that I should not know to which form to refer them with certainty. This is especially the case with the Mexican birds; and I have two examples which seem to be changing the blue-black crest of *C. macrolopha* for the brighter blue one of *C. diademata*.

1. Cyanocitta cristata.

The Blew Jay, *Catesby, Nat. Hist. Carol.* i. p. 13, pl. 13.
The Blue Jay, *Edwards, Gleanings N. H.* v. p. 60, pl. 239.
Le Geay bleu de Canada, *Briss. Orn.* ii. p. 55; *Montb. Pl. Enl.* pl. 529.
Corvus cristatus, *Linn. S. N.* i. p. 157; *Wils. Am. Orn.* i. p. 11, pl. 1. fig. 1; *Aud. Orn. Biogr.* ii. p. 11; *id. B. Amer.* pl. cii.
Le geai bleu, *Levaill. Ois. Parad.* pl. 45.
Garrulus cristatus, *Vieill. N. Dict. d'Hist. Nat.* xii. p. 477; *id. Gal. Ois.* i. pl. 102; *Sw. & Richards. Faun. Bor.-Amer. Birds*, p. 293; *Audub. B. Amer.* roy. 8vo, iv. p. 110, pl. ccxxi.
Pica cristata, *Wagler, Syst. Av.* Pica, sp. 8.
Cyanurus cristatus, *Swains. Faun. Bor.-Amer. Birds*, p. 495; *Baird, B. N. Am.* p. 495; *Gray, Hand-l. B.* ii. p. 3; *Coues, Key N.-Am. B.* p. 165; *Allen, Bull. Harv. Coll.* ii. p. 297; *Baird, Brewer, & Ridgway, N.-Amer. B.* p. 273, pl. 42. fig. 3; *Coues, B. N.West*, p. 214.
Cyanocorax cristatus, *Bp. Comp. List B. Eur. & N. Amer.* p. 27; *Gray, Gen. B.* ii. p. 307.
Cyanocitta cristata, *Strickl. Ann. N. H.* 1845, xv. p. 261; *Cab. Mus. Hein.* Th. i. p. 221; *Sclater, Cat. Am. B.* p. 142.
Cyanogarrulus cristatus, *Bp. Consp.* i. p. 376.

Adult. Above greyish purple, rather bluer on the hind neck, and more distinctly grey on the sides of the neck; head very strongly crested, greyish purple, with somewhat of a lilac lustre, the feathers above the eye lilac-grey; nasal bristles, sides of the face and entire throat whitish, glossed with lilac, the feathers with narrow black shaft-lines; round the nape a band of black feathers, running in a narrower line behind the ear-coverts (the hindermost of which are tipped with black) down the sides of the neck to the fore neck, which is black, forming a large patch glossed slightly with purple both above and below; a narrow band across the forehead, loral plumes, and a narrow line of feathers through the eye blackish; rest of under surface of body pale lilac-brown, inclining to white on the lower abdomen and under tail-coverts; under wing-coverts black, the outermost shaded with purplish blue; the median series tipped with white, forming a transverse bar; a spot at the edge of the wing white; wings above purplish blue, glossed with cobalt, the least coverts resembling the back, the greater series brighter and more cobalt, transversely barred with black, and broadly tipped with white, forming a bar across the wing; quills ashy black; the primaries externally bright blue, the secondaries inclining to purple externally and barred with black, the innermost bright cobalt, transversely barred with black and broadly tipped with white; tail bright blue, all but the two centre feathers broadly tipped with white, these two barred across with black, these bars only seen on the outer webs of the other tail-feathers, and disappearing entirely on the outermost; "bill and feet brownish black; iris brown" (*Audubon*). Total length 11·5 inches, culmen 1·2, wing 5·5, tail 5·4, tarsus 1·5.

Young. Similar to the adult, the colours much duller, but distributed in the same manner; the ring round the nape and the fore neck brown instead of black, as in the old bird; under wing-coverts greyer than in the adult, the bluish white tips to the median series very large and distinct.

Hab. North America.

a. Ad. st.	North America.	J. Gould, Esq.
b. Ad. sk.	North America.	
c. Ad. sk.	North America.	Lord Odo Russell [P.].
d. Juv. sk.	North America.	Dr. Ridgway [P.].
e. Ad. sk.	Toronto, Canada.	Mr. James Whitely [C.].
f. Juv. sk.	Toronto, Canada.	Mr. James Whitely [C.].
g, *h*. Ad. sk.	Ann Arbor, Michigan.	Prof. J. B. Steere [P.].
i, *k*, *l*. Ad. sk.	49th parallel (*Dawson*).	N.-A. Boundary Commission [C.].
m. Skeleton.	North America.	R. B. Sharpe, Esq.

2. Cyanocitta stelleri.

Steller's Crow, *Lath. Gen. Syn.* i. p. 389 (1781); *Penn. Arct. Zool.* ii. p. 249, pt. 1.
Corvus stelleri, *Gm. Syst. Nat.* i. p. 370; *Pall. Zoogr. Rosso-Asiat.* i. p. 393; *Audub. Orn. Biogr.* iv. p. 453; *id. B. Amer.* pl. ccclxii.
Garrulus stelleri, *Vieill. N. Dict.* xii. p. 481; *Bp. Amer. Orn.* ii. p. 44, pl. xii. fig. 1; *Sw. & Richards. Faun. Bor.-Am. Birds*, p. 294 (*nec* pl. 54); *Audub. B. Amer.* roy. 8vo, iv. p. 107, pl. ccxxx.
Cyanurus stelleri, *Swains. Faun. Bor.-Am. Birds*, p. 495; *Baird, B. N. Amer.* p. 581; *Lord, Natural. Vanc. Isl.* p. 73; *Dall & Bann. Tr. Chicago Acad.* i. p. 286; *Gray, Hand-l. B.* ii. p. 4; *Cooper, Orn. Calif.* i. p. 298; *Coues, Key N.-Amer. B.* p. 165; *Ridgw. Am. Journ. Sci. & Arts* (3), v. p. 43; *Baird, Brewer, & Ridgw. N.-Amer. B.* ii. p. 227, pl. 39. fig. 1; *Coues, B. N. West*, p. 214.

Pica stelleri, *Wagler, Syst. Av.* Pica, sp. 10.
Cyanocorax stelleri, *Gray, Gen. B.* ii. p. 307; *Finsch, Abhandl. nat. Ver. Bremen*, iii. p. 40.
Cyanocitta stelleri, *Strickl. Ann. N. Hist.* 1845, xv. p. 261; *Cab. Mus. Hein.* Th. i. p. 221.
Cyanogarrulus stelleri, *Bp. Consp.* i. p. 377.
Cyanura stelleri, var. frontalis, *Ridgw. Amer. Journ. Sci. & Arts*, v. p. 43; *Baird, Brewer, & Ridgw. N.-Amer. B.* ii. p. 279, pl. 39. fig. 2; *Coues, B. N. West*, p. 215.

Adult. Head and crest sooty black, including the sides of the face, sides of neck, entire throat and fore neck; the face and crown rather deeper black, with a few silvery blue streaks on the former; mantle and upper back sooty black, with an imperceptible shade of greyish blue; the lower back, rump, and upper tail-coverts verditer-blue, brighter on the latter; tail ultramarine, with indistinct dusky cross bars under certain lights; wings ultramarine, the inner webs of the quills black, the secondaries rather brighter, the inner ones barred externally with black; the primaries externally verditer, as also are the adjacent wing-coverts; the greater coverts with indistinct dusky blackish bars; throat and fore neck sooty black, deeper in shade than on the mantle, the chin and upper throat streaked with greyish; all the rest of the under surface of the body verditer-blue, including the under tail-coverts; under wing-coverts and thighs black, the latter slightly shaded with blue, as also the edge of the wing; bill and legs black; iris hazel. Total length 12 inches, culmen 1·2, wing 5·8, tail 6, tarsus 0·8.

Obs. The throat-marking varies extremely, even in specimens from the same locality, being very plain in some, almost obsolete in others.

Mr. Ridgway separates a race as *frontalis*, and says that "all references to Steller's Jay as occurring in California, excepting on the northern coast-range, relate to this bird." From his synopsis we gather that this race has the following characters:—"The whole head, neck, jugulum, and dorsal region plumbeous umber" instead of "sooty black;" the "forehead conspicuously streaked with blue, and the crest washed with the same." The blue colour is "of two different shades, the wings and tail being deep indigo, the body and tail-coverts greenish cobalt-blue; whereas in true *C. stelleri* the blue everywhere is of a uniform dull greenish blue shade." Although the blue of the back and wings is deeper in the latter bird, it is true there is considerable contrast in the two tints, and the Californian Blue Jay does not show much difference, at least as far as I have examined specimens. None of them exhibited the least shade of blue on the crest; but all were much browner on the head and neck. For this reason I do not admit *C. frontalis* as a distinct race.

Hab. Eastern North America, west to the Missouri. North-eastern Texas, north to Red River and Moose Factory (*Ridgway*).

a. Ad. sk.	North-west America.	Purchased.
b. Ad. st.	Sitka.	G. Barclay, Esq. [P.].
c. Ad. sk.	Sitka.	Purchased.

d. Jun. sk.	Sitka.	Purchased.
e. Juv. sk.	49th parallel (*Dawson*).	N.-A. Boundary Commission [C.].
f. Ad. sk.	San Juan de Fuca.	Capt. Kellett and Lieut. Wood [P.].
g. Ad. sk.	"Saturna Island."	Dr. Lyall [P.].
h, i, k. Ad. sk.	British Columbia.	J. K. Lord, Esq. [C.].
l. Ad. sk.	California.	Zoological Society.
m, n. Ad. sk.	California (*Lorquin*).	Purchased.
o. Juv. sk.	California, June 24, 1871 (*Dr. Finsch*).	R. B. Sharpe, Esq. [P.].

3. Cyanocitta macrolopha.

Garrulus stelleri, *Swains. & Rich. Faun. Bor.-Amer. Birds*, pl. 54 (*nec* p. 294).
Cyanocorax stelleri, *Woodhouse in Sitgreave's Report Zuni & Color. Rivers*, p. 77.
Cyanocitta macrolopha, *Baird, Pr. Acad. N. Sci. Philad.* 1854, p. 118; *Sclater, Cat. Am. B.* p. 143; *Sclater & Salv. Nomencl. Av. Neotr.* p. 39.
Cyanura macrolopha, *Baird, B. N. Amer.* p. 582; *Coues, Am. Nat.* v. p. 770; *Elliot, B. N. Amer.* i. pl. 17.
Cyanurus macrolophus, *Gray, Hand-l. B.* ii. p. 4.
Cyanurus stelleri, var. macrolophus, *Allen. Bull. Harv. Coll.* ii. p. 178; *Coues, Key N.-Amer. B.* p. 165, fig. 107; *Baird, Brewer, & Ridgw. N.-Amer. B.* ii. p. 281, pl. 29. fig. 3; *Coues, B. N. West*, p. 214.
Cyanura coronata, var. macrolopha, *Ridgw. Am. Journ.* (3) v. p. 43.

Adult male. Head and neck all round, including the mantle and scapulars on the upper surface, and extending as far as the chest below, sooty black, with a slight wash of grey on the mantle and scapulars; the head, crest, and face blacker than the back, the fore part of the crown streaked with cobalt-blue; over the eye a spot of silvery white; the throat also sooty grey; lower back, rump, and upper tail-coverts dull cobalt-blue, brighter on the latter; tail deep ultramarine, brighter on the margins of the feathers, which are waved under certain lights with obsolete bars of black; wings deep ultramarine, the primaries externally cobalt, the secondaries brighter blue, the innermost barred with black, as also are the greater coverts; the median series likewise with a few faint indications of terminal black bars; under surface of body from the chest downwards dull cobalt; thighs and under wing-coverts black, shaded with blue; bill and legs black. Total length 12 inches, culmen 1·25, wing 6, tail 6·05, tarsus 1·75.

Young. Has the bright blue of the body and the black of the head replaced by a dull slate; the head unvaried (*Ridgway*).

Hab. Central line of Rocky Mountains, from the northern border of the United States to tablelands of Mexico; Arizona (*Ridgway*).

a. Ad. st.	[North America.]	Hudson's Bay Co. [P.].
b, c. ♂ ♀ ad. sk.	West side of Rocky Mountains.	J. K. Lord, Esq. [P.].
d. ♂ ad. sk.	Fremont County, Colorado, March 25, 1876.	Purchased.
e. ♀ ad. sk.	Arapahoe County, Colorado, Nov. 24, 1872.	Purchased.

4. Cyanocitta coronata.

Garrulus coronatus, *Swains. Phil. Mag.* i. p. 437; *Jard. & Selby, Ill. Orn.* ii. pl. 64.
Pica coronata, *Wagler, Isis*, 1829, p. 750.
Cyanurus coronatus, *Swains. Faun. Bor.-Am. Birds*, p. 495; *Baird, B. N. Amer.* p. 583; *Scl. P. Z. S.* 1859, p. 381; *Scl. & Salv. Ibis*, 1859, p. 22; *Gray, Hand-l. B.* ii. p. 4.
Cyanocorax coronatus, *Gray, Gen. B.* ii. p. 307; *Bp. P. Z. S.* 1837, p. 115.
Cyanogarrulus coronatus, *Bp. Consp.* i. p. 377.
Cyanocitta coronata, *Strickl. Ann. N. H.* 1845, p. 261; *Cab. Mus. Hein.* Th. i. p. 222; *Sclater, Cat. Am. B.* p. 142; *Scl. & Salv. Nomencl. Av. Neotr.* p. 39.
Cyanocitta coronata, var. coronata, *Ridgw. Am. Journ.* v. p. 42.
Cyanura stelleri, var. coronata, *Baird, Brewer, & Ridgw. N.-Amer. B.* ii. p. 272, pl. 39. fig. 4; *Coues, B. N. West*, p. 215.

Adult. Above purplish grey on the back and scapulars, the lower back, rump, and upper tail-coverts cobalt-blue, brighter on the latter; wings ultramarine, brighter on the inner secondaries, which are barred with black, as also the greater series of wing-coverts, the bars on the latter very indistinct; tail ultramarine, inclining to purplish brown on the inner webs; head all round and the fore neck sooty black, the crest shaded with deep blue, all the feathers of the crown also tipped with deep cobalt, inclining to silvery white on the frontal plumes; throat ashy white, the feathers slightly washed with blue; rest of under surface clear blue, inclining to purple on the breast; thighs and under wing-coverts ashy black, shaded with blue; bill and legs black. Total length 12·5 inches, culmen 1·1, wing 5·75, tail 5·8, tarsus 1·7.

Obs. This bird is scarcely distinguishable from *C. diademata* when the latter is blackish-crested, and cannot be considered a very well pronounced species. The back is purplish grey, of a different tint from the blue of the other, while both are distinct from the true *C. macrolopha*, which has the back blackish.

Hab. Highlands of Mexico.

a. Ad. sk.	[Mexico.]	Purchased.
b. Ad. sk.	Mexico.	Purchased.
c, d. Ad. sk.	Mexico.	A. Boucard, Esq. [C.].

5. Cyanocitta diademata.

Cyanogarrulus diadematus, *Bp. Consp.* i. p. 377.
Lophocorax diademata, *Kaup, J. f. O.* 1854, p. lv.
Cyanocitta diademata, *Sclater, Cat. Am. B.* p. 143; *Sclater & Salvin, Nomencl. Av. Neotr.* p. 39.
Cyanurus diadematus, *Gray, Hand-l. B.* ii. p. 4.
Cyanura coronata, var. diademata, *Ridgw. Amer. Journ.* (3) v. p. 42.
Cyanura stelleri, var. diademata, *Baird, Brewer, & Ridgw. N.-Amer. B.* ii. p. 272; *Coues, Key N.-Amer. B.* p. 215.

Adult male. General colour above greenish blue, brighter and more cobalt on the rump and upper tail-coverts; the tail darker blue and more ultramarine, all the feathers very plainly crossed with numerous dusky bars of blackish; wings deeper blue than the back, all the greater series of coverts very plainly barred with black, as also the inner secondaries, all these latter quills shining cobalt, the primaries externally slightly greenish blue; head and crest ultramarine blue, the forehead silvery cobalt, and the fore part of the crown brighter blue; nasal bristles, lores, and sides of face black, the cheeks and ear-coverts shaded with blue; above the eye a distinct spot of white, and a smaller one below the eye; chin greyish white; all the rest of the under surface light cobalt-blue, rather more purple on the throat and chest; thighs purplish; under wing-coverts ashy black, shaded with blue; bill and legs black. Total length 11·3 inches, culmen 1·1, wing 5·6, tail 5·6, tarsus 1·6. (Oaxaca, *Rébouch.*)

Some birds, probably females or younger individuals, approach *C. coronata* in having a blackish or blue-black crest, while the region of the throat and the sides of the face are blackish, but always shaded with blue, which in the adult of the present species extends all over the under surface, excepting on the grey throat.

Hab. Mexico.

a. Ad. sk.	[Mexico.]	Purchased.
b, c. Jun. sk.	[Mexico.]	A. Boucard, Esq. [C.].
d, e. Ad. sk.	Sierra de Oaxaca.	A. Boucard, Esq. [C.].
f, g. ♂ ad. sk.	Parada, Oaxaca.	M. Rébouch [C.].

28. APHELOCOMA.

Type.

Aphelocoma, *Cab. Mus. Hein.* Th. i. p. 221, note ...... A. californica.

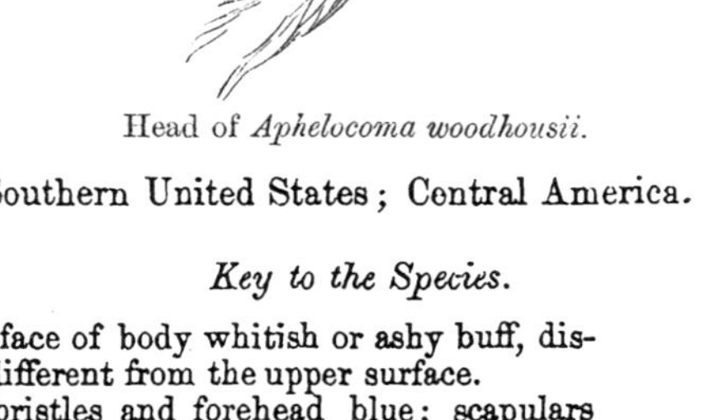

Head of *Aphelocoma woodhousii.*

Range. Southern United States; Central America.

Key to the Species.

a. Under surface of body whitish or ashy buff, distinctly different from the upper surface.
  a'. Nasal bristles and forehead blue; scapulars ashy, like the back; back much darker than the lower parts *.

* Much of the information contained in the above 'Key' is derived direct from the 'North-American Birds' of Messrs Baird, Brewer, and Ridgway, as the Museum wants specimens of some of the races mentioned by them. Their synoptic tables also are so complete as to leave very little to add.

a″. Over the eye a distinct line of whitish feathers.
a.‴ Under tail-coverts whitish; ear-coverts dusky blackish, with a very faint blue wash.
a⁴. Eyebrow very strongly marked...... *californica*, p. 113.
b⁴. Eyebrow very faintly indicated...... *sumichrasti*, p. 114.
b‴. Under tail-coverts bluish; ear-coverts also bluish .... *woodhousii*, p. 114.
b″. No white superciliary streak.
c‴. Tail nearly square.................. { *ultramarina*, p. 115. *couchi*, p. 116.
d‴. Tail much rounded .... { *sordida*, p. 116. *arizonæ*, p. 117.
b′. Nasal bristles and forehead hoary white, running into a distinct eyebrow of the same colour; scapulars blue, like the wings; interscapulary region as lightly coloured as the lower parts.......................... *floridana*, p. 117.
b. Under surface of body blue, like the upper.
c′. Size much larger: throat blue, like the breast; forehead and eyebrow blue, like the crown.. *unicolor*, p. 118.
d′. Size much smaller: throat bluish white, contrasting with the breast; forehead and eyebrow bluish white, lighter than the rest of the crown .... *nana*, p. 118.

1. Aphelocoma californica.

Garrulus californicus, *Vigors in Zool. Beechey's Voyage*, p. 21, pl. 5.
Cyanocitta californica, *Strickl. Ann. N. H.* 1845, xv. p. 342; *Bp. Consp.* i. p. 377; *Baird, B. N. Amer.* p. 584; *Scl. Cat. Am. B.* p. 143; *Cooper, Orn. Calif.* i. p. 302; *Baird, Brewer, & Ridgw. N.-Amer. B.* p. 288.
Cyanocitta superciliosa, *Strickl. Ann. N. H.* xv. p. 260.
Corvus ultramarinus, *Audub. Orn. Biogr.* iv. p. 456; *id. B. Amer.* pl. ccclxii.
Garrulus ultramarinus, *Audub. B. Amer.* iv. p. 116, pl. 232 (*nec Bp.*).
Cyanocorax californicus, *Gamb. Pr. Acad. Nat. Sci. Philad.* 1847, p. 201.
Aphelocoma californica, *Cab. Mus. Hein.* Th. i. p. 221; *Bp. C. R.* xxxii. p. 828.
Cyanurus californicus, *Gray, Hand-l. B.* ii. p. 4.
Aphelocoma floridana, var. californica, *Coues, Key N.-Amer. B.* p. 166; *id. B. N. West*, p. 219.

Adult. Head and neck bright cobalt-blue, including also the sides of the latter; back drab-brown, fading off into grey on the rump; the upper tail-coverts pale cobalt, like the tail; wings pale blue, the primary-coverts and the primaries brown towards the tips; lores dusky; ear-coverts and hinder part of cheeks blackish, shaded with blue; over the eye a long and distinct eyebrow; fore part of cheeks dull white, streaked with blue; under surface of body ashy white, inclining to pure white on the under tail-coverts, the throat and chest rather lighter, and streaked slightly with brown, with which colour some of the feathers are edged; the sides of the breast more or less blue, forming a crescentic mark; under wing-coverts ashy, shaded with blue; bill and legs black. Total length 11·8 inches, culmen 1·2, wing 4·8, tail 5·7, tarsus 1·6.

Young. Lighter brown than the adult, the head only a little darker and shaded with blue, not blue entirely; upper tail-coverts white; sides of face and of neck dull ashy, running down each side of the breast and enclosing the throat, which is paler than the rest of the under surface.

Hab. Pacific province of United States.

a. Ad. sk.	San Francisco, California.	G. Barclay, Esq. [P.].
b. Ad. sk.	San Francisco, California.	Capt. Kellett and Lieut. Wood [P.].
c. Ad. sk.	California.	Purchased.
d. Ad. sk.	California.	M. Lorquin [C.].
e. Juv. sk.	California.	M. Lorquin [C.].
f. Ad. sk.	California.	Purchased.

2. **Aphelocoma sumichrasti.**

Cyanocitta californica, *Scl. & Salv. Nomencl. Av. Neotr.* p. 39 (*nec Vig.*).
Cyanocitta sumichrasti, *Ridgw. Rep. Geol. Expl. 40th par.*
Cyanocitta californica, var. sumichrasti, *Baird, Brewer, & Ridgw. N.-Amer. B.* ii. p. 283.
Cyanocitta floridana, var. sumicrasti, *Coues, B. N. West*, p. 219.

Adult. Similar to *A. californica*, but with the white eyebrow obsolete or scarcely distinguishable. Total length 12·5 inches, culmen 1·3, wing 5·8, tail 6·3, tarsus 1·7.

Hab. Mexico.

a. Ad. sk.	Mexico.	Purchased.
b. ♂ ad. sk.	Atlisko, Mexico.	A. Boucard, Esq. [C.].
c. Ad. sk.	Oaxaca, Mexico.	A. Boucard, Esq. [C.].

3. **Aphelocoma woodhousii.**

Cyanocorax californica, *Woodh. in Sitgr. Report Zuni & Color. Rivers*, p. 77 (*nec Vig.*).
Cyanocitta woodhousii, *Baird, Birds N. Amer.* p. 585, pl. lix.; *id. Report Mexican Bound.* ii. *Birds*, p. 20, pl. xxi.; *Cooper, B. Calif.* i. p. 304; *Allen, Bull. Harv. Coll.* ii. p. 299, note.
Cyanurus woodhousei, *Gray, Hand-l. B.* ii. p. 4.
Cyanocitta californica, var. woodhousei, *Baird, Brewer, & Ridgw. in N.-Am. B.* ii. p. 291.
Aphelocoma floridana, var. woodhousei, *Coues, B. N. West*, p. 219.

Head and hind neck greyish cobalt, lores and feathers round the eye dusky blackish, as also the ear-coverts, the latter being shaded with blue; over the eye a narrow streak of silvery-tipped feathers, and below the eye a few more plumes tipped with silvery white; sides of face and neck and sides of neck dull blue; fore part of the cheeks spotted with white; mantle and scapulars drab-brown, the latter washed externally with dull blue; wings greyish cobalt; the quills blackish, blue only on the outer web; lower back and rump greyish; the upper tail-coverts and tail cobalt, some of the tail-feathers blackish along the edge of their inner webs; throat and fore neck white, somewhat streaked with ashy, many of the feathers being margined with this colour, those of the fore neck being edged with bluish ash-colour; rest of the under surface of the body light drab-brown, fading into whitish on the lower abdomen; under wing- and tail-coverts light greyish blue; bill and legs black. Total length 10·5 inches, culmen 1·5, wing 4·85, tail 5·05, tarsus 1·55.

Female. Similar to the male. Total length 10·5, culmen 1·15, wing 5, tail 5·4, tarsus 1·5.

Hab. Rocky Mountains and middle province of United States.

a.	♂ ad. sk.	Apache, Arizona (*H. W. Henshaw*).	Smithsonian Institution [P.].
b.	♀ ad. sk.	Arizona (*Newberry*).	Smithsonian Institution [P.].
c.	♂ ad. sk.	El Paso County, Colorado, March 9, 1873.	Purchased.
d.	♂ ad. sk.	El Paso County, Colorado, Oct. 22, 1873.	Purchased.
e.	♀ ad. sk.	El Paso County, Colorado, Nov. 22, 1873.	Purchased.
f.	♂ juv. sk.	Apisha-Pau Creek, Colorado.	Purchased.

4. Aphelocoma ultramarina.

Garrulus ultramarinus, *Bp. Journ. Acad. N. Sci. Philad.* 8vo, iv. p. 386; *Temm. Pl. Col.* ii. pl. 439.

Cyanocitta ultramarina, *Strickl. Ann. & Mag. N. H.* xv. p. 260; *Bp. Consp.* i. p. 378; *Baird, Brewer, & Ridgw. N.-Amer. B.* ii. p. 284.

Cyanurus ultramarinus, *Gray, Hand-l. B.* ii. p. 4.

Incumbent setaceous feathers of the base of the bill partly black and partly blue; whole plumage above, including the wings and tail, bright azure, most vivid and somewhat sericeous on the head and tail-coverts, duller and slightly intermixed with dusky on the back; inner webs and tips of the quill-feathers dusky; shafts of the wing- and tail-feathers black; lores black; cheeks dull blackish blue; chin whitish, intermixed with black bristles; whole inferior surface dirty whitish, more tinged with cinereous on the anterior parts, and becoming purer towards the vent; inferior wing- and tail-coverts slightly intermixed with blue; inferior surface of the wings and of the tail dusky grey, the latter darker; wings, when closed, reaching almost to the middle of the tail, which is seven inches long, and perfectly even at tip. Length thirteen inches; bill one inch and a half long, strongly notched at tip, and with the feet black. (*Bonaparte, l. c.*)

Hab. Mexico.

Obs. This species is not at present represented in the Museum; and I am only able to judge of its distinctness by the observations given by the authors of the 'North-American Birds.' They do not seem to be well acquainted with the species either, and separate it from *A. sordida* principally on account of its square tail. The dimensions given (*l. c.* p. 284) are copied from Bonaparte.

Subsp. *a*. Aphelocoma couchi.

Cyanocitta couchi, *Baird, B. N. Amer.* p. 588, pl. 60. fig. 2; *id. Mex. B.* ii. *Birds*, p. 21, pl. xxii.
Cyanocitta ultramarina, var. couchi, *Baird, Brewer, & Ridgway, B. N. Amer.* ii. p. 284.

Tail rounded, but little graduated; lateral feather about a quarter of an inch shortest. Wings longer than the tail; when closed, reaching nearly to its middle. Above and on sides of head and neck bright blue; the lores blackish; the middle of the back slightly duller, the tips of some of the feathers dark brown. Beneath brownish ash, paler on the chin and towards the anal region, which, with the crissum, is white. No trace of white or black on the sides of the head, nor of any streaks or collar on the breast. Length (fresh) 11·50 inches; wing 6·00; tail (dried) 5·40; tarsus 1·50. (*Baird, Brewer, and Ridgway, l. c.*)

Hab. South side of the valley of Rio Grande, near the coast, and southward. (*Iid. l. c.*)

Obs. The above-mentioned authors separate *A. couchi* from *A. ultramarina* on account of its smaller size &c. (*l. c.* p. 284).

5. Aphelocoma sordida.

Garrulus sordidus, *Swains. Phil. Mag.* i. p. 437 (1827); *id. Zool. Illustr.* 2nd ser. pl. 86.
Pica sieberi, *Wagl. Syst. Av.* Pica, sp. 33.
Cyanurus sordidus, *Swains. Faun. Bor.-Am. Birds*, p. 495.
Cyanocorax sieberi, *Gray, Gen. B.* ii. p. 307.
Cyanocitta sieberi, *Bp. Consp.* i. p. 378.
Aphelocoma sieberi, *Cab. Mus. Hein.* i. p. 221, note.
Cyanocitta sordida, *Sclater, Cat. Am. B.* p. 142; *Sclater & Salvin, Nomencl. Av. Neotr.* p. 39.
Cyanurus sieberi, *Gray, Gen. B.* ii. p. 307.

Adult. General colour above dull blue, rather ashy on the back, the head, rump, and upper tail-coverts brighter and more cobalt; wings and tail dull blue, the inner webs of the quills brown, those of the tail-feathers inclining to purplish brown; nasal bristles, lores, and sides of face and of neck dull blue, blacker on the ear-coverts; entire under surface of body brownish ash-colour, inclining to ashy white on the abdomen and under tail-coverts; thighs and under wing-coverts ashy grey, the edge of the wing shaded with blue: bill and legs black. Total length 12·5 inches, culmen 1·3, wing 7·15, tail 6·8, tarsus 1·75.

Hab. Mexico.

a. Ad. sk.	Mexico.	Major-Gen. Hardwicke.
b. Ad. st.	Mexico.	J. Gould, Esq.
c. Ad. sk.	Mexico.	A. Boucard, Esq. [C.].

Subsp. α. Aphelocoma arizonæ.

Cyanocitta sordida, *Baird, B. N. Amer.* p. 587, pl. lx. fig. 1 (*nec Swains.*); *id. Mex. B.* ii. *Birds*, p. 21, pl. xxii. fig. 1; *Cooper, Orn. Calif.* i. p. 305 (1870).
Cyanocitta sordida, var. arizonæ, *Ridgw. Rep. U. S. Geol. Expl. 40th par.*; *Baird, Brewer, & Ridgw. N.-Amer. B.* ii. p. 292.

Bill short, thick, half as high as long; wings considerably longer than the tail, which is slightly graduated (·50 of an inch). Upper surface, including whole side of head to the throat, light sky-blue, the whole dorsal region inclining to pure bluish ashy. Beneath fine uniform pale ash for anterior half (including the throat), this gradually fading into white on the posterior portions (including whole abdomen), the lower tail-coverts being pure white; lores blue. Length 13 inches, wing 6·20, tail 5·70, culmen 1·30, depth of bill ·40, tarsus 1·50, middle toe ·97. Fourth, fifth, and sixth quills equal and longest, second shorter than ninth, first 2·20 shorter than longest. (*Baird, Brewer, and Ridgway, l. c.*)

Immature. The blue, except that of the wings and tail, replaced by dull ash, the blue feathers appearing in scattered patches. (*Iid. l. c.*)

Hab. Arizona.

Obs. The differences between this race and true *A. sordida* are given as follows by the above-mentioned authors (p. 284):—" Graduation of the colours as in *A. sordida*; but the blue, instead of being a bright ultramarine, is very much paler and duller and with a greenish cast, the whole dorsal region decidedly ashy; ash of the pectoral region much paler, and throat similar, instead of decidedly whitish, in contrast; pure white of posterior lower parts covering whole abdomen instead of being confined to crissum."

a.	Juv. sk.	Near Crittenden, Arizona, Aug. 29th (*H. W. Henshaw*).	Smithsonian Institution [P.].
b.	♀ ad.	Fort Bayard, near Mexico.	Purchased.

6. Aphelocoma floridana.

Corvus floridanus, *Bartr. Trav. S. Carol.* p. 291; *Audub. Orn. Biogr.* i. p. 444; *id. B. Amer.* pl. lxxxvii.
Garrulus floridanus, *Bp. Amer. Orn.* ii. p. 59, pl. xiv. fig. 1; *Audub. B. Amer.* roy. 8vo, iv. p. 118, pl. ccxxxiii.
Cyanurus floridanus, *Swains. Faun. Bor.-Am. Birds*, p. 495; *Gray, Hand-l. B.* ii. p. 4.
Cyanocorax floridanus, *Bp. Comp. List B. Eur. & N. Amer.* p. 27.
Cyanocitta floridana, *Strickl. Ann. & Mag. N. Hist.* xv. p. 261; *Bp. Consp.* i. p. 377; *Baird, B. N. Am.* p. 586; *Allen, Bull. Harv. Coll.* ii. p. 298; *Baird, Brewer, & Ridgway, N. Am. B.* i. p. p. 285.
Aphelocoma floridana, *Cab. Mus. Hein.* Th. i. p. 221, note; *Coues, Key N.-Amer. B.* p. 165, fig. 108.
Aphelocoma floridana, var. floridana, *Coues, B. N. West*, p. 219.

Adult (Florida; *G. C. Taylor*). Nasal bristles and forehead hoary grey, extending backwards over the eye, and forming an indistinct eyebrow; lores and ear-coverts blackish, the latter with a slight wash of blue; fore part of cheeks hoary; crown of head, nape, hind neck, and sides of the latter dull cobalt-blue, more dingy on the crown; entire mantle drab-brown, shading off into blue on the lower back and rump; scapulars inwardly like the mantle, but blue externally, so as to resemble the wings; the latter blue, the quills all black on the inner web, the inner secondaries all shaded with blue; upper tail-coverts brighter cobalt; tail deep blue, with indistinct wavy lines of dusky under certain lights; under surface of body pale drab-brown, with a faint shade of lilac; the throat white, shaded with lilac on the lower portion, and encompassed with an indistinct circlet of dull blue across the jugular region, which has a slight lilac shade; thighs darker drab-colour, washed with blue near the tarsal joint; under tail-coverts cobalt; under wing-coverts greyish drab, slightly washed with cobalt, those near the edge of the wing brighter blue; bill and legs black. Total length 10·6 inches, culmen 1·05, wing 4·5, tail 5·5, tarsus 1·45. (*Mus. Salvin and Godman.*)

Hab. Florida.

7. **Aphelocoma unicolor.**

Cyanocorax unicolor, *Du Bus, Bull. Acad. R. Brux.* xiv. pt. 2, p. 103; *id. Esq. Orn.* pl. 17; *Scl. P. Z. S.* 1859, p. 365; *Schl. Mus. P.-B.* Coraces, p. 49.
Cyanocorax concolor, *Cass. Pr. Acad. Nat. Sci. Philad.* iv. p. 26.
Cyanocitta unicolor, *Bp. Consp.* i. p. 378; *Scl. P. Z. S.* 1857, p. 204; *id. Cat. Amer. B.* p. 143; *Scl. & Salv. Nomencl. Av. Neotr.* p. 39.
Cyanurus unicolor, *Gray, Hand-l. B.* ii. p. 5.

Adult. Above and below ultramarine blue, with a certain gloss of brighter blue under certain lights; wings and tail uniform with the back, the quills blackish on their inner web, the tail-feathers inclining to purplish brown on their inner web; feathers in front of the eye blue-black. Total length 12·5 inches, culmen 1·2, wing 6·5, tail 6·8, tarsus 1·55.

Hab. Central America, from Mexico to Guatemala.

a. Ad. sk.	Jalapa, Mexico.	R. de Oca [C.].
b. Ad. sk.	Mexico.	Purchased.
c. Ad. sk.	Guatemala.	J. Gould, Esq.
d, e. Ad. sk.	Vera Paz, Guatemala.	O. Salvin, Esq. [C.].

8. **Aphelocoma nana.**

Cyanocorax nanus, *Du Bus, Bull. Acad. R. Brux.* xiv. pt. 2, p. 103; *id. Esq. Orn.* pl. 25; *Scl. P. Z. S.* 1857, p. 204, 1859, p. 365; *Schl. Mus. P.-B.* Coraces, p. 49; *Scl. & Salv. Nomencl. Av. Neotr.* p. 39.
Cyanocitta nana, *Bp. Consp.* i. p. 378; *Scl. P. Z. S.* 1859, p. 381; *id. Cat. Amer. B.* p. 143.
Cyanurus nanus, *Gray, Hand-l. B.* ii. p. 5.

Adult. General colour above smalt-blue, brighter and inclining to cobalt on the head and tail; wings coloured like the back, the inner webs of all but the innermost secondaries black; nasal bristles, lores, feathers round the eye, cheeks, and ear-coverts black; forehead and a faintly defined eyebrow greyish cobalt; entire throat whitish shaded with purplish blue, with a small spot of black at the base of the chin, the gular patch margined below with a shade of blue-black; all the rest of the under surface of body smalt-blue, the thighs and under wing-coverts rather more ashy; bill and legs black. Total length 9·6 inches, culmen 0·95, wing 4·55, tail 4·9, tarsus 1·25.

Hab. Mexico.

a. Ad. sk.	Cordova, Mexico.	Purchased.

29. CYANOCORAX*.

	Type.
Cyanocorax, *Boie, Isis*, 1826, p. 975	C. chrysops.
Cyanurus, *Swains. Faun. Bor.-Am. Birds*, p. 495 (1831)	C. azureus?
Coronideus, *Cab. Mus. Hein.* Th. i. p. 224 (1850)	C. violaceus.

Head of *Cyanocorax chrysops.*

Range. Central and South America.

---

* *Obs.* I agree with Messrs. Sclater and Salvin that *Cyanurus* of Swainson must be considered a synonym of *Cyanocorax* of Boie; but I arrive at this conclusion in a totally different manner from that adopted by these authors. *Cyanurus* was a composite genus, and included ten species, viz.:—1. *cristatus*; 2. *stelleri*; 3. *sordidus*; 4. *floridanus*; 5. *coronatus*; 6. *cyanopogon*; 7. *pileatus*; 8. *azureus*; 9. *formosus*; 10. *cristatellus*. Of these *C. pileatus* (vel *chrysops*) ought not to have been included, as it is the sole type of Boie's genus *Cyanocorax* (1826). In 1841, Mr. G. R. Gray removed *C. formosus* as the type of his genus *Calocitta*; and in 1845 Strickland instituted his genus *Cyanocitta*, with *C. cristata* as its type, but including also Nos. 2, 4, and 5 of Swainson's list. Of these latter *C. floridanus* as well as *C. sordidus* were removed by Cabanis, in 1850, into his genus *Aphelocoma*; and therefore in the year 1850 there remained but three of Swainson's original *Cyanuri*, viz. *C. cyanopogon*, *C. azureus* (vel *cæruleus*, Vieill.), and *C. cristatellus* (vel *cyanoleucus*, Max.). This last-named bird was chosen by Cabanis in 1850 as the type of his genus *Argurocitta* (ex Bp.); and therefore the type of *Cyanurus* must be either *C. cyanopogon* or *C. azureus*. Both of these birds I keep in the same genus as *C. chrysops*; but should the very different form of the crest-feathers ever induce any systematist to separate these two birds generically, it would be better to keep *C. azureus* as the type of *Cyanurus*, not only on account of the inapplicability of the name to *C. cyanopogon*, but on account of the closer relationship of the latter to *C. chrysops*.

Key to the Species.

a. Middle tail-feathers tipped with white; breast white or yellowish; a blue or white spot above the eye, another below the latter, and a third on the fore part of the cheeks.
  a'. Outer tail-feather resembling the others, violet with a white tip.
    a''. Hinder neck purplish blue; nape and spot above the eye rich cobalt; occipital crest thick and velvety .... { *chrysops*, p. 120. *diesingi*, p. 121. }
    b''. Hinder neck and nape purplish, with no velvety crest; spot above the eye purplish cobalt.... { *affinis*, p. 121. *sclateri*, p. 122. }
    c''. Hinder neck and nape whitish; spots above and below the eyes and cheek-spot also whitish shaded with blue .... { *cayanus*, p. 122. *intermedius*, p. 123. }
  b'. Outer tail-feather resembling the others, black with a white tip; hinder neck bluish white; crest very long, equal to the culmen in length.. *cyanopogon*, p. 123.
  c'. Outer tail-feather entirely white, the middle ones blue, tipped with white .... *mystacalis*, p. 124.
b. Tail uniform; no white tips to any of the feathers.
  d'. Forehead uniform with the crown; no white line across the sinciput.
    d''. Back brown, the hind neck uniform.
      a'''. Throat and fore neck brown .... *cyanomelas*, p. 124.
      b'''. Throat and fore neck black........ *chilensis*, p. 125.
    e''. Back pale violet, as also the under surface; hind neck lighter violet .... *violaceus*, p. 125.
    f''. Back purplish blue, as also the under surface; head and neck all round black .... { *cæruleus*, p. 126. *heckeli*, p. 126. }
  e'. Forehead black, succeeded by a white line across the sinciput.
    g''. Throat blackish.
      c'''. Size larger: wing 5·35 inches; hinder crown and nape bright cobalt, separated from forehead by a broad line of white, which runs also along the sides of the crown .... *ornatus*, p. 127.
      d'''. Size smaller: wing 4·55 inches; hinder crown and nape deep blue, scarcely at all darker than the back .... *pumilo*, p. 127.
    h''. Throat silvery greyish white, tinged with blue *argentigula*, p. 128.

1. Cyanocorax chrysops.

Acahé, *Azara, Apunt.* i. p. 253.
Pica chrysops, *Vieill. N. Dict. d'Hist. Nat.* xxvi. p. 124 (*ex Azara*); *id. Gal. Ois.* i. p. 157, pl. 101.
Corvus pileatus, *Temm. Pl. Col.* ii. pl. 58 (*ex Illig.*).
Cyanocorax pileatus, *Boie, Isis,* 1826, p. 975; *Gray, Gen. B.* ii. p. 307; *Burm. Th. Bras.* ii. p. 284; *Heine, J. f. O.* 1860, p. 114; *Schl. Mus. P.-B.* Coraces, p. 52; *Pelz. Orn. Bras.* p. 189; *Scl. & Salv. Nomencl. Av. Neotr.* p. 39.

Pica pileata, *Wagler, Syst. Av.* Pica, sp. 28.
Cyanurus pileatus, *Swains. Faun. Bor.-Amer. Birds*, p. 495; *Gray, Hand-l. B.* ii. p. 5.
Uroleuca pileata, *Bp. Consp.* i. p. 379.

Adult male. General colour above blackish, with a very slight purple gloss, deepening into blackish purple on the rump and upper tail-coverts; nasal bristles, forehead, and crown black, all the feathers dense, and those of the front erect, the plumes of the crown forming a thick crest; nape silvery or greyish blue, deepening into purple on the hind neck; above the eye a spot of silvery cobalt; below the eye a second spot, and a patch at the base of the lower mandible, purplish cobalt, the eye-spot slightly tinged with silvery blue on its upper margin; sides of face, throat, and chest rich velvety brownish black; rest of under surface creamy yellow; wings like the back, shaded with deeper purple on the quills; tail-feathers deep purple, tipped with creamy yellow, more broadly on the outermost feathers; "bill and legs black; iris yellow" (*Alan Peel, MS.*). Total length 14 inches, culmen 1·3, wing 6·3, tail 7·1, tarsus 1·9.

Hab. Southern Brazil; Uruguay; Paraguay.

a. ♂ ad. sk.	Uruguay, Dec. 16, 1875.	Alan Peel, Esq. [C.].
b, c. Ad. st.	Brazil.	Purchased.
d. Skeleton.		Zoological Society.

2. Cyanocorax diesingi.

Cyanocorax diesingi, *Pelzeln, Sitz. Akad. Wien*, xx. p. 164; *Heine, J. f. O.* 1860, p. 115; *Pelzeln, Orn. Brasil.* p. 189.
Cyanurus diesingi, *Gray, Hand-l. B.* ii. p. 5.

Head crested; the sides of the neck, the throat, and jugulum black; a spot above and behind the eye and a streak at the base of the lower mandible bluish violet; nape white, shaded with violet, changing into violet towards the back; the back, the rump, the outer aspect of the wings, and the upper surface of the tail violet; the wings inwardly and the under surface of the tail blackish brown; the lower wing-coverts, the breast, abdomen, vent, and the tips of the tail-feathers white; the bill and feet blackish. Total length 13 inches, wing 6. (*Pelzeln, l. c.*)

Hab. Borba, Brazil (*Natterer*).

3. Cyanocorax affinis.

Cyanocorax affinis, *Pelz. Sitz. Akad. Wien*, xx. p. 164; *Scl. Cat. Am. B.* p. 145; *Schl. Mus. P.-B.* Coraces, p. 53; *Scl. & Salv. Nomencl. Av. Neotr.* p. 39.
Cyanurus affinis, *Gray, Hand-l. B.* ii. p. 5.

Adult. Above rich purplish brown, the upper tail-coverts and tail deep purple, the latter broadly tipped with yellowish white; wings deep purple, the inner webs dark brown; lores and frontal plumes, which are recurved, black, as also are the crown and sides of the head, throat, and entire fore neck; the occiput and nape purple, the shade extending down the sides of the neck; above and below the eye a patch of rich cobalt feathers, as well as a longitudinal malar patch; rest of the under surface light yellow from the chest downwards, including the under wing- and tail-coverts. Total length 13·2 inches, culmen 1·4, wing 6·3, tail 6·5, tarsus 1·75.

Hab. Columbia; Panama; Veragua.

a. Ad. sk.	Bogotá.	Purchased.
b. ♂ ad. sk.	Chepo, Panama.	O. Salvin, Esq.
c. ♂ ad. sk.	Isthmus of Panama.	J. M'Cleannan, Esq. [C.].
d. Ad. sk.	[South America.]	

4. Cyanocorax sclateri.

Cyanocorax sclateri, *Heine, J. f. O.* 1860, p. 115.
Cyanurus sclateri, *Gray, Hand-l. B.* ii. p. 5.

Above violet-brown; the whole head, cheeks, throat, and fore part of neck black; lores, spots near the eye, and nape blue; wing-coverts and quills dusky, margined with blue; tail-feathers blue, underneath dusky, the outer ones broadly, the middle ones more narrowly white at the tip; underneath white. Total length 13″ 9‴, culmen 1″, wing 6″ 3‴, tail 7″, tarsi 1″ 10‴. (*Heine, l. c.*)

Hab. Cartagena.

5. Cyanocorax cayanus.

Le Geay de Cayenne, *Briss. Orn.* ii. p. 52, pl. iv. fig. 1; ? *Montb. Pl. Enl.* iii. pl. 373 *.
Corvus cayanus, *Linn. S. N.* i. p. 157 (*ex Briss.*).
Pica albicapilla, *Vieill. N. Dict. d'Hist. Nat.* xxvi. p. 125.
Pica larvata, *Wagler, Syst. Av.* Pica, sp. 29.
Cyanocorax cayanus, *Gray, Gen. B.* ii. p. 307; *Cab. in Schomb. Reis. Guian.* iii. p. 683; *id. Mus. Hein.* Th. i. p. 224; *Sclater, Cat. Amer. B.* p. 145; *Schl. Mus. P.-B.* Coraces, p. 51; *Pelz. Orn. Bras.* p. 190; *Scl. & Salv. Nomencl. Av. Neotr.* p. 39.
Uroleuca cayana, *Bp. Consp.* i. p. 379.
Cyanurus cayanus, *Gray, Hand-l. B.* ii. p. 5.

Adult. General colour above lavender-brown, with a distinct shade of purple on the scapulars and under tail-coverts, which are darker brown; wings dark brown, externally purplish; tail dull purple, with a broad white band at the tip, slightly shaded with lilac; frontal plumes stiff and recurved, black, as also the feathers above and round the eye; the rest of the sides of the face, sides of neck, throat, and fore neck blackish brown; hinder part of head and hind neck

* This figure is apparently made up from Brisson's, and coloured from his description. The result is a tolerably good representation, *not* of *C. cayanus*, but of *C. bellus* of Schlegel (= *C. mystacalis*, Geoffr.). It is only the fact of the base of the outer tail-feathers being blue that prevents it from representing *C. mystacalis* with accuracy; and if *C. bellus* were distinct from the latter, which we do not believe, then Montbeillard's plate ought, apparently, to be referred to it.

white, shaded with blue near the margin of the black on the crown and sides of the head; above the eye a small spot of cobalt-blue; below the eye another spot of silvery white feathers, shaded with lilac and cobalt; anterior part of cheeks white, shaded with lilac; sides of breast and rest of under surface of body yellowish white, including the thighs and under wing- and tail-coverts. Total length 14 inches, culmen 1·4, wing 6·4, tail 6·6, tarsus 2.

Hab. Guiana.

a. Ad. sk.	British Guiana.	Purchased.
b. Ad. sk.	[South America.]	

6. Cyanocorax intermedius.

Cyanocorax intermedius, *Heine, J. f. O.* 1860, p. 116.
Cyanurus intermedius, *Gray, Hand-l. B.* ii. p. 5.

Above violet; forehead, sinciput, cheeks, and throat black; lores and spots near the eye whitish; nape and hind neck dull bluish white; wing-coverts and quills dusky, slightly margined with blue; tail-feathers dusky, washed with bluish, underneath dusky, the outer feathers more broadly, the inner ones more narrowly tipped with white; under surface of body, including the whole of the breast and the abdomen, whitish. Total length 12″ 3‴, culmen 1″, wing 5″ 9‴, tail 5″ 9‴, tarsus 1″ 9‴. (*Heine, l. c.*)

Hab. Venezuela.

7. Cyanocorax cyanopogon.

Corvus cyanopogon, *Max. Beitr. Naturg. Bras.* iii. p. 1247; *Temm. Pl. Col.* ii. pl. 169.
Cyanurus cyanopogon, *Swains. Faun. Bor.-Amer. Birds*, p. 495; *Gray, Hand-l. B.* ii. p. 5.
Cyanocorax cyanopogon, *Gray, Gen. B.* ii. p. 307; *Cab. Mus. Hein.* Th. i. p. 224; *Burm. Th. Bras.* ii. p. 285; *Scl. Cat. Am. B.* p. 145; *Schl. Mus. P.-B.* Coraces, p. 51; *Pelz. Orn. Bras.* pp. 190, 430; *Scl. & Salv. Nomencl. Av. Neotr.* p. 39.
Uroleuca cyanopogon, *Bp. Consp.* i. p. 379.

Adult. General colour above purplish brown, with a decided shade of purple on the lower back, rump, and upper tail-coverts; wings dark brown, the quills externally purplish, of a brighter shade than the back; tail purple, browner on the inner webs, broadly banded with white at the tip, extending further on the outer feathers than on the centre ones; head and sides of face, sides of neck, throat, and chest blackish brown; the feathers of the crown of a velvety texture, and forming a closely set crest; over the eye a large spot of cobalt-blue, fading off on the upper margin into silvery blue; a spot behind the lower part of the eye and a broad malar patch deep purple; nape and hind neck silvery white, with a purplish shade, deepening into purple on the hind neck; rest of under surface of body white, including the thighs and the under wing- and tail-coverts; some of the lower chest-feathers black, tipped with white; "bill black; legs blackish grey; iris citron-yellow" (*Neuwied*). Total length 13·5 inches, culmen 1·25, wing 6, tail 6·8, tarsus 1·8.

Hab. Brazil.

a. Ad. sk.	Brazil.	
b. Ad. sk.	Brazil.	M. Claussen [P.].
c, d. Ad. sk.	Brazil.	Purchased.
e. Ad. sk.	Brazil.	Purchased.
f. Skeleton.	Brazil.	Zoological Society.

8. Cyanocorax mystacalis.

Pica mystacalis, *Geoff. St.-Hilaire, Mag. de Zool.* 1835, pl. 34.
Cyanocorax mystacalis, *Gray, Gen. B.* ii. p. 307; *Scl. & Salv. Nomencl. Av. Neotr.* p. 39.
Cyanocorax uroleucus, *Heine, J. f. O.* 1860, p. 115.
Cyanocorax bellus, *Schl. Mus. P.-B.* Coraces, p. 50.
Cyanurus mystacalis, *Gray, Hand-l. B.* ii. p. 5.

Adult. Above purplish blue, much brighter and inclining to ultramarine on the wings; the quills greyish black, externally blue, the inner secondaries entirely of the latter colour; tail pure white, the two centre feathers ultramarine blue, brighter than the back, tipped with white, the extreme end shaded with lavender-grey, the next two feathers with more or less of a blue shade near the base of the tail; frontal plumes recurved, black, as also the crown of the head, which is shaded with blue on the hinder margin of the black cap; occiput, nape, hinder part of the neck, and upper part of mantle snowy white; lores and feathers below the eye, ear-coverts, and sides of neck, entire throat and fore neck black; a small patch of feathers above the eye, another below the hinder part of the latter, and cheeks (forming a broad moustachial band) silvery white slightly shaded with cobalt; rest of under surface of body, including the under wing- and tail-coverts, pure white. Total length 12·3 inches, culmen 1·45, wing 5·65, tail 5·75, tarsus 1·7.

Hab. Ecuador.

a. Ad. sk.	Ecuador.	Capt. Kellett and Lieut. Wood [P.].
b, c. Ad. sk.	Ecuador.	J. Gould, Esq.

9. Cyanocorax cyanomelas.

Arraca morada, *Azara, Apunt.* i. p. 256.
Pica cyanomelas, *Vieill. N. Dict. d'Hist. Nat.* xxvi. p. 127.
Pica cyanomelæna, *Wagler, Syst. Av.* Pica, sp. 24.
Garrulus cyanomelas, *Lafr. & D'Orb. Syn.* p. 9.
Cyanocorax cyanomelas, *Gray, Gen. B.* ii. p. 307; *Hartl. Ind. Azara,* p. 4; *Bp. Consp.* i. p. 379; *Schl. Mus. P.-B.* Coraces, p. 47; *Pelz. Orn. Bras.* pp. 190, 430; *Scl. & Salv. Nomencl. Av. Neotr.* p. 40.
Coronoideus cyanomelas, *Burm. Th. Bras.* ii. p. 287, note.
Cyanurus cyanomelas, *Gray, Hand-l. B.* ii. p. 6.
Corvus ænas, *Licht. in Mus. Berol.*

Adult female. General colour above and below violet- or drab-brown, the upper tail-coverts purplish blue; frontal plumes of a velvety texture, and lores blackish; the ear-coverts, sides of face, and throat also darker brown than the rest of the body; over the whole of the breast and abdomen a shade of violet, deepening into violet blue on the under tail-coverts; under wing-coverts also shaded with violet, like the breast; wings above pale brown, like the back, with a very faint shade of violet; tail deep purple. Total length 12·5 inches, culmen 1·4, wing 7·05, tail 6·3, tarsus 1·6; (*Mus. Salvin and Godman.*)

Hab. Brazil.

a. Ad. sk.	Brazil.	M. Claussen [C.].

10. Cyanocorax chilensis.

Psilorhinus chilensis, *Bp. Consp.* i. p. 381 *; *Gray, Hand-l. B.* ii. p. 6.

Cyanocorax nigriceps, *Scl. & Salv. P. Z. S.* 1876, p. 354.

Adult (type of *C. nigriceps*). General colour above dark violet- or drab-brown, with a shade of violet-blue, very faintly indicated on the back, but becoming deeper in colour on the rump; the upper tail-coverts and tail deep purple; wings a little darker than the back, with a distinct shade of violet on the outer coverts and inner secondaries; head rather darker than the back; lores, forehead, sides of face and throat blackish; rest of under surface violet-brown, deepening into purple on the thighs and under tail-coverts; under wing-coverts resembling the breast; bill and legs black. Total length 13·5 inches, culmen 1·45, wing 7·55, tail 6·7, tarsus 1·85. (*Mus. Salvin and Godman.*)

Hab. Bolivia.

a. Ad. st.	Bolivia.	Mr. Brydges [C.].
b. Ad. sk.	Bolivia.	Mr. Brydges [C.].

11. Cyanocorax violaceus.

Cyanocorax violaceus, *Du Bus, Bull. Acad. Brux.* xiv. pt. 2, p. 103; *id. Rev. Zool.* 1848, p. 243; *id. Esq. Zool.* pl. 30; *Bp. Consp.* i. p. 379; *Scl. P. Z. S.* 1855, p. 158, 1858, pp. 72, 456; *id. Cat. Am. B.* p. 145; *Schl. Mus. P.-B.* Coraces, p. 46; *Pelz. Orn. Bras.* p. 190; *Scl. & Salv. Nomencl. Av. Neotr.* p. 40.

Cyanocorax harrisii, *Cass. Pr. Acad. Nat. Sci. Philad.* iv. p. 26.

Cyanocorax hyacinthinus, *Cab. in Schomb. Reis. Guian.* iii. p. 683.

Coronideus hyacinthinus, *Cab. Mus. Hein.* Th. i. p. 225; *Burm. Th. Bras.* ii. p. 287, note.

Cyanurus hyacinthinus, *Gray, Hand-l. B.* ii. p. 6.

Adult. General colour above greyish purple, the wings deeper-coloured, the quills with somewhat of an ultramarine shade; tail purple, glossed with ultramarine; head, sides of face, sides of neck, entire throat, and fore neck blackish brown, all the feathers of the forehead and region of the eye of a velvety texture, with a very full

* *Cf.* Schl. Mus. P.-B. *Coraces*, p. 47.

crest springing from the centre of the crown; across the occiput a line of white drawn from the hinder region of the eye above the ear-coverts, and slightly shaded with purple; the nape and hind neck greyish purple, paler than the back; rest of under surface of body from the chest downwards light greyish purple, richer on the chest and on the thighs; lower wing-coverts greyish black, slightly washed with purple; inner lining of quills greyish black. Total length 15 inches, culmen 1·4, wing 8·15, tail 6·9, tarsus 2.

Hab. Amazonia, Columbia, and Guiana.

a. Ad. st.	Rio Javarri.	H. W. Bates, Esq. [C.].
b. Ad. sk.	Peru.	Purchased.
c. Ad. sk.	New Granada.	Purchased.
d. Ad. sk.	British Guiana.	Sir R. Schomburgk [C.].

12. **Cyanocorax cæruleus.**

Urraca celeste, *Azara, Apunt.* i. p. 259.
Pica cærulea, *Vieill. N. Dict. d'Hist. Nat.* xxvi. p. 126 (*ex Azara*).
Corvus azureus, *Temm. Pl. Col.* ii. pl. 168.
Pica azurea, *Wagler, Syst. Av.* Pica, sp. 25.
Cyanurus azureus, *Swains. Faun. Bor.-Amer. Birds*, p. 495.
Cyanocorax azureus, *Gray, Gen. B.* ii. p. 307; *Hartl. Ind. Azara*, p. 4; *Bp. Consp.* i. p. 379; *Pelz. Orn. Bras.* pp. 190, 433; *Scl. & Salv. Nomencl. Av. Neotr.* p. 40.
Cyanocorax cyanescens, *Licht. Nomencl.* p. 40.
Coronideus cæruleus, *Burm. Th. Bras.* ii. p. 287, note.
Cyanocorax cæruleus, *Schl. Mus. P.-B.* Coraces, p. 46.
Cyanurus cæruleus, *Gray, Hand-l. B.* ii. p. 6.

Adult. General colour above deep smalt-blue, shading off into brighter blue on the rump, the upper tail-coverts more cobalt; wings like the back, the inner webs of the quills black, the secondaries only at the base, the innermost of the latter like the back; tail deep blue, the inner webs blackish along the edge; head all round, including the throat and fore neck, black, the hinder crown shaded with blue, and the feathers of the fore neck tipped with blue where they adjoin the breast, which, with the rest of the under surface, is rich smalt-blue, as also the thighs and under wing- and tail-coverts; bill and legs black. Total length 15·5 inches, culmen 1·6, wing 8·15, tail 6·9, tarsus 2·05. (*Mus. Salvin & Godman.*)

Hab. Southern Brazil.

13. **Cyanocorax heckeli.**

Cyanocorax heckelii, *Pelz. Sitz. k. Akad. Wien*, xx. p. 163; *id. Orn. Brasil.* p. 191.
Cyanurus heckelii, *Gray, Hand-l. B.* ii. p. 6.

Frontal plumes erect and rigid, directed forwards; the entire head, throat, and jugulum brownish black, without any lustre; the wings inwardly, as well as their lower surface and that of the tail, blackish brown; the rest of the body intense violet-blue; bill and feet black. Length 1 ft. $4\frac{3}{4}$ inches. (*Pelzeln, l. c.*)

Hab. Rio Boraxudo, near Paranagua (*Natterer*).

14. Cyanocorax ornatus.

Pica ornata, *Less. Rev. Zool.* 1839, p. 41.
Cyanocorax ornatus, *Gray, Gen. B.* ii. p. 307; *Scl. P. Z. S.* 1857, p. 204, 1859, p. 365; *Schl. Mus. P.-B.* Coraces, p. 47.
Cyanocitta ornata, *Bp. Consp.* i. p. 144; *Scl. & Salv. Ibis,* 1859, p. 21; *Scl. P. Z. S.* 1859, p. 381; *id. Cat. Amer. B.* p. 143; *Scl. & Salv. Nomencl. Av. Neotr.* p. 39.
Cyanocitta cyanicollis, *Licht. Nomencl.* p. 9.
Cyanurus ornatus, *Gray, Hand-l. B.* ii. p. 4.

Adult. General colour above deep ultramarine, the tail rather brighter, and having a slight cast of verditer-blue under certain lights; wings coloured like the back, the quills blackish on the inner web, excepting the inner secondaries, which are deep ultramarine like the back; hind neck and mantle black, occupying also the sides of the neck and joining underneath the fore neck, which is also black like the entire throat; rest of under surface of body deep ultramarine, the under wing-coverts black, shaded with blue; lores and nasal bristles black, as also the entire sides of the face, feathers round the eye and frontal band of plumes, which are of a velvety nature; hinder head and nape cobalt-blue, separated from the black fore part of the head by a narrow band of silvery white, which follows the margin of the black down the sides of the occiput to behind the ear-coverts, separating the black from the blue all the way. Total length 12·5 inches, culmen 1·35, wing 5·35, tail 6·1, tarsus 1·75.

Hab. Central America from Mexico to Veragua.

a, b. Ad. sk.	Jalapa, Mexico.	R. de Oca [C.].
c. Ad. st.	Mexico.	Purchased.
d. Ad. sk.	Mexico.	Purchased.
e. Ad. sk.	Guatemala.	Purchased.
f. Ad. sk.	Veragua.	M. E. Arcé [C.].

15. Cyanocorax pumilo.

Cyanocorax nanus?, *Strickl. Contr. Orn.* 1849, p. 122, pl. 33.
Cyanocorax pumilo, *Strickl. t. c.* p. 122.
Cyanocitta pumilo, *Bp. Consp.* i. p. 378; *Scl. & Salv. Ibis,* 1859, p. 21; *id. Nomencl. Av. Neotr.* p. 39.
Cyanurus pumilio, *Gray, Hand-l. B.* ii. p. 5.

Adult. General colour deep smalt-blue; quills blackish on inner web, externally blue like the back, the inner secondaries entirely blue; tail deep blue, the inner web inclining to purplish brown; head ultramarine, forming a cap which contrasts with the blue of the back; lores and forehead, feathers round the eye, sides of face black; the black forehead separated from the crown by a tiny line of silvery white feathers, in the form of a row of dots; throat blackish with a blue shade; rest of under surface of body deep smalt-blue, including the under tail- and wing-coverts; the lower series of the latter ashy black, like the inner lining of the wing. Total length 9·5 inches, culmen 1, wing 4·55, tail 4·75, tarsus 1·35.

Hab. Guatemala.

a. Ad. sk.	Guatemala.	Purchased.
b. Ad. sk.	Coban, Guatemala.	Purchased.

16. **Cyanocorax argentigula.**

Cyanocitta argentigula, *Lawr. Ann. Lyc. N. Y.* xi. p. 88; *Scl. & Salv. P. Z. S.* 1876, p. 268.

Female. The feathers of the front, lores, with those above and below the eye, and of the chin are rigid and of an intense black; crown, hind neck, upper part of back, sides of neck, and the breast black, lower part of back and rump of a dull deep blackish blue; the abdomen, sides, and under tail-coverts are also dull deep blue, but brighter than the back, especially on the sides and under tail-coverts; on the front part of the crown is a crescentic mark of bluish white, which becomes narrower along each side of the crown, and extends over the ear-coverts as far as the occiput; the feathers of the crown are elongated, forming a short crest; the throat is marked with a sharply defined patch of silvery greyish white, with just a tinge of blue, and is in strong contrast with the black that surrounds it; this patch widens out a little at its lower extremity, which is on the upper part of the breast; the wings and exposed portions of the quill-feathers are of a rich dark cobalt-blue; the inner webs and under surface of the quill-feathers are glossy black; the tail above is coloured like the wings, underneath it is glossy black; bill and feet black. Length $10\frac{3}{4}$ inches, wing $4\frac{3}{4}$, tail $5\frac{1}{4}$, bill 1, tarsus $1\frac{5}{16}$. (*Lawrence.*)

Hab. Costa Rica, Salamanca.

30. **XANTHURA.**

	Type.
Xanthoura, *Bp. Consp.* i. p. 380 (1850)	X. yncas.
Cissolopha, *Bp. t. c.* p. 380 (1850)	X. sanblasiana.
Cyanolyca, *Cab. Mus. Hein.* i. p. 223 (1851)	X. armillata.
Xanthocitta, *Cab. t. c.* p. 223 (1851, *ex Bp. MS.*)	X. yncas.

Range. Central and South America.

Key to the Species.

a. Outer tail-feathers yellow.
  a'. Hind part of the neck and crown of the head yellowish white .... *yncas*, p. 129.
  b'. Crown and nape bright cobalt.
    a''. Breast and under tail-coverts bright yellow, without any shade of green; frontal crest-plumes erect, and measuring 0·6 inch in length .... *cæruleocephala*, p. 130.

b''. Underneath yellow, washed with green on the sides of the body; under tail-coverts bright yellow; frontal crest-plumes short, not exceeding 0·3 inch in length ...... *cyanocapilla*, p. 131.
c''. Underneath greenish, including the under tail-coverts .... *luxuosa*, p. 132.
b. Outer tail-feathers blue above, black beneath.
c'. Throat black.
d''. Belly black.
a'''. Frontal plumes forming a thin elongated black crest* .... *sanblasiana*, p. 132.
b'''. Frontal plumes short, erect, but not forming a crest.
a^4. Larger: back bright blue; nostrils entirely covered by the frontal plumes .... *beecheii*, p. 133.
b^4. Smaller: back greenish blue; nostrils half-covered by the frontal plumes.. *yucatanica*, p. 133.
e''. Belly blue .... *melanocyanea*, p. 134.
d'. Throat blue, encircled by a white necklace across the fore neck.
f''. Crown blue, with a narrow band of white behind the black forehead .... *viridicyanea*, p. 134.
g''. Crown white .... *jolyæa*, p. 135.
e'. Throat blue, encircled by a black necklace.
h''. Lower back, rump, tail, and wings greenish cobalt.
c'''. Fore part of crown silvery whitish; throat silvery or turquoise-blue ...... *turcosa*, p. 135.
d'''. Fore part of crown rich cobalt, like the rest of the head; throat rich cobalt *quindiuna*, p. 135.
i''. Lower back, rump, wings, and tail deep cobalt.
e'''. Purplish above, the head only slightly lighter; throat-patch very little lighter than the under surface, which is deep purple .... *meridana*, p. 136.
f'''. Deep rich cobalt-blue above and below; the throat-patch brighter cobalt, and the head inclining to silvery blue .... *armillata*, p. 136.

1. Xanthura yncas.

Geai du Pérou, *Montb. Pl. Enl.* iii. pl. 625.
Peruvian Jay, *Lath. Gen. Syn.* i. pt. 1, p. 391.
Corvus yncas, *Bodd. Tabl. Pl. Enl.* p. 38 (*ex Montb.*).
Corvus peruvianus, *Gm. S. N.* i. p. 373 (*ex Lath.*).
Le Geai Péruvien, *Levaill. Ois. Parad.* pl. 44.
Pica chloronota, *Wagler, Syst. Av.* Pica, sp. 12.
Pica luteola, *Less. Traité*, p. 331.
Garrulus peruvianus, *Lafr. & D'Orb. Syn. Av.* p. 9.
Cyanocorax peruvianus, *Cab. in Tschudi's Faun. Peruan.* p. 36.

---

* The characters of these three birds are copied from Messrs. Sclater and Salvin's paper (*l. c.*).

Cyanocorax peruanus, *Cab. t. c.* p. 232.
Cyanocorax yncas, *Gray, Gen. B.* ii. p. 307; *Schl. Mus. P.-B.* Coraces, p. 53.
Xanthoura peruviana, *Bp. Consp.* i. p. 380.
Xanthocitta peruviana, *Cab. Mus. Hein.* Th. i. p. 223.
Cyanocorax incas, *Scl. P. Z. S.* 1854, p. 114, 1858, p. 72; *id. Cat. Amer. B.* p. 144; *Scl. & Salv. Nomencl. Av. Neotr.* p. 39.
Cyanurus yncas, *Gray, Hand-l. B.* ii. p. 5.
Xanthura peruviana, *Taczan. P. Z. S.* 1874, p. 524.
Cyanocorax yncas, var. cyanodorsalis, *Dubois, Bull. Acad. Brux.* xxxviii. p. 492.
Xanthoura incas, var. incas, *Baird, Brewer, & Ridgw. N.-Amer. B.* ii. p. 295.

Adult. Above bright green, slightly yellowish; the wings rather darker green than the back, the quills bronzy towards the tips of the outer webs; two centre tail-feathers dark green, with a shade of indigo, all the rest of the feathers bright yellow; frontal plumes (recurved) and nasal bristles deep cobalt, as also a small spot above the eye, the feathers below the latter and the cheeks also deep cobalt; the rest of the sides of the face, lores, throat, and fore neck black, with a slight shade of blue on the upper chin; crown of the head, from the fore part of the head to the hind neck, ivory-yellow, with a dash of brighter yellow and a shade of lilac under certain lights, the hinder neck bright yellow where it adjoins the green back; rest of under surface of body bright yellow, some of the plumes of the lower throat black, tipped with yellow; under wing-coverts bright yellow, as also the inner webs of the quills. Total length 11·3 inches, culmen 1·15, wing 5·05, tail 6·2, tarsus 1·65.

Obs. Ecuadorian specimens are smaller than Brazilian, and have quite a distinct shade of lilac-blue on the head and hind neck, especially near the mantle. Total length 10·5 inches, culmen 1, wing 4·6, tail 5·8, tarsus 1·45. Dr. Dubois has given a name to this form; but it is the true *C. yncas* (*cf.* Pl. Enl. pl. 625), and the character is by no means strongly pronounced in a large series.

Hab. Western South America, from Peru to Ecuador.

a, b. Ad. sk.	South America.	Purchased.
c. Ad. sk.	West Coast of South America.	Capt. Kellett and Lieut. Wood [P.].
d, e. Ad. sk.	Quito.	J. Gould, Esq.
f. Ad. sk.	New Granada.	J. K. Salmon, Esq. [C.].
g. Ad. sk.	Huiro, Valle de Santa Ana, July 1874.	H. Whitely, Esq. [C.].

2. Xanthura cæruleocephala.

Cyanocorax yncas, var. γ. cæruleocephala, *Dubois, Bull. Acad. Roy. Belg.* xxxviii. p. 493.

Adult. General colour above green, washed with bluish green on the upper tail-coverts; wings green, like the back, the inner webs of the quills dusky brown, the primaries somewhat tinged with yellow towards the tips of the outer webs; four centre tail-feathers dark green, the four outer ones on each side bright yellow; crown of head and nape deep cobalt, the frontal plumes rather stiff and erect, and separated from the rest of the crown by a line of white; lores and sides of face black, a spot above and below the eye and a large patch at the base of the lower mandible deep cobalt; entire throat and fore neck black; rest of under surface of body pure bright yellow, without any shade of green anywhere, the under wing-coverts yellow like the breast. Total length 11·8 inches, culmen 1·05, wing 4·6, tail 6, tarsus 1·5.

Obs. The Venezuelan species seems to be quite distinct, and will, I believe, be found to bear Dr. Dubois's name of *cæruleocephala*. At the same time the characters given by him are not the true distinctive ones of the species; for the bird has very long frontal plumes, not *short*, as in the Guatemalan species; and, again, there is a distinct line of white across the crown, which Dr. Dubois states to be absent in his typical specimen. This may be due to defective preparation of the skin; but the pure yellow abdomen ("abdomen flavissimum," *Dubois*), being such a peculiar character of the Venezuelan bird, obliges me to refer the specimens in the Museum to *X. cæruleocephala*, notwithstanding the discrepancies in the description.

Hab. Venezuela and Trinidad.

a. Ad. st.	Trinidad.	Mrs. Wright [P.].
b. Ad. sk.	Venezuela (*Rozas*).	Purchased.

3. Xanthura cyanocapilla.

? Pica chloronota, *Wagler, Isis,* 1829, p. 749.
Cyanocorax cyanocapillus, *Cab. in Tschudi, Faun. Peruan.* p. 233, note; *Scl. & Salv. Nomencl. Av. Neotr.* p. 39.
Xanthoura guatemalensis, *Bp. Consp.* i. p. 380.
Xanthocitta cyanocapilla, *Cab. Mus. Hein.* i. p. 223.
Cyanocorax guatemalensis, *Scl. & Salv. Ibis,* 1859, p. 22, 1860, p. 113; *Scl. Cat. Amer. B.* p. 144.
Cyanurus cyanocapillus, *Gray, Hand-l. B.* ii. p. 5.
Xanthoura incas, var. guatemalensis, *Baird, Brewer, & Ridgw. N.-Amer. B.* p. 295.

Adult. Above dark green, as also the wings and four centre tail-feathers, the rest of the tail bright yellow; head and hind neck deep cobalt-blue, shading off into the back and not forming a distinct cap; nasal bristles and frontal plumes very large and erect, bright blue in colour; feathers in front of eye, sides of face, throat, and fore neck black, the lower feathers of the latter tipped with yellow; fore part of cheeks and a small spot above and below the hinder part of the eye blue; rest of under surface of body bright yellow, including the thighs and under wing- and tail-coverts. Total length 10 inches, culmen 1·1, wing 4·55, tail 5·8, tarsus 1·45.

Hab. Guatemala.

a. Ad. sk.	[Guatemala.]	Purchased.
b, c. Ad. sk.	Guatemala.	Purchased.

4. Xanthura luxuosa.

Pica chloronota, *Wagler*, *Isis*, 1829, p. 750 (♂ juv.).
Garrulus luxuosus, *Less. Rev. Zool.* 1839, p. 100.
Cyanocorax luxuosus, *Du Bus*, *Esq. Orn.* pl. 18; *Cass. B. Calif. & Texas*, p. 1, pl. i.; *Sclater*, *Cat. Am. B.* p. 144; *Sclater & Salvin*, *Nomencl. Av. Neotr.* p. 39.
Xanthoura luxuosa, *Bp. Consp.* i. p. 380; *Baird*, *B. N. Amer.* p. 589; *Lawr. Mem. Bost. Soc. N. H.* ii. p. 285.
Xanthocitta luxuosa, *Cab. Mus. Hein.* Th. i. p. 224.
Cyanocorax peruvianus, *Cass. Pr. Acad. Nat. Sci. Philad.* iv. p. 89.
Cyanocorax yncas, *Lawr. Ann. Lyc. N. Y.* 1851, p. 115.
Cyanurus luxuosus, *Gray*, *Hand-l. B.* ii. p. 5.
Xanthoura incas, var. luxuosa, *Baird, Brewer, & Ridgw. N. Amer. B.* ii. p. 295.

Adult. Above green, with a slight tinge of blue on the back and upper tail-coverts; the wings green, like the back, the secondaries darker; four centre tail-feathers green, the two middle ones washed with indigo-blue; all the other feathers bright yellow, shaded with green near the base of those next to the middle ones; head deep cobalt, as also the nasal bristles and frontal plumes, the latter separated from the crown by a frontal bar of white feathers; plumes in front of and round the eye, sides of face, and entire fore neck black, the region of the eye velvety, the black feathers of the lower neck tipped with green; fore part of cheeks and a small patch of feathers above and below the hinder part of the eye bright blue like the head; rest of under surface pale green from the chest to the under tail-coverts, the ground-colour being yellow, only slightly apparent on the lower abdomen; under wing-coverts bright yellow; bill black; feet lead-colour. Total length 11 inches, culmen 1·25, wing 4·55, tail 5·55, tarsus 1·55.

Hab. Texas southwards to Mexico.

a, b. Ad. st.	Mexico.	Purchased.
c, d. Ad. sk.	Jalapa, Mexico.	R. de Oca [C.].

5. Xanthura sanblasiana.

Geai de San Blas, *Néboux*, *Rev. Zool.* 1840, pp. 290, 323.
Pica sanblasiana, *Lafr. Rev. et Mag. de Zool.* 1842, Ois. pl. 28.
Cyanocorax sanblasianus, *Gray*, *Gen. B.* ii. p. 307.
Cissolopha sanblasiana, *Bp. Consp.* i. p. 380; *Lawr. Mem. Bost. Soc. N. H.* ii. p. 284.
Cyanocorax de San-Blas, *Prev. et Desmurs*, *Voy. 'Vénus,'* v. p. 200.
Cyanurus sanblasianus, *Gray*, *Hand-l. B.* ii. p. 4.
Cyanurus geoffroyi, *id. l. c.* p. 4.
Cyanocitta sanblasianus, *Sclater & Salvin*, *P. Z. S.* 1876, p. 269.

Adult. General colour above deep cobalt-blue, with a shade of ultramarine, especially on the quills; tail ultramarine-blue; head and neck all round and the entire under surface of body black; frontal feathers velvety, the anterior ones forming an erect crest; under wing- and tail-coverts black, washed with blue; bill brownish black. Total length 8·5 inches, culmen 1·25, wing 5·25, tail 6·1, tarsus 1·6.

Hab. Western Mexico.

a. Ad. st. [Mexico.]

6. Xanthura beecheii.

Pica beecheii, *Vigors, Zool. Journ.* iv. p. 353; *id. in Beechey's Voyage, Birds*, p. 22, pl. 6.
Cyanocitta crassirostris, *Bp. Consp.* i. p. 378.
Cyanocorax geoffroyi, *Bp. C. R.* xxxi. p. 564.
Cyanocorax beecheyi, *Finsch, Abhandl. nat. Ver. Bremen*, ii. p. 333.
Cyanocitta beecheii, *Sclater & Salvin, Nomencl. Av. Neotr.* p. 39; *Lawr. Mem. Bost. Soc. N. H.* ii. p. 283; *Sclater & Salvin, P. Z. S.* 1876, p. 270.

Adult male. Head all round, nape, and hind neck black; the lores stiff, and directed forwards, so as to hide the nostrils and base of the bill; upper surface of body purplish blue, rather brighter on the rump and upper tail-coverts; wing-coverts exactly like the back; quills black, externally purplish blue, the innermost secondaries and the tail entirely of the latter colour; entire under surface of body glossy black, the thighs and under wing- and tail-coverts purplish blue; bill black; legs yellow; iris yellow. Total length 16·5 inches, culmen 1·65, wing 6·75, tail 8·5, tarsus 2. (*Mus. Salvin and Godman.*)

Adult female. Similar to the male, but having the bill yellow and the iris grey (*Grayson*). Total length 15 inches, wing 6·6, tail 7·9, tarsus 1·9. (*Mus. Salvin and Godman.*)

Hab. N.W. Mexico.

7. Xanthura yucatanica.

Cyanocitta beachii, *Bp. Consp.* i. p. 378 (*nec Vigors*).
Corvus (Pica) beecheii, *Eyd. et Gerv. Mag. de Zool.* 1836, pl. 72; *iid. Voyage 'Favorite,'* pl. 20.
Cyanocitta crassirostris, *Pucher. Rev. Zool.* 1858, p. 196; *Salv. Ibis*, 1861, p. 353; *Moore, P. Z. S.* 1859, p. 57; *Lawr. Ann. Lyc. N. Y.* ix. p. 201; *Scl. & Salv. Nomencl. Av. Neotr.* p. 39.
Cyanurus crassirostris, *Gray, Hand-l. B.* ii. p. 5.
Cyanurus beecheii, *Gray, Hand-l. B.* ii. p. 4.
Cyanocitta yucatanica, *Du Bois, Bull. Acad. Brux.* xl. p. 797.
Cyanocitta germana, *Scl. & Salv. P. Z. S.* 1876, p. 270.

Adult male. Above verditer-blue or greenish, inclining to purer cobalt on the upper tail-coverts; wing-coverts resembling the back, as do also the inner secondaries; quills blackish, externally of the same colour as the back; tail brighter and more cobalt-blue above, black below; head and neck all round and entire under surface of body, including the under wing- and tail-coverts, black; thighs black, washed with blue; lower surface of quills ashy, inclining to whitish near base of the inner web; bill black; feet yellow. Total length 13 inches, culmen 1·5, wing 5·75, tail 7·4, tarsus 1·75.

Adult female. Similar to the male, but smaller, having the bill yellow and the tail-feathers tipped with white. Total length 12 inches, culmen 1·4, wing 5·35, tail 6·6, tarsus 1·7.

Hab. Honduras and Yucatan.

a, b. [♀] ad. sk.	Honduras.	Mr. Dyson [C.].
c. [♂] ad. sk.	Honduras.	Mr. Leyland [C.].
d. [♀] ad. sk.	[Central America.]	Purchased.

8. **Xanthura melanocyanea.** (Plate VI.)

Garrulus melanocyaneus, *Hartl. Rev. Zool.* 1844, p. 215.
Cyanolyca melanocyanea, *Cab. Mus. Hein.* Th. i. p. 223.
Cyanocitta melanocyanea, *Bp. Consp.* i. p. 378; *Sclater & Salvin, Ibis,* 1859, p. 21; *Sclater, Cat. Am. B.* p. 144; *Sclater & Salvin, Nomencl. Av. Neotr.* p. 39.
Cyanocorax melanocyaneus, *Sclater, P. Z. S.* 1858, p. 359; *Schlegel, Mus. P.-B.* Coraces, p. 45.
Cyanurus melanocyaneus, *Gray, Hand-l. B.* ii. p. 4.

Adult. General colour above verditer-blue, with a strong lustre of greenish cobalt under certain lights, especially on the wings; tail purple, washed with cobalt, the latter being the prevailing shade; head and neck all round, including the upper part of mantle above and the breast below, black; rest of under surface of body very dull verditer-blue, including the under wing- and tail-coverts; bill and legs black; iris red. Total length 13 inches, culmen 1·3, wing 5·4, tail 6·5, tarsus 1·45.

Hab. Guatemala.

a. Ad. st.	Guatemala.	
b. Ad. sk.	Coban, Vera Paz.	Purchased.
c. Ad. sk.	[Central America.]	

9. **Xanthura viridicyanea.**

Garrulus viridicyaneus, *D'Orb. et Lafr. Syn.* p. 9; *D'Orb. Voy. Am. Mérid. Ois.* pl. 53. fig. 1.
Cyanocorax viridicyaneus, *Gray, Gen. B.* ii. p. 307.
Cyanocitta viridicyanea, *Bp. Consp.* i. 378; *Sclater & Salvin, Nomencl. Av. Neotr.* p. 39; *iid. P. Z. S.* 1873, p. 185.
Cyanurus viridicyaneus, *Gray, Hand-l. B.* ii. p. 4.

Adult. General colour above dark verditer-blue, brighter on the wings and tail, the quills internally purplish brown, externally blue, the inner secondaries entirely of the latter colour; tail rather brighter blue than the back, the feathers internally purplish brown; frontal plumes, lores, and nasal bristles, feathers round the eye, cheeks, ear-coverts, and sides of neck black; the black frontal line succeeded by a broad band of white, which crosses the fore part of the vertex, and extends backwards on each side of the head, forming an eyebrow, but not very distinct, being mixed with blue; the rest of the head blue, like the back, but rather brighter, especially where it adjoins the white band; throat and fore neck dull blue, succeeded by a narrow crescentic line of white, the rest of the under surface of body verditer-blue, duller on the thighs and under wing- and tail-coverts; "bill and feet black; iris blackish." (*D'Orbigny.*) Total length 12·5 inches, culmen 1·05, wing 5·45, tail 6·9, tarsus 1·55.

Hab. Bolivia and Peru.

a. Ad. st.	Bolivia.	Mr. Brydges [C.].
b. Ad. sk.	Bolivia.	Mr. Brydges [C.].
c, d. Ad. sk.	Huasampilla, Peru, March 1872.	Henry Whitely, Esq. [C.].

10. **Xanthura jolyæa.**

Cyanocitta jolyæa, *Bp. J. f. O.* 1853, p. 47; *Scl. & Salv. Nomencl. Av. Neotr.* p. 39; *Taczan. P. Z. S.* 1874, p. 524; *Scl. & Salv. P. Z. S.* 1876, p. 271.

Cyanurus joliæa, *Gray, Hand-l. B.* ii. p. 5.

Similar to *C. viridicyanea*, but of a more splendid blue colour; forehead and cheeks broadly black; crown and a narrow collar on the throat whitish blue; throat blue, somewhat ashy (not black). (*Bonaparte, l. c.*)

Hab. Junin district of Peru.

11. **Xanthura turcosa.** (Plate VII.)

Cyanocitta turcosa, *Bp. C. R.* xxxvii. p. 830; *Scl. P. Z. S.* 1858, p. 552, 1860, p. 58; *id. Cat. Amer. B.* p. 144; *Scl. & Salv. Nomencl. Av. Neotr.* p. 39.

Cyanocorax turcosus, *Schl. Mus. P.-B.* Coraces, p. 48; *Scl. P. Z. S.* 1857, p. 18.

Cyanurus turcosus, *Gray, Hand-l. B.* ii. p. 5.

Adult. General colour above greenish cobalt, of a purer blue on the mantle and scapulars; wings greenish cobalt, the inner webs black, the inner secondaries entirely blue like the back; tail uniform greenish cobalt, of a more decided greenish tinge than the back; nasal bristles, forehead, feathers above and below the eye, cheeks, and ear-coverts black; fore part of crown silvery cobalt, shading off into bright cobalt on the nape and hind neck; chin black; entire throat light turquoise-blue, girdled by a black band across the fore neck; rest of under surface of body greenish cobalt, including the thighs and under wing- and tail-coverts; bill and legs black. Total length 13 inches, culmen 1·35, wing 5·45, tail 6·6, tarsus 1·7.

Hab. Ecuador, Columbia.

a. Ad. sk.	Ecuador.	C. Buckley, Esq. [C.].
b, c. Ad. sk.	Quito.	J. Gould, Esq.
d. Ad sk.	Ecuador.	Purchased.
e. Ad. sk.	Bogotá.	Purchased.

12. **Xanthura quindiuna.** (Plate VIII.)

Cyanocitta armillata, γ. quindiuna, *Scl. & Salv. P. Z. S.* 1876, p. 272.

Adult male. Head, neck, and mantle rich purplish cobalt, shading off on the back into greener blue, the rump and upper tail-coverts being decidedly greenish cobalt; tail greenish cobalt, the edges of the inner webs blackish; wing-coverts resembling the back, the quills black, externally greenish cobalt, the innermost secondaries being entirely of this colour; forehead, lores, sides of face, and chin black; the throat rich cobalt, encircled by a collar of black; rest of the under surface greenish cobalt, inclining to purplish blue on the thighs, breast, and under wing-coverts; bill and legs black. Total length 14·4 inches, culmen 1·3, wing 5·65, tail 7·8, tarsus 1·7.

Hab. Cordillera of Quindiu, New Granada.

a, b, c, d. Ad. sk.	Quindiu.	T. K. Salmon, Esq. [C.].

13. **Xanthura meridana.**

Cyanocitta armillata, *Scl. & Salv. P. Z. S.* 1870, p. 788.
Cyanocitta armillata, *a.* meridana, *Scl. & Salv. P. Z. S.* 1876, p. 271.

General colour above rich purplish blue, the tail scarcely any darker than the back; wing-coverts coloured like the back, as also are the inner secondaries; quills black, externally purplish blue, like the back; head and neck very little darker than the back; the fore part of the crown rather brighter purple; forehead, nasal bristles, feathers round the eye, and sides of the face black; under surface of body deep purple, including the under wing- and tail-coverts; chin black; the throat bluish purple, surrounded on all sides by a black collar; bill and legs black. Total length 12·8 inches, culmen 1, wing 5·4, tail 6·6, tarsus 1·55. (*Mus. Salvin and Godman.*)

Hab. Andes of Merida, Venezuela.

14. **Xanthura armillata.**

Cyanocorax armillatus, *Gray & Mitch. Gen. B.* ii. p. 307, pl. 74; *Scl. P. Z. S.* 1855, p. 153; *Schl. Mus. P.-B.* Coraces, p. 48.
Cyanocitta armillata, *Bp. Consp.* i. p. 378; *Scl. Cat. Amer. B.* p. 144; *Scl. & Salv. Nomencl. Av. Neotr.* p. 39; *iid. P. Z. S.* 1876, p. 271.
Cyanurus armillatus, *Gray, Hand-l. B.* ii. p. 4.

Adult female (Pamplona, *Wyatt*). General colour above deep smalt-blue, the head and hind neck very much lighter and of a greyish cobalt-blue; forehead, lores, feathers above the eye, entire sides of face (running down the sides of the throat and round the fore neck, so as to encircle the throat) black; throat entirely azure-cobalt, the chin black; rest of under surface of body deep smalt-blue, including the under wing- and tail-coverts; wings above resembling the back, the quills blackish, externally of the same blue as the back, the innermost secondaries and tail-feathers blue, not quite so bright as the back; bill and legs black. Total length 12 inches, culmen 1·1, wing 5·4, tail 6·8, tarsus 1·55. (*Mus. Salvin & Godman.*)

Hab. New Granada.

a. Ad. sk.	Bogotá.	Purchased.	(Type of species.)
b, *c*. Ad. sk.	Bogotá.	Purchased.	
d, *e*. Ad. sk.	Bogotá.	Purchased.	

31. UROLEUCA.

Type.

Uroleuca, *Cab. Mus. Hein.* Th. i. p. 225 (*ex Bp. Consp.* i. p. 379, descr. nullâ, typo haud indicato) .... U. cyanoleuca.
Argurocitta, *Heine, J. f. O.* 1860, p. 117 .... U. cyanoleuca.

Head of *Uroleuca cyanoleuca.*

Range. Confined to Brazil.

1. Uroleuca cyanoleuca.

Corvus cyanoleucus, *Max. Reis. Bras.* ii. p. 190.
Corvus tricolor, *Mikan, Del. Faun. Bras.* pl. 16 (*ex Natt. MS.*).
Corvus splendidus, *Licht. Verz. Doubl.* p. 21.
Corvus cristatellus, *Temm. Pl. Col.* ii. pl. 193; *Max. Beitr. Naturg. Bras.* iii. p. 1251.
Pica cristatella, *Wagler, Syst. Av.* Pica, sp. 26.
Cyanurus cristatellus, *Swains. Faun. Bor.-Amer. Birds*, p. 495.
Cyanocorax cyanoleucus, *Gray, Gen. B.* ii. p. 307; *Schl. Mus. P.-B.* Coraces, p. 50; *Pelz. Orn. Bras.* pp. 189, 430; *Scl. & Salv. Nomencl. Av. Neotr.* p. 39.
Uroleuca cyanoleuca, *Bp. Consp.* i. p. 379; *Cab. Mus. Hein.* Th. i. p. 225.
Uroleuca cristatella, *Burm. Th. Bras.* ii. p. 286.
Argurocitta cyanoleuca, *Heine, J. f. O.* 1860, p. 117.
Cyanurus cyanoleucus, *Gray, Hand-l. B.* ii. p. 6.

Adult male. Head and neck all round brown; frontal plumes (which are developed into a long recurved crest), lores, feathers round the eye and cheeks, blacker; mantle and back violet-brown, deepening into rich violet on the lower back, rump, and upper tail-coverts; wing-coverts deep violet, as also the quills on their outer webs, the inner webs blackish brown, the inner secondaries entirely violet like the wing-coverts, the primaries externally edged with dull cobalt; tail deep violet for the basal half, the apical half pure white; under surface of body from the fore neck downwards creamy white, the lower feathers of the fore neck brown terminally, edged with white; under wing-coverts white, the edge of the wing blue; bill and legs black; "iris greyish brown" (*Max.*). Total length 13·5 inches, culmen 1·5, wing 7·95, tail 6·2, tarsus 1·85.

Hab. The same as that of the genus.

a. Ad. st. Brazil.

32. GYMNOKITTA.

	Type.
Gymnorhinus, *Max. Reis. N.-Amer.* ii. p. 21 (1841) *	G. cyanocephala.
Cyanocephalus, *Bp. Oss. Stat. Zool. Eur. Vertebr.* 1840–41, p. 17 (1842) †	G. cyanocephala.
Gymnokitta, *Bp. Consp.* i. p. 382 (1850)	G. cyanocephala.

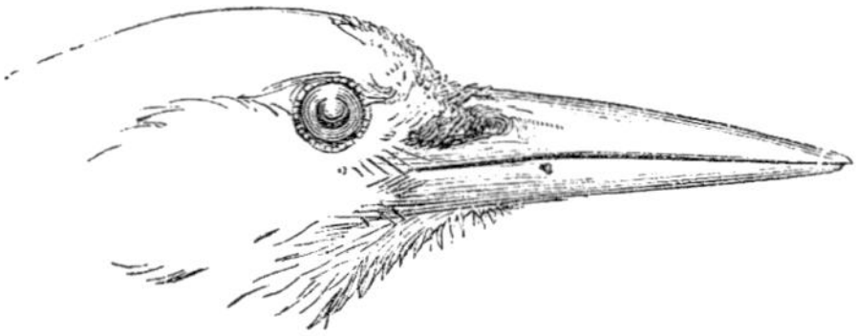

Head of *Gymnokitta cyanocephala.*

Range. Rocky Mountains of Colorado to Cascade Mountains of California and Oregon; south to Mexico and Arizona.

1. Gymnokitta cyanocephala.

Gymnorhinus cyanocephalus, *Max. Reis. N.-Amer.* ii. p. 21.
Cyanocephalus wiedi, *Bp. Oss. Stat. Zool. Eur. Vert.* 1840–41, p. 17; *Salvin, Ibis,* 1873, p. 103.
Gymnokitta cyanocephala, *Bp. Consp.* i. p. 382; *Cass. B. Calif. and Texas,* p. 165, pl. 28; *Baird, B. N. Amer.* p. 574; *Coues, Ibis,* 1872, p. 152; *id. Key N.-Amer. B.* p. 163; *Cooper, Orn. Calif.* i. p. 292; *Baird, Brewer, & Ridgw. N.-Amer. B.* ii. p. 260, pl. 38. fig. 2: *Coues, B. North-west,* p. 210.
Psilorhinus cyanocephalus, *Gray, Gen. B.* iii. App. p. 14.
Cyanocorax cassini, *McCall, Pr. Acad. N. Sci. Philad.* 1851, p. 216.
Nucifraga cyanocephala, *Gray, Hand-l. B.* ii. p. 10.

General colour above pale ashy blue, rather lighter and more cobalt on the upper tail-coverts; head deep cobalt, including the sides of the face; the forehead, cheeks, and ear-coverts rather brighter; lores and feathers round the eye blackish; wing-coverts coloured like the back; the quills blackish, externally greyish blue, the inner secondaries entirely of the latter colour, and resembling the back; tail pale greyish blue, with a very narrow whitish edging at the tip of the outermost feathers; under surface of body greyish blue, including the under wing- and tail-coverts, and rather lighter on the lower abdomen and thighs; throat whitish, streaked with greyish blue, with which most of the feathers are rather broadly

* Nec *Gymnorhina,* G. R. Gray, 1840. † Preoccupied in Botany.

edged; bill and legs black. Total length 10·3 inches, culmen 1·4, wing 5·75, tail 5·6, tarsus 1·5.

Hab. The same as that of the genus.

a. ♀ ad. sk.	Tierra Amarella, New Mexico, Sept. 15, 1874 (*C. E. Aiken*).	Smithsonian Institution [P.].
b. ♀ ad. sk.	Colorado, March 31, 1875.	Purchased.
c. ♂ ad. sk.	Fort Garland, July 16, 1875.	Purchased.

33. PSILORHINUS.

Type.

Psilorhinus, *Rüppell, Mus. Senckenb.* 1837, p. 188...... P. mexicanus.

Range. Central America, ranging into the extreme south of North America.

Key to the Species.

a. Cheeks brown.
  a'. Tail brown, not white-tipped; abdomen brownish .... *morio*, p. 139.
  b'. Tail brown, broadly tipped with white; abdomen creamy white .... *mexicanus*, p. 140.
b. Cheeks bluish, as also a spot above and below the hinder part of the eye .... *cyanogenys*, p. 140.

1. Psilorhinus morio.

Pica morio, *Wagler, Isis*, 1829, p. 751.
Corvus morio, *Eyd. & Gerv. Voy. Favorite, Zool.* p. 54.
Pica fuliginosa, *Less. Traité*, p. 333.
Psilorhinus morio, *Gray, Gen. B.* ii. p. 308; *Bp. Consp.* i. p. 381; *Cab. Mus. Hein.* i. p. 226; *Baird, B. N. Amer.* p. 592, pl. lxviii. figs. 1, 2; *Scl. P. Z. S.* 1859, pp. 57, 365; *Scl. & Salv. Ibis*, 1859, p. 22; *Scl. Cat. Amer. B.* p. 146; *Gray, Hand-l. B.* ii. p. 6; *Scl. & Salv. Nomencl. Av. Neotr.* p. 40; *Baird, Brewer, & Ridgw. N.-Amer. B.* ii. p. 304.

Adult. General colour brown, blacker on the head, neck, and mantle; the frontal plumes stiff, erect, the front ones directed forwards, so as to partially hide the nostrils; sides of face, throat, and chest blackish brown, shading off into paler brown on the breast, the rest of the under surface pale ashy on the abdomen; thighs and under wing- and tail-coverts darker brown; wings and tail above brown like the back, all the feathers distinctly shaded with ashy grey on the outer webs; bill blackish. Total length 17·5 inches, culmen 1·75, wing 8·75, tail 9, tarsus 2·05.

Some specimens, probably females, are smaller, the abdomen paler ashy, and the bill yellow instead of black. Total length 14·5 inches, culmen 1·6, wing 8·1, tail 7·8, tarsus 1·85.

Hab. Mexico.

a, b. [♂ ♀] ad. sk.	Mexico.	Purchased.
c. Ad. sk.	Orizaba, Mexico.	A. Boucard, Esq. [C.].
d. Juv. sk.	Orizaba, Mexico.	A. Boucard, Esq. [C.].

2. Psilorhinus mexicanus.

Pica morio ♀, *Wagler, Isis,* 1829, p. 751.
Psilorhinus mexicanus, *Rüpp. Mus. Senckenb.* 1837, p. 189, pl. 11. fig. 2; *Gray, Hand-l. B.* ii. p. 6; *Scl. & Salv. Nomencl. Av. Neotr.* p. 40.

Adult. General colour above umber-brown, darker on the head, and inclining to blackish on the lores, sides of face, and throat, becoming paler brown again on the upper breast, the sides of which are ashy brown; rest of the under surface creamy white, slightly washed with ashy brown on the breast, thighs, and on the under tail-coverts; under wing-coverts light ashy brown, darker near the edge of the wing; wings and tail above brown, all but the two centre tail-feathers broadly tipped with white. Total length 14·8 inches, culmen 1·85, wing 7·55, tail 7·2, tarsus 1·9.

Hab. Central America: from Mexico to Guatemala; Honduras; Costa Rica.

a, b. Ad. sk.	Honduras.	Mr. Dyson [C.].
c. Ad. sk.	W. Mexico.	A. Boucard, Esq. [C.].
d. Ad. sk.	Costa Rica.	Mr. Van Patten [C.].
e, f. Ad. st.	Mexico.	Purchased.

3. Psilorhinus cyanogenys. (Plate IX.)

Psilorhinus cyanogenys, *Gray, Hand-l. B.* ii. p. 6 (descr. nullâ).

Similar to *Ps. mexicanus*, but differs in having a distinct greyish blue malar patch, as well as a spot of the same colour above and below the hinder part of the eye. Total length 16 inches, culmen 1·6, wing 7·6, tail 8·5, tarsus 1·85.

Hab. Mosquito coast.

Obs. This species seems to me to be quite distinct, and shows a tendency to the markings on the face of a *Xanthura* or *Cyanocorax*.

a. Ad. sk. Pearl-Bay Lagoon, Mosquito. Henry Wickham, Esq. [P.].

34. STRUTHIDEA.

	Type.
Struthidea, *Gould, P. Z. S.* 1836, p. 143	S. cinerea.
Brachystoma, *Swains. Classif. B.* ii. p. 267	S. cinerea.
Brachyprorus, *Cab. Mus. Hein.* Th. i. p. 217	S. cinerea.

Range. Confined to Australia.

1. Struthidea cinerea.

Struthidea cinerea, *Gould, P. Z. S.* 1836, p. 143; *id. Syn. B. Austr.* i. pl. 9; *id. B. Austr.* iv. pl. 17; *Reich. Vög. Neuholl.* p. 148; *Gray, Gen. B.* ii. p. 309, pl. 75. fig. 3; *Bp. Consp.* i. p. 368; *Reich. Syst. Av.* Taf. lxxx. fig. 7; *Gould, Handb. B. Austr.* i. p. 472; *Gray, Hand-l. B.* ii. p. 7.
Brachystoma cinerea, *Swains. Classif. B.* ii. p. 267; *id. An. in Menag.* p. 729.
Brachyprorus cinereus, *Cab. Mus. Hein.* Th. i. p. 217.

Adult male. General colour grey, both above and below, rather browner on the back, the feathers of the head, sides of face, throat, and breast rather scaly in appearance, by reason of their lighter greyish tips; wings pale brown, inclining to darker brown on the inner webs of all the quills and on the centres of the wing-coverts; tail glossy black, with a slight greenish lustre on the outer webs; under wing-coverts and lining of wing pale ashy brown; bill and legs black; iris pearly white. Total length 12·7 inches, culmen 0·95, wing 5·9, tail 6·4, tarsus 1·6.

Adult female. Similar to the male. Total length 12·8 inches, wing 6, tail 6·9, tarsus 1·6.

Hab. The same as that of the genus.

a. Ad. st.	New South Wales.	J. Gould, Esq. [P.].
b, c. ♂ ♀ ad. sk.	N. Australia.	Sir T. Mitchell [P.].
d. ♀ ad. sk.	Gilbert River, Sept. 23, 1856.	J. R. Elsey, Esq. [P.].
e, f. Ad. sk.	N.W. Australia.	J. R. Elsey, Esq. [P.].

35. PICATHARTES.

	Type.
Galgulus, *Wagler, Syst. Av.* Pica, note (1827, *nec Briss.*)	P. gymnocephalus.
Picathartes, *Less. Man. d'Orn.* i. p. 375 (1828)	P. gymnocephalus.

Range. Gold Coast.

1. Picathartes gymnocephalus.

Corvus gymnocephalus, *Temm. Pl. Col.* 327.
Tufted Grakle, *Lath. Gen. Hist.* iii. p. 162.
Galgulus gymnocephalus, *Wagler, Syst. Av. art.* Pica, note.
Picathartes gymnocephalus, *Less. Man. d'Orn.* i. p. 375; *Gray, Gen. B.* ii. p. 316; *Bp. Consp.* i. p. 387; *Hartl. Orn. W.-Afr.* p. 114; *Schl. Mus. P.-B.* Coraces, p. 54; *Gray, Hand-l. B.* ii. p. 14; *Sharpe, Ibis,* 1872, p. 71; *id. Cat. Afr. B.* p. 57; *Ussher, Ibis,* 1874, p. 67, pl. ii.
Pica gymnocephala, *Schinz, Naturg. u. Abbild. Vög.* p. 44, Taf. 16. fig. 3.

Adult. Head bare and of a bright yellow colour all over, excepting on the hinder part of the skull, where there is a circular area of bare skin quite black and separated on the occiput by a narrow band of yellow skin; hind neck yellowish, the skin feathered with silky white plumes; chin bare; the throat, sides of neck, and entire under surface of body silky white, with somewhat of a yellowish tinge; general colour of upper surface slaty grey, inclining to black on the upper part of the mantle; wing-coverts slaty like the back, the greater series as well as the quills brown; tail brown, slightly shaded with grey on the centre feathers; under wing-coverts and inner lining of quills dark brown. Total length 16·5 inches, culmen 1·4, wing 5·95, tail 7·5, tarsus 2·6.

Nestling. Of the same colour and markings as the adult, the bare parts, however, not being so bright.

Hab. Forests of Denkera, in the interior of the Gold Coast, West Africa.

a. Ad. st.	Western Africa.	J. Whitfield, Esq. [P.].
b. Ad. sk.	Denkera (*Ussher*).	R. B. Sharpe, Esq.
c. Pull. sk.	Denkera (*Aubinn*).	R. B. Sharpe, Esq. [P.]. (Specs. fig. Ussher, *l. c.*)
d. Skeleton.	Denkera (*Aubinn*).	R. B. Sharpe, Esq.

36. GLAUCOPIS.

	Type.
Glaucopis, *Gm. S. N.* i. p. 363 (1788)	G. cinerea.
Callæas, *Forster, Descr. Anim.* p. 74 (1844)	G. cinerea.

Range. Confined to New Zealand.

Key to the Species.

a. Bill larger; wattles yellow, blue at base .... *cinerea*, p. 142.
b. Bill smaller; wattles blue (grey in skin), with no yellow; a faint eyebrow of clear grey .... *wilsoni*, p. 143.

1. Glaucopis cinerea.

Cinereous Wattle-bird, *Lath. Gen. Syn.* i. p. 364, pl. xiv.
Glaucopis cinerea, *Gm. S. N.* i. p. 363; *Finsch, J. f. O.* 1872, p. 167, 1874, p. 191; *Buller, B. N. Zealand*, p. 155, pl. 15. fig. 2; *Sharpe, Voy. Ereb. & Terror, Birds*, App. p. 27.
Cryptorhina callæas, *Wagler, Syst. Av.* Cryptorhina, sp. 5.
Callæas cinerea, *Forster, Descr. Anim.* p. 74; *Gray, Voy. Ereb. & Terror, Birds*, p. 8; *id. Hand-l. B.* ii. p. 7.

Adult male. General colour slaty grey all over; quills browner on their inner web; tail blackish at tip; forehead, lores, and feathers in front of the eye velvety black, as well as the chin and a spot at the base of the lower mandible; wattles rich orange, blue at the base; bill and legs black; iris blackish brown. Total length 16·5 inches, culmen 1·35, wing 6·4, tail 7·1, tarsus 2·7.

Adult female. Exactly similar to the male. Total length 15 inches, culmen 1·35, wing 5·95, tail 6·7, tarsus 2·6.

Young. More dusky than the adults, with the orange wattles much smaller, but having no tinge of olive-brown either above or below.

Obs. This species represents *G. wilsoni* in the South Island, and seems to have always a very much stronger bill. Neither the female nor the young ever have a tinge of brown, as is the case with *G. wilsoni*, from which it is also distinguished, in all plumages, by its orange wattles.

Hab. South Island, New Zealand.

a, b. ♂ ♀ ad. sk.	Port Cooper, May 4, 1849 (*F. Strange*).	Captain Stokes [P.].
c. Jun. sk.	Pigeon Bay, N. Z.	Captain Stokes [P.].
d. Jun. sk.	New Zealand.	Colonial Museum [C.].
e, f. Ad. sk.	New Zealand.	Colonial Museum [C.].
g. Skeleton.		

2. Glaucopis wilsoni.

Glaucopis wilsoni, *Bp. Consp.* i. p. 368; *Finsch, J. f. O.* 1872, p. 167, 1874, p. 172; *Buller, B. N. Zeal.* p. 152, pl. 16. fig. 1; *Sharpe, Voy. Ereb. & Terror, Birds*, App. p. 27.
Callæas wilsoni, *Gray, Ibis*, 1862, p. 227; *id. Hand-l. B.* ii. p. 7.
Callæas olivascens, *Pelz. Verh. z.-b. Wien*, 1867, p. 317, note; *Gray, Hand-l. B.* ii. p. 7.
Glaucopis olivascens, *Finsch, J. f. O.* 1870, p. 324.

Adult male. General colour dark bluish grey, very slightly inclining to brown on the lower back, rump, upper and under tail-coverts; forehead, lores, chin, and feathers round the eye velvety black, these black parts margined with a narrow line of pale grey, which separates it from the darker grey of the back; wing-coverts slaty brown, with a slight greyish wash on the outer wing-coverts and primaries; bill and legs black; wattles at gape bright ultramarine (in life); iris dark brown. Total length 14 inches, culmen 1·25, wing 6·1, tail 6·5, tarsus 2·75.

Adult female. Similar to the male, but more deeply tinged with brown on the lower part of the back, rump, and abdomen. (*Buller.*)

Young. Similar to the adult female, but more deeply washed with brown, this colour extending over the entire upper surface; wattles smaller and of a paler blue. (*Buller.*)

Obs. Dr. Buller's account of the sexes and young birds seems to be quite satisfactory, and is borne out by the specimens in the Museum. Von Pelzeln's *Callæas olivascens* is doubtless nothing but the young bird.

Hab. North Island, New Zealand.

a. Juv. sk.	River Hokianga.	Miss R. Stone [P.].
b, c. Ad. sk.	North Island, N. Z.	Colonial Museum [E.].
d. Juv. sk.	North Island, N. Z.	Colonial Museum [E.].
e. Ad. sk.	North Island, N. Z.	Purchased.
f. Juv. sk.	North Island, N. Z.	The Admiralty [P.].

37. HETERALOCHA.

	Type.
Neomorpha, *Gould, P. Z. S.* 1836, p. 144	H. acutirostris.
Heteralocha, *Cab. Mus. Hein.* Th. i. p. 218 (1850)	H. acutirostris.

Range. Confined to New Zealand.

1. Heteralocha acutirostris.

Neomorpha acutirostris, *Gould, P. Z. S.* 1836, p. 144 (♀).
Neomorpha crassirostris, *Gould, t. c.* p. 145 (♂).
Neomorpha gouldi, *Gray, List Gen. B.* 1841, p. 15; *id. Voy. Ereb. & Terror, Birds*, p. 2.
Heteralocha gouldi, *Cab. Mus. Hein.* th. i. p. 218.
Heteralocha acutirostris, *Buller, B. N. Zeal.* p. 63, pl. 7; *Finsch, J. f. O.* 1872, p. 105, 1874, p. 192; *Sharpe, Voy. Ereb. & Terror, Birds*, App. p. 26.

Adult male. Entirely black above and below, with somewhat of a bottle-green lustre; tail black, broadly tipped with white; at the gape a large pendent wattle "of a rich orange colour in life; bill ivory-white, darkening to blackish grey at the base; tarsi and toes bluish grey, the claws light horn-colour" (*Buller*). Total length 19 inches, culmen 2·5, wing 8·3, tail 8, tarsus 3·25.

Adult female. Similar to the male in colour, but distinguished by its long curved bill. Total length 19·5 inches, culmen 3·9, wing 7·95, tail 8, tarsus 3·05.

Obs. "In some examples the white at the end of the tail is tinged more or less with rufous, while in others the under tail-coverts are tipped with white. It should be noted also that the brightness of the fleshy wattles depends, in some measure, on the health or condition of the bird; for during sickness they change to lemon-yellow. The palate and soft parts of the throat are bright yellow. The tongue is horny at the tip, bifurcate, and furnished with minute barbs." (*Buller, l. c.*)

It appears to me that the white markings on the under tail-coverts are a sign of immaturity, as both the birds so marked in the collection are young females with the bills not fully grown.

Hab. North Island, New Zealand.

a, b, c. ♂ ad. sk.	Wainuiomata Valley, Rimutaka range, Province of Wellington (*Dr. Hector*).	Colonial Museum [C.].
d–g. ♀ ad. and jun. sk.	Wainuiomata Valley.	Colonial Museum [C.].
h. ♂ ad. st.	New Zealand.	Percy Earl, Esq. [C.].
i. ♀ ad. st.	New Zealand.	Sir George Grey [P.].

38. CREADION.

	Type.
Creadion, *Vieill. N. Dict. d'Hist. Nat.* viii. p. 390 (1817)	C. carunculatus.
Oxystomus, *Swains. Classif. B.* ii. p. 270 (1837)	C. carunculatus.

Range. New Zealand.

1. Creadion carunculatus.

Wattled Stare, *Lath. Gen. Syn.* iii. p. 9, pl. 36.
Sturnus carunculatus, *Gm. S. N.* i. p. 805.
Creadion pharoides, *Vieill. N. Dict. d'Hist. Nat.* viii. p. 390.
Icterus rufusater, *Less. Voy. Coquille*, i. p. 649, pl. xxiii. fig. 1.
Xanthornus carunculatus, *Quoy et Gaim. Voy. de l'Astrol.* i. p. 212, pl. 12. fig. 4.
Oxystomus carunculatus, *Swains. Classif. B.* ii. p. 270.
Creadio carunculatus, *Gray, Gen. B.* ii. p. 338; *id. Voy. Ereb. & Terror, Birds*, p. 8; *Bp. Consp.* i. p. 389; *Gray, Hand-l. B.* ii. p. 22; *Finsch, J. f. O.* 1872, p. 167, 1874, p. 192; *Buller, B. N. Zeal.* p. 149, pl. 15; *Sharpe, Voy. Ereb. & Terror, Birds*, App. p. 26.
Creadio carunculatus, *Cab. Mus. Hein.* Th. i. p. 218.
Creadion cinereus, *Buller, Essay N.Z. Orn.* p. 10; *Gray, Hand-l. B.* ii. p. 22.

Adult male. Head and neck all round, sides of face and of neck, glossy jet-black; mantle, scapulars, and entire back deep chestnut;

tail black; wing-coverts chestnut like the back, the greater series black at base, chestnut at tip; primary-coverts and all the quills entirely black; under tail-coverts chestnut; at the gape a wattle, "varying in tint from a clear yellow to a bright vermilion, being apparently affected by physical conditions, such as the health of the bird or the temperature of the weather; bill and legs black; irides blackish brown" (*Buller*). Total length 8·5 inches, culmen 1·35, wing 3·8, tail 3·5, tarsus 1·5.

Adult female. Similar to the male. Total length 7·8 inches, culmen 1·35, wing 3·35, tail 3·1, tarsus 1·35.

Young male. Brown above and below, the rump and upper tail-coverts dull chestnut, the back slightly marked with the same; wing-coverts like the back, the median and greater coverts externally chestnut; underneath olivaceous brown; under tail-coverts chestnut.

Hab. The same as that of the genus.

a. Ad. sk.	Bay of Islands, N. Z.	Antarctic Expedition [C.].
b. ♂ juv. sk.	Tiapo, New Zealand, Sept. 1873.	Dr. Buller [P.].
c, d. ♂ ♀ ad. sk.	New Zealand.	Capt. Stokes, R.N. [P.].
e. Ad. sk.	New Zealand.	

39. FALCULIA.

Type.

Falculia, *Isid. Geoff. St.-Hilaire, Mag. de Zool.* 1836, pl. 49. F. palliata.

Range. Confined to Madagascar.

1. Falculia palliata.

Falculia palliata, *Isid. Geoffr. St.-Hilaire, Mag. de Zool.* 1836, pl. 49; *Reichenb. Syst. Av.* pl. xli. fig. 3; *id. Handb. Orn.* p. 322, Taf. dcxiii. fig. 4097; *Hartl. Faun. Madag.* p. 33; *Schl. P. Z. S.* 1866, p. 421; *Schl. & Poll. Faun. Madag.* p. 105, pl. 33; *Schl. Dierk.* p. 188, pl. iii. fig. 40; *Gray, Hand-l. B.* i. p. 105.

Adult. Above black with a deep steel-blue gloss, the feathers being edged with the latter colour, and having conspicuous white basal markings, very plain on the disarrangement of the plumage, especially on the lower back; across the rump a band of greyish white; wings and tail entirely black, with somewhat more of a greenish gloss; all the rest of the plumage, including the head and neck all round and the under surface pure white; under wing-coverts and axillaries white, the outermost and lower series of coverts blackish, like the inner lining of the quills, which are black below, with a small white spot at the base of the primaries. Total length 9·5 inches, culmen 2·35, wing 6, tail 4·2, tarsus 1·25.

Hab. Madagascar.

a. Ad. sk.	Madagascar.	Purchased.
b. Ad. sk.	Madagascar.	

Subfam. II. FREGILINÆ.

Key to the Genera.

a. Wings long, falling short of the tip of the tail by less than the length of the tarsus.
  a'. Wings very long and pointed, the distance between the tips of the primaries and secondaries nearly double the length of the tarsus.
    a''. Bill long and curved, the distance from the eye to the tip of bill longer than the tarsus; base of lower mandible feathered .... 40. GRACULUS, p. 146.
    b''. Bill short, the distance between the eye and the tip of bill shorter than the tarsus; base of lower mandible bare . 41. PYRRHOCORAX, p. 148.
  b'. Wings shorter and more rounded, the distance between the tips of the primaries and the tips of the secondaries not so long as the tarsus .... 42. CORCORAX, p. 149.
b. Wings short, falling short of the tip of the tail by more than the length of the tarsus 43. PODOCES, p. 150.

40. GRACULUS*.

	Type.
Coracia, *Brisson, Orn.* ii. p. 3 (1760)	G. graculus.
Graculus, *Koch, Syst. baier. Zool.* p. 91 (1816)	G. graculus.
Fregilus, *Cuvier, Règn. Anim.* i. p. 416 (1817)	G. graculus.

Range. Europe and Northern Africa, North-east Africa, eastwards through Central Asia and the Himalaya Mountains to China and Eastern Siberia.

1. Graculus graculus.

Le Coracias, *Briss. Orn.* ii. p. 3, pl. 1. fig. 1.
? Le Coracias hupé, *Briss. t. c.* p. 6.
Corvus graculus, *Linn. S. N.* i. p. 158; *Naum. Vög. Deutschl.* ii. pl. 57. fig. 2; *Schl. u. Susem. Vög. Eur.* Taf. 8. fig. 2; *Yarr. Br. B.* ii. p. 56.
Gracula pyrrhocorax, *Scop. Ann.* i. p. 42 (*nec L.*).
Coracias des Alpes, *Montb. Pl. Enl.* 255.
Red-legged Crow, *Lath. Gen. Syn.* i. pt. 1, p. 401.
? Hermit Crow, *Lath. t. c.* p. 403.
? Corvus eremita, *Gm. S. N.* i. p. 377 (*ex Lath.*).
Graculus eremita, *Koch, Syst. baier. Zool.* p. 91.
Fregilus graculus, *Cuv. Règne Anim.* 1817, p. 406; *Gould, B. Eur.* pl. 219; *Macgill. Brit. B.* i. p. 587; *Blyth, Cat. B. Mus. A. S B.* p. 9; *Reichenb. Syst. Av.* Taf. lxxxv. fig. 3; *Bp. Consp.* i. p. 388; *Cab. Mus. Hein.* i. p. 228; *Horsf. & Moore, Cat. B. Mus. E.I. Co.*

* It seems better to reinstate this name for the Chough. *Coracia* of Brisson is too like *Coracias* of Linnæus to be retained; and the next generic name is *Graculus* of Koch. The only objection to the adoption of this one would seem to be that it had been applied by Linnæus to the Cormorants. But this was in 1735, a date not recognized by those ornithologists who follow the rules of the British Association; and for the last-named birds the generic name of *Phalacrocorax* of Brisson appears to be the right one.

p. 549; *Fritsch, Vög. Eur.* tab. 28. fig. 4; *Linderm. Vög. Griechenl.* p. 72; *Blanf. Geol. & Zool. Abyss.* p. 395; *Salvad. Ucc. Ital.* p. 169; *Heugl. Orn. N.O.-Afr.* p. 495.
Coracia erythrorhamphos, *Vieill. N. Dict. d'Hist. Nat.* viii. p. 2; *id. Gal. Ois.* pl. 103; *Roux, Orn. Prov.* pl. 137.
Pyrrhocorax graculus, *Temm. Man.* i. p. 122; *Werner, Atlas,* Omnivores, pl. 12; *Gould, B. Gt. Brit.* iii. pl. 62; *Schleg. Mus. P.-B.* Coraces, p. 55; *Irby, B. Gibr.* p. 128; *Dresser, B. Eur.* pt. xxxvii.
Fregilus europæus, *Less. Tr. d'Orn.* p. 324.
Pyrrhocorax rupestris, *Brehm, Vög. Deutschl.* p. 175.
Fregilus erythropus, *Swains. Classif. B.* ii. p. 268.
Coracia gracula, *Gray, Gen. B.* ii. p. 321; *Degl. & Gerbe, Orn. Eur.* i. p. 205; *Doderl. Avif. Sicil.* p. 64; *Gray, Hand-l. B.* ii. p. 15.
Fregilus himalayanus, *Gould, P. Z. S.* 1862, p. 125; *Jerd. B. Ind.* ii. p. 318.
Fregilus graculus, var. orientalis, *Dybowski, J. f. O.* 1868, p. 332.
Fregilus graculus, var. brachypus, *Swinh. P. Z. S.* 1871, p. 383.

Adult male. General colour above glossy purplish black, of a silky texture; wings black, externally dull green; tail dull greenish black; entire under surface of body glossy purplish black, shading into greenish on the under tail-coverts; bill deep vermilion; legs vermilion; iris brown. Total length 15 inches, culmen 2·1, wing 12, tail 6·7, tarsus 2·3.

Adult female. Similar to the male.

Young. Dusky black instead of glossy black; bill and legs orange-yellow.

Obs. The Himalayan Chough, separated on account of its larger size by Mr. Gould, is so thoroughly approached in a series, that it cannot be specifically recognized, and at most can only be regarded as a larger Eastern form. The following are the measurements of the series in the Museum:—

		Total length. in.	Culm. in.	Wing. in.	Tail. in.	Tarsus. in.
a.	Lizard Point	16·0	2·05	11·0	5·8	2·25
b.	Cornwall (*Mitchell*)	15·0	2·15	10·85	5·8	2·15
c.	Cornwall (*Mitchell*)	15·5	2·2	11·2	5·8	2·25
d. ♂.	Wales (*Mitford*)	14·5	1·9	11·0	5·9	2·1
e.	Italy	15·0	1·95	10·9	6·0	2·1
f. ♀.	Hautes-Pyrénées (*Brooke*)	15·0	2·0	11·2	5·9	2·05
g. ♀.	Smyrna (*Merlin*)	15·0	2·05	11·3	6·2	2·05
h.	Granada (*Saunders*)	15·0	1·9	10·8	5·7	2·05
i.	Tangier (*Olcese*)	15·0	1·95	10·8	5·85	2·15
k.	Tangier (*Olcese*)	16·8	2·4	12·2	6·4	2·3
l. ♀.	Lemien (*Petherick*)	17·5	2·45	12·4	6·4	2·2
m.	Alatau Alps	14·0	1·9	11·2	6·4	1·75
n.	Snowy range, Himalayas	16·5	2·15	13·2	7·2	2·25
o. ♂.	Valley of Aster, Cashmere	15·0	2·05	13·25	7·0	2·0
p. ♀.	Valley of Aster, Cashmere	15·0	2·05	12·5	6·5	2·05
q. ♂.	N.W. Himalayas (*Pinwill*)	17·0	2·2	12·7	7·2	2·0
r.	Darjiling (*Hodgson*)	14·0	2·2	11·0	5·8	2·15
s.	Kumaon (*Strachey*)	14·5	2·3	13·0	6·9	2·4
t.	Abadah, Persia (*Blanford*)	14·0	2·05	11·4	6·0	2·15
u.	Thibet (*Gifford*)	14·5	2·1	12·4	6·8	2·3
v.	Altai Mountains	14·0	1·95	12·0	6·9	1·8

Mr. Swinhoe has kindly lent me some specimens of his *Graculus brachypus* from China. The general bulk of the Chinese bird is no doubt smaller than those either from Europe or India, and the toes are proportionately shorter; but considering that a little difference in size, more or less, is all that can be found to separate the Red-billed Choughs from various parts of the world, I do not propose to divide them, even as races.

Hab. The same as that of the genus.

a. Ad. st.	Great Britain.	Col. Montagu [P.].
b, c. Ad. st.	Cornwall.	D. W. Mitchell, Esq. [P.].
d. Ad. sk.	Lizard Point.	Mrs. Rouchette [P.].
e. ♂ ad. sk.	Wales (*Mitford*).	J. Gould, Esq.
f. Juv. sk.	Monmouthshire, July 1874.	Rev. A. Morgan [P.].
g. Ad. sk.	Italy.	J. Gould, Esq.
h. ♀ ad. sk.	Hautes-Pyrénées, April 18, 1872.	A. B. Brooke, Esq. [C.].
i. ♂ ad. sk.	Granada (*H. Saunders*).	R. B. Sharpe, Esq. [P.].
k, l. Ad. sk.	Tangier.	M. Olcese [C.].
m. ♀ ad. st.	Athens.	C. W. L. Merlin, Esq.
n. Ad. sk.	Smyrna.	C. W. L. Merlin, Esq. [P.].
o. Ad. sk.	Kytch Territory, west of the Nile, Jan. 1862 (*Petherick*).	J. Gould, Esq.
p. Ad. sk.	Abadah, north of Shiraz, Persia, June 1869.	Capt. St. John [C.].
q. Ad. st.	Altai Mountains.	Moscow Museum [E.].
r. Ad. sk.	Alatau Alps, June 1841.	Moscow Museum [E.].
s, t. ♂ ♀ ad. sk.	Valley of Aster, Cashmere, 14,000 feet, Sept. 1875.	Purchased.
u. ♂ ad. sk.	N.W. Himalayas.	Capt. Stackhouse Pinwill [P.].
v. Ad. sk.	Snow-range, Himalayas.	J. Gould, Esq.
w. Ad. sk.	Kumaon (*Strachey*).	J. Gould, Esq. (Type of *G. himalayensis.*)
x. Ad. sk.	Darjiling.	B. H. Hodgson, Esq. [P.].
y. Ad. sk.	Thibet.	Earl of Gifford [P.].
z. Ad. sk.	Shanghai, China.	Shanghai Museum [P.].

41. PYRRHOCORAX.

Type.

Pyrrhocorax, *Vieill. N. Dict. d'Hist. Nat.* p. 568 (1816) . . P. alpinus.

Range. Mountain regions of Southern Europe, eastwards to the Himalayas.

1. Pyrrhocorax alpinus.

Le Choucas des Alpes, *Briss. Orn.* ii. p. 30, pl. 1. fig. 2; *Montb. Pl. Enl.* pl. 531.

Corvus pyrrhocorax, *Linn. S. N.* i. p. 158; *Schl. & Susem. Vög. Eur.* Taf. 8. fig. 1.

Alpine Crow, *Lath. Gen. Syn.* i. pt. 1, p. 381.

Pyrrhocorax alpinus, *Vieill. N. Dict. d'Hist.* vi. p. 568; *Naum. Vög. Deutschl.* ii. pl. 57; *Vieill. Gal. Ois.* i. pl. 104; *Roux, Orn. Prov.* pl. 138; *Gould, B. Eur.* iii. pl. 218; *Gray, Gen. B.* ii. p. 320, pl. 378. fig. 2; *Blyth, Cat. B. Mus. A. S. B.* p. 90; *Bp. Consp.* i. p. 388; *Cab. Mus. Hein.* Th. i. p. 227; *Reich. Syst. Av.* pl. lxxxv. fig. 1; *Horsf. & Moore, Cat. B. Mus. E.I. Co.* ii. p. 549; *Jaub. et Barth.-Lapomm. Rich. Orn.* p. 96; *Linderm. Vög. Griechenl.* p. 72; *Jerd. B. Ind.* ii. p. 319; *Filippi, Viagg. Pers.* p. 350; *Fritsch, Vög. Eur.* tab. 28. fig. 5; *Degl. et Gerbe, Orn. Eur.* i. p. 204; *Doderl. Avif. Sicil.* p. 64; *Salvad. Faun. Ital. Ucc.* p. 170; *Gray, Hand-l. B.* ii. p. 15; *Shelley, B. Egypt*, p. 161; *Heugl. Orn. N.O.-Afr.* p. 496; *Henders. & Hume, Lahore to Yark.* p. 249; *Dresser, B. Eur.* part xxxviii.; *Irby, B. Gibr.* p. 128; *Severtz. Turkest. Jevotn.* p. 64; *Dresser, Ibis*, 1875, p. 237; *Scully, Str. F.* 1876, p. 162; *Blanf. East. Persia*, p. 263.

Pyrrhocorax pyrrhocorax, *Temm. Man. d'Orn.* i. p. 121; *Werner, Atlas*, Omnivores, pl. 9.

Pyrrhocorax alpinus, var. digitatus, *Hempr. & Ehr. Symb. Phys.* fol.

Pyrrhocorax montanus, *Brehm, Vög. Deutschl.* p. 176.

Fregilus pyrrhocorax, *Swains. Classif. B.* ii. p. 268.

Pyrrhocorax planiceps, *Brehm, Naum.* 1855, p. 275.

Pyrrhocorax forsythi, *Stoliczka, Str. F.* 1874, p. 462.

Adult. General colour glossy black all over; "bill yellow, with an orange tinge; legs dark vermilion-red; iris brown" (*Dresser*). Total length 16·5 inches, culmen 1·4, wing 10·75, tail 6·8, tarsus 2.

Hab. The same as that of the genus.

a.	Ad. sk.	Smyrna.	C. W. L. Merlin, Esq. [P.].
b.	Ad. sk.	Pass of Afka, Lebanon.	Canon Tristram [C.].
c.	Ad. sk.	N.W. Himalayas.	Captain Stackhouse Pinwill [P.].
d.	Ad. sk.	Between Simla and Mussoorie.	Captain Stackhouse Pinwill [P.].
e.	Ad. sk.	Northern India (*Major Hay*).	J. Gould, Esq.
f.	Ad. sk.	Ladakh.	Purchased.
g.	Skeleton.		Zoological Society.

42. CORCORAX.

	Type.
Corcorax, *Less. Traité*, p. 325 (1831)	C. melanorhamphus.
Cercoronus, *Cab. Arch. f. Naturg.* 1847, p. 335..	C. melanorhamphus.

Range. Confined to Australia.

1. Corcorax melanorhamphus.

Coracias melanorhamphos, *Vieill. N. Dict.* viii. p. 2.

Pyrrhocorax leucopterus, *Temm. Man.* i. p. 121.

Fregilus leucopterus, *Vigors & Horsf. Tr. Linn. Soc.* xv. p. 265.

Corcorax australis, *Less. Tr. d'Orn.* p. 325.

Corvus leucopterus, *Wagl. Syst. Av.* Corvus, sp. 14; *Gray, Hand-l. B.* ii. p. 15.

Corcorax leucopterus, *Gray, List Gen. B.* 1841, p. 52; *id. & Mitch. Gen. B.* ii. pl. 78; *Gould, B. Austr.* iv. pl. 16.

Corcorax melanorhynchus, *Gray, Gen. B.* ii. p. 321; *Bp. Consp.* i. p. 388.

Cercoronus melanorhynchus, *Cab. Arch. f. Naturg.* 1847, p. 335.

Corcorax melanorhamphos, *Reich. Syst. Av.* pl. lxxxv. fig. 5; *Gould, Handb. B. Austr.* i. p. 470.

Cercoronus melanorhamphus, *Cab. Mus. Hein.* i. p. 227.

Pyrrhocorax melanorhynchus, *Schleg. Mus. P.-B.* Coraces, p. 56.

Adult male. General colour sooty black above and below, the upper parts with a slight purplish gloss, less distinct on the lower surface, the quills externally glossed with dull green; all the inner webs of the primaries white, excepting the bases and the tips, which are black; bill black; feet black; iris scarlet. Total length 16 inches, culmen 1·7, wing 9·9, tail 9·2, tarsus 2·5.

Adult female. Similar to the male. Total length 15·5 inches, culmen 1·7, wing 9·7, tail 9·2, tarsus 2·45.

Young. Everywhere much browner than the adults, including the feathers of the wings and tail.

Hab. The same as that of the genus.

a. Juv. sk.	S. Australia.	Sir G. Grey [P.].
b, c. ♂ ♀ ad. sk.	Murray river.	Sir G. Grey [P.].
d. ♀ juv. sk.	Murray river.	Sir G. Grey [P.].
e. Imm. sk.	Australia.	C. D. E. Fortnum, Esq. [P.].

43. PODOCES.

Type.

Podoces, *Fischer, Mém. Soc. Imp. Nat. Moscou,* v. pl. 21 (1823) .... P. panderi.

Range. Central Asia, from Buchara to Turkestan and Thibet.

Key to the Species.

a. Tail blue-black.
  a'. With a pectoral band .... *panderi,* p. 150.
  b'. With no pectoral band .... *hendersoni,* p. 151.
b. Tail creamy white .... *biddulphi,* p. 151.
c. Tail white or creamy white, excepting the centre feathers; entire colour above and below sandy .. *humilis,* p. 152.

1. Podoces panderi.

Podoces panderi, *Fischer, Mém. Soc. Imp. Nat. Moscou,* v. pl. 21; *Bp. Consp.* i. p. 388; *Sharpe, P. Z. S.* 1870, p. 334; *Cab. J. f. O.* 1873, p. 63, Taf. 3. figs. 37, 38; *Severtz. Turkest. Jevotn.* p. 64; *Gould, B. Asia,* part xxvii.
Corvus panderi, *Eversm. Reise Buchara,* p. 126.
Pica panderi, *Wagler, Syst. Av.* Pica, sp. 17.
Garrulus panderi, *Gray, Hand-l. B.* ii. p. 3.

Adult female. Above grey, rather darker on the head, and inclining to isabelline colour on the lower scapulars, lower back, and rump; upper wing-coverts clear grey, the greater series shading off into black, and broadly tipped with white, forming a duplicate bar across the wing; primary-coverts black, broadly edged with white; primaries pure white, black at extreme base and at tip, the shafts also blackish; the secondaries black, white for the apical half, the black gradually increasing in extent towards the inner secondaries, which are entirely black, the innermost grey like the scapulars; upper tail-coverts black; tail glossy purplish black; nasal bristles isabelline; a line of feathers over the eye white; between the eye and the base of the bill a small spot of black feathers; sides of face light grey, the throat whitish; on the fore neck a large patch of black; rest of under surface vinaceous isabelline-colour, shading off into white on the middle of the abdomen, thighs, and under tail-coverts; under wing-coverts pure white, the black bases to the primaries forming a bar across the wing; bill black; feet lead-colour. Total length 10 inches, culmen 1·2, wing 4·45, tail 3·9, tarsus 1·6.

Hab. Buchara and Turkestan.

a. ♀ ad. sk.	Wells of Tambai-Kasgan, Desert of Kysil-Kum, north of the Lower Oxus, July 13th, 1874.	Dr. N. Severtzoff [C.].

2. Podoces hendersoni.

Podoces hendersoni, *Hume, Ibis,* 1871, p. 408; *Hume & Henders. Lahore to Yark.* p. 244, pl. 22; *Gould, B. Asia,* pt. xxvii.

Adult male (type of species). General colour above sandy, distinctly inclining to vinous on the scapulars, lower back, rump, and upper tail-coverts; lesser and median wing-coverts vinaceous sandy colour; the rest of the coverts and the quills glossy blue-black; bastard-wing blue-black externally, the inner web white; primaries pure white, except at the base and at the tip, where they are black; the secondaries entirely blue-black like the wing-coverts; tail blue-black, the outermost feathers narrowly edged with sandy colour; nasal plumes sandy, the lower ones blue-black; crown of the head glossy blue-black, the lateral feathers tipped with white, forming a cap; lores, eyebrow, sides of face and of neck, and entire under surface of body sandy colour, paler and more whitish on the throat; under wing-coverts black, excepting those near the edge of the wing, which are white; bill black; feet black. Total length 12 inches culmen 1·7, wing 5·8, tail 4·75, tarsus 1·65.

Hab. Yarkand.

a. ♂ ad. sk.	Oi-tograk, Yarkand.	Dr. Henderson [C.]. (Type of species.)

3. Podoces biddulphi.

Podoces biddulphi, *Hume, Str. F.* ii. pp. 503, 529; *Gould, B. Asia,* pt. xxvii.

Adult male. General colour above rich creamy brown, including the entire back, scapulars, and least wing-coverts, the outermost of the latter shaded with silvery white; greater wing-coverts black, glossed with purple, the bastard quills pure white; the primary-coverts purplish blue, the outermost externally white; primaries white, dusky at base, and black for the terminal third, decreasing on the inner primaries, the last one pure white; secondaries purplish black, tipped with white, the innermost almost entirely white along the inner web and towards the tip of the outer web; upper tail-coverts and centre tail-feathers creamy buff, the latter with a broad central streak of black, rest of the tail-feathers white with black shafts; nasal bristles creamy buff; crown of head purplish black from the forehead to the nape; lores, feathers round the eye, sides of face and of neck, creamy buff; cheeks and fore part of the ear-coverts black, with cream-coloured edges to the feathers; throat black, entirely obscured by cream-coloured margins; rest of under surface of body rich creamy brown like the back, shading off into whitish on the abdomen, vent, and under tail-coverts; under wing-coverts and axillaries black, the outermost along the edge of the wing white; bill and legs black. Total length 11·6 inches, culmen 2, wing 5·95, tail 4·3, tarsus 1·85. (*Coll. F. Stoliczka*).

Hab. Maralbash, E. Turkestan.

Obs. The above description is taken from a skin procured by the late Dr. Stoliczka, for the permission to describe which I am indebted to Mr. Hume, who has forwarded the specimen to England for illustration in the forthcoming work 'The Zoology of Yarkand.'

4. Podoces humilis.

Podoces humilis, *Hume, Ibis*, 1871, p. 408; *Hume & Henders. Lahore to Yark.* p. 247, pl. 23; *id. Str. F.* 1874, p. 452; *Gould, B. Asia*, part xxvii.

Adult male. General colour above sandy brown, slightly more rufescent on the head and neck; round the latter an indistinct collar of buffy white; lores and feathers over the front of the eye fulvescent; a few blackish feathers between the bill and the eye; nasal bristles sandy brown; sides of face and ear-coverts pale sandy, browner on the upper margin of the latter; entire under surface of body whitish, inclining to isabelline on the abdomen and flanks; under wing-coverts isabelline; upper wing-coverts coloured like the back, and terminally shading off into sandy brown, giving the appearance of faintly indicated bars across the wing, the outer small coverts blackish; quills brown, with paler margins of sandy brown, the secondaries paler and more resembling the back; tail-feathers white, slightly shaded externally with isabelline, this colour increasing towards the centre feathers, which are blackish brown, margined all round with sandy brown, the next pair being blackish on the inner web and isabelline on the outer; bill and feet black. Total length 7·5 inches, culmen 0·1, wing 3·8, tail 2·85, tarsus 1·15.

Adult female. Similar to the male, but with shorter bill. Total length 7 inches, culmen 0·8, wing 3·6, tail 2·8, tarsus 1·2.

Hab. Yarkand and Thibet.

a, b. ♂ ♀ ad. sk.	Sanju Pass, Yarkand, August 9th, 1870.	Dr. G. Henderson [C.]. (Types of species.)

Family PARADISEIDÆ.

a. Bill elongated and slender, the culmen longer than the tarsus ...... *Subfam. EPIMACHINÆ*, p. 153.
b. Bill short and more or less stout, the culmen not so long as the tarsus. *Subfam. PARADISEINÆ*, p. 163.

Subfam. I. EPIMACHINÆ.

Key to the Genera.

a. Tail not so much elongated, not equal in length to the body of the bird.
  a'. With a jugular shield of metallic plumes occupying the throat and fore neck .. 1. PTILORHIS, p. 153.
  b'. With no jugular shield; fore neck without metallic plumes.
    a''. Flank-feathers produced, yellowish like the breast, the shafts elongated into six threads; tail square .... 2. SELEUCIDES, p. 159.
    b''. Flank-feathers produced into a brown fan-like shield, all the feathers tipped with purple; tail graduated. 3. DREPANORNIS, p. 160.
b. Tail enormously elongated, much longer than the body of the bird .... 4. EPIMACHUS, p. 161.

1. PTILORHIS.

	Type.
Ptilorhis, *Swains. Zool. Journ.* i. p. 481 (1825)	P. paradisea.
Craspedophora, *Gray, List Gen. B.* 1841, p. 15........	P. magnifica.

Range. Australia and New Guinea.

Key to the Species.

Adult males.

a. Metallic shield on breast extending from the lower throat and occupying the fore neck; long silky plumes attached to the flanks, not reaching beyond the under tail-coverts; breast purple below the pectoral shield .... { *paradisea*, p. 154. *victoriæ*, p. 155. }
b. Metallic shield extending from the chin to the upper breast, and followed by a double band of purple and oily golden green; long flank-plumes reaching beyond the tail.
  a'. Breast washed with oily green, the abdomen and flanks with purple, in female ashy brown on the head and back .... *alberti*, p. 156.
  b'. Breast below the double pectoral band purple *magnifica*, p. 157.

Adult females.

a. Head rufous like the rest of the back .... *magnifica*, p. 157.
b. Head ashy brown, contrasting with the rufous of the wings.
a'. Underneath dull white, thickly barred across with black, from the lower throat downwards .... *alberti*, p. 156.
b'. Underneath white tinged with fawn-colour, and marked with narrow black bars of a horseshoe-shape...................... *paradisea*, p. 154.
c'. Underneath fawn-colour, with a few indistinct spots and bars of dusky brown.......... *victoriæ*, p. 155.

1. Ptilorhis paradisea.

Ptiloris paradisea, *Swains. Zool. Journ.* i. 481; *Less. H. N. Ois. Parad.* pp. 25, 213, pls. 29, 30; *Gould, B. Austr.* iv. pl. 100; *Gray, Gen. B.* i. p. 94; *Bp. Consp.* i. p. 412; *Cab. Mus. Hein. Th.* i. p. 214; *Reichenb. Handb. spec. Orn.* Scansoriæ, p. 328, Taf. dclx. figs. 3083–85; *Gould, Handb. B. Austr.* i. p. 591; *Gray, Hand-l. B.* i. p. 105; *Elliot, P. Z. S.* 1871, p. 582; *id. Monogr. Parad.* pl. xxv.

Epimachus brisbani, *Wilson, Ill. Zool.* pl. 9; *Less. Man. d'Orn.* ii. p. 6.

Epimachus regius, *Less. Voy. Coquille, Zool.* p. 667, pl. 28; *id. Cent. Zool.* pl. 3.

Epimachus paradiseus, *Schl. J. f. O.* 1861, p. 386; *id. Mus. P.-B.* Coraces, p. 97.

Adult male. General colour above velvety black, changing to deep fiery purple when viewed away from the light; wings black, the coverts deep purple like the back and scapulars, the quills bluish purple at tip, the inner secondaries shaded with violet and bluish purple; tail-feathers velvety black, the two centre feathers burnished steel-green, the next ones on each side glossed with steel-blue near the base; crown of head burnished coppery green, fringed on the hind neck with metallic steel-blue; behind the eye, above the ear-coverts, a longitudinal patch of velvety purple; sides of face, sides of neck and throat black, shaded purple like the back when held away from the light; from the lower throat a large triangular patch of burnished steel-green, extending to the fore neck; breast velvety purple, each feather with a mesial shade of purplish blue, the plumes of the lower breast edged with olive-green; the rest of the under surface entirely of the latter colour with coppery purple bases; under wing- and tail-coverts black; bill and legs black. Total length 11·5 inches, culmen 2, wing 6·2, tail 4·4, tarsus 1·45.

Adult female. General colour above ashy brown; the head also of this colour, all the feathers narrowly streaked with buffy white; over the eye a long streak of dull white, forming a distinct eyebrow; lores and sides of face brown, minutely streaked with buffy white, these streaks being also visible on the sides of the neck; cheeks and throat pale ochraceous buff, with whitish shaft-streaks;

rest of under surface ochraceous buff, mottled with narrow black cross markings of irregular pattern, in the form of bars of different shapes; the lower abdomen, flanks, and under tail-coverts transversely barred with blackish; under wing-coverts orange-chestnut, with which colour also the quills are lined on the inner web; upper wing-coverts ashy brown like the back and scapulars, the greater series and the primaries externally washed with orange-rufous; quills brown, externally olivaceous brown, shading into orange-rufous on the edge of the quills; tail warm brown, slightly shaded with olive and edged with orange-rufous along the inner web. Total length 11·2 inches, culmen 2, wing 5·9, tail 4, tarsus 1·45.

Hab. South-eastern Australia, ranging as for north as Queensland.

a, b. ♂ ♀ ad. st.	Australia.	
c, d. ♂ ♀ ad. sk.	New South Wales.	J. E. Harting, Esq. [P.].

2. Ptilorhis victoriæ.

Ptilorhis victoriæ, *Gould, P. Z. S.* 1849, p. 111, pl. 12; *id. B. Austr. Suppl.* pl.; *id. Handb. B. Austr.* i. p. 593; *Reich. Handb.* Scansoriæ, p. 329, Taf. dcx. figs. 4086–88; *Elliot, P. Z. S.* 1871, p. 582; *id. Monogr. Parad.* pl. xxvi.

Ptilornis victoriæ, *Bp. Consp.* i. p. 412; *Rosenb. J. f. O.* 1864, p. 123; *Gray, Hand-l. B.* i. p. 105.

Adult male (type of species). Above velvety black, appearing rich purple against the light; wing-coverts coloured like the back; quills black, externally glossed with deep purple, the secondaries entirely of this colour; tail rich velvety purple, the two centre feathers metallic steel-green; head metallic steel-green with a coppery gloss, inclining to steel-blue on the nape; sides of face, sides of neck and throat velvety purple, with a strong gloss of fiery copper on the chin and sides of the throat; a triangular shield of metallic green slightly shaded with purple extending from the upper part of the throat and occupying all the fore neck; entire fore neck and chest rich velvety purple, the lower plumes tipped with oily green; rest of under surface of body oily green, all the feathers with concealed velvety black bases; under wing-coverts purple, the lower surface of the wings and tail blue-black; bill and legs black. Total length 10 inches, culmen 1·5, wing 5·3, tail 3·3, tarsus 1·4.

Adult female. General colour above ashy brown, the head rather more dusky, all the feathers narrowly streaked with fulvous, as also the sides of the neck; lores and ear-coverts also brown, narrowly streaked with whitish; from above the fore part of the eye a broad fulvous line of feathers running above the ear-coverts and forming a distinct eyebrow; cheeks and chin buffy white, the throat and underparts of the body fawn-colour, spotted on the breast and barred on the flanks with dark brown; under wing- and tail-coverts uniform fawn-colour; wing-coverts above ashy brown like the back, the outermost of the greater series and the primary-coverts externally washed with orange-chestnut; quills brown, externally washed with olive, inclining to orange-chestnut towards the ends of the quills; tail olive-brown. Total length 9 inches, culmen 1·5, wing 4·9, tail 3·2, tarsus 1·4.

Obs. Although from the description and plate in Mr. Elliot's 'Monograph' this Rifle-bird would not be considered a very distinct species, an examination of the types shows that it is so. It is a very much smaller bird than *P. paradisea*; and instead of the purple shade of the throat the latter has a distinct reflection of burnished copper; the gular shield is much smaller and has a coppery gloss; the purple is confined to the chest; and the rest of the under surface is oily green. In *P. paradisea* the upper breast is purple as well as the chest, and the shade of the underparts is olive-green. The females of the two birds are quite different.

Hab. Barnard Islands, North-eastern Australia.

a. ♂ ad. sk.	Barnard Islands.	J. Macgillivray, Esq. [C.]. (Type of species.)
b. ♀ ad. st.	Barnard Islands.	J. Macgillivray, Esq. [C.].
c. ♀ ad. sk.	Barnard Islands.	J. Macgillivray, Esq. [C.].

3. Ptilorhis alberti *.

Ptilorhis magnificus, *Gould, B. Austr.* v. *Suppl.* pl. 51.
Craspedophora magnifica, *Gould, Handb. B. Austr.* i. p. 595.
Ptilorhis alberti, *Elliot, P. Z. S.* 1871, p. 583; *id. Monogr. Parad.* pl. xxiv.

Adult male. General colour velvety black, appearing purple when held away from the light, with somewhat of a bluish purple gloss on some of the feathers, this latter shade being especially distinct on the outer wing-coverts, which are otherwise like the back; quills blue-black, of a velvety texture, the inner secondaries glossed with purplish blue; tail velvety black, the two centre feathers metallic

---

PTILORHIS WILSONI.

Ptiloris wilsoni, *Ogden, Pr. Acad. Nat. Sci. Philad.* 1875, p. 451, pl. 25.

Top of head, occiput, throat, upper part of the breast extending around beneath the eyes, of a shining, metallic, purplish-green colour; the lower margin of the metallic-coloured feathers of the breast is marked with a line of purple and light green feathers, the latter blending with the purple-coloured feathers of the abdomen; entire upper parts velvety green, with purple reflections; the two central tail-feathers metallic green, the rest black, with the outer margin shaded with purple; the wings black, with the outer border purplish. Total length (mounted specimen) 14″, bill 2″ 3‴, wing 5″ 4‴, tail 4″ 2‴, tarsus 1″ 4‴, middle toe 1″ 1‴.

Hab. New Guinea.

"This species is larger than either *P. magnificus* or *P. albertii.* The metallic-coloured feathers are larger, and extend over a greater surface of the throat and breast; the lower border is rounded. The tarsi are shorter and the toes not so stout as in the above two species."

Dr. Ogden, who describes this bird, does not give very trenchant characters for its separation; and the distinctions of size do not appear to be well founded. If the plumage is correctly described, the "velvety green" colour of the upper surface and the "light green" chest-band would be better specific characters.

steel-green, the next one on each side also glossed with steel-green towards the base; crown of head and nape metallic steel-green with a slight coppery lustre; lores, feathers above the eye and sides of face, as well as the chin and the sides of the throat, glossy purple; from the chin a shield of metallic steel-blue plumes spreads out onto the chest, and is followed by a double band, the first velvety black with a burnished coppery lustre, the second golden green with an olivaceous lustre; rest of under surface olivaceous, shaded with reddish purple, especially on the abdomen and flank-feathers, the latter being elongated into silky plumes, which reach beyond the tail; under wing- and tail-coverts and vent velvety blue-black; bill and legs black; iris brown. Total length 12·2 inches, culmen 2·2, wing 6·65, tail 4·25, tarsus 1·65.

Adult female. General colour ashy brown on the head and neck, shading into olivaceous brown on the back and rump, slighly tinged with reddish on the upper tail-coverts; lores, feathers round the eye, and ear-coverts dusky brown, with slightly indicated shaft-streaks of ochraceous, these being also faintly perceptible on the feathers of the crown; over the eye a broad line of white, forming a distinct eyebrow; cheeks white, as also the throat, which is separated from the cheeks by a broad malar streak of blackish; rest of under surface of body dull white, mottled with cross bars of blackish, much wider apart on the abdomen and flanks, the feathers of the latter being somewhat elongated; under wing-coverts like the breast, the cross bars less distinct; upper wing-coverts like the back, the greater series orange-chestnut, this being also the colour of the quills, which are brown on the inner webs. Total length 11·2 inches, culmen 1·95, wing 6, tail 4·3, tarsus 1·6.

Obs. Mr. Elliot was quite right in separating this species from the New-Guinea bird. The male is not only smaller, but the breast in the latter has not the same olivaceous wash, the reddish purple shade appearing to begin immediately below the golden-green collar. The females are extremely different, that of the New-Guinea bird being nearly uniform rufous above.

Hab. North-eastern Australia.

a, *b*. ♂ ♀ ad. sk.	Port Albany.	Purchased.
c. ♂ ad. st.	Cape York.	J. Macgillivray, Esq. [P.].
d, *e*. ♀ ad. st.	Cape York.	J. Macgillivray, Esq. [P.].
f. ♂ juv. st.	Cape York.	Capt. Stanley [P.].
g. ♂ imm. st.	Australia.	Purchased.

4. Ptilorhis magnifica.

Le Proméfil, *Levaill. Ois. de Parad.* p. 36, pl. 16; *Less. Mus. Zool.* pl. 29.

Falcinellus magnificus, *Vieill. N. Dict. d'Hist. Nat.* xxviii. p. 167, pl. G 80. no. 3.

Epimachus splendidus, *Steph. Gen. Zool.* xiv. p. 77.

Epimachus magnificus, *Wagler, Syst. Av.* Epimachus, sp. 10; *Cuvier, Règne Anim.* 1829, p. 440; *Less. Cent. Zool.* p. 22, pls. 4, 5; *id. Ois. Parad. Syn.* p. 27; *id. H. N. Ois. Parad.* p. 218, pls. 32–34;

Bp. Consp. i. p. 412; *Gray, P. Z. S.* 1859, p. 155; *Schl. J. f. O.* 1861, p. 386; *id. Mus. P.-B.* Coraces, p. 96.
Craspedophora magnifica, *Gray, List Gen. B.* 1841, p. 15; *Reichenb. Handb.* Scansoriæ, p. 330, Taf. dcxi. figs. 4089–91; *Wall. P. Z. S.* 1862, p. 160; *Rosenb. J. f. O.* 1864, p. 128; *Salvad. Ann. Mus. Civic. Genov.* ix. p. 191.
Epimachus paradiseus, *Gray, Gen. B.* ii. pl. xxxii. (*nec Swains.*).
Ptilornis magnificus, *Gray, Hand-l. B.* i. p. 105.
Ptilorhis magnificus, *Elliot, P. Z. S.* 1871, p. 583; *id. Monogr. Parad.* pl. xxiii.; *Salvad. Ann. Mus. Civic. Genov.* vii. p. 785.
Ptiloris superbus, *Beccari, Ann. Mus. Civic. Genov.* vii. p. 713.

Male. Top of head and occiput, centre of throat, and entire upper part of breast shining bluish green, purple in certain lights; entire upper parts deep velvety black, with rich dark purple reflections; primaries black, with green reflections; a narrow line of green, red in some lights, beneath the metallic of the breast; flanks and abdomen purple; side plumes also purple, basal half and filamentary ends black; two centre tail-feathers shining green; remainder velvety black, with green reflections on their outer webs; bill, feet, and legs stout, black. (*Elliot, l. c.*)*

Female. Above cinnamon-rufous, the wings and tail entirely of the same colour as the back, the inner webs browner; over the eye a narrow streak of white; lores and sides of face dusky brown, the former washed with rufous, the ear-coverts minutely streaked with rufous along the shafts of the feathers; cheeks white, the feathers somewhat scaly in appearance; a malar streak of dark brown on each side of the throat; throat white, slightly mottled with minute dusky cross markings; rest of under surface of body dull white, very numerously and thickly barred across with dusky blackish. Total length 12 inches, culmen 1·95, wing 6·6, tail 4·45, tarsus 1·65. (*S.E. New Guinea: coll. O. C. Stone.*)

Young male. Similar to the adult female, but of a deeper rufous, the head and neck rather dingier than the back; a tolerably well-defined white eyebrow, the feathers edged with brown; lores and ear-coverts dusky chocolate-brown, with very few ochraceous shaft-streaks; under surface of body dirty white, very thickly barred across with black; the flank-plumes elongated, but barred exactly like the breast; from the base of the lower mandible a malar streak of dusky black, continued down the sides of the throat onto the sides of the chest. Total length 14·5 inches, culmen 2·3, wing 6·95, tail 4·9, tarsus 1·65. (*Dorey: A. R. Wallace.*)

Hab. New Guinea.

a, b. ♂ ad. sk.	New Guinea.	A. R. Wallace, Esq. [C.].
c. Ad. st.	New Guinea.	[Native skin.]
d. ♂ juv. sk.	Dorey.	A. R. Wallace, Esq.

* The above description is taken from Mr. Elliot's work, as the specimens in the British Museum are imperfect.

2. SELEUCIDES.

	Type.
Seleucides, *Less. Hist. Nat. Ois. Parad. Syn.* p. 29	S. niger.
Nematophora, *Gray, List Gen. B.* 1840, p. 12	S. niger.

Range. Confined to New Guinea.

1. Seleucides niger *.

Le Manucode à douze filets, *Audeb. et Vieill. Ois. Dor.* ii. p. 29, pl. 13.
Le Nébouleux, *Levaill. Ois. de Parad.* i. pls. 16, 17.
Le Promérops multifil, *Levaill. H. N. Promér. et Guêp.* pl. 17.
Paradisea nigra, *Shaw, Gen. Zool.* vii. pt. 2, p. 488, pl. 60 (1809).
Paradisea alba, *Blumenb. Abbild. nat. Gegenst.* pl. 96; *Schl. J. f. O.* 1861, p. 386.
Paradisea nigricans, *Shaw, t. c.* p. 489.
Paradisea violacea, *Bechst. Kurze Uebers.* p. 133.
Paradisea vaillanti, *Shaw & Nodd. Nat. Misc.* xxiv. pl. 1025.
White-winged Paradise-bird, *Lath. Gen. Hist.* iii. p. 197.
Twelve-wired Paradise-bird, *Lath. t. c.* p. 199, pl. 48.
Le Promérops à douze filets, *Vieill. Gal. Ois.* pl. 185.
Epimachus albus, *Wagler, Syst. Av.* 1827, Epimachus, sp. 9; *Gray, Gen. B.* i. p. 94; *id. P. Z. S.* 1861, p. 433; *Wall. P. Z. S.* 1862, p. 160; *Schl. Mus. P.-B.* Coraces, p. 95; *Gray, Hand-l. B.* i. p. 105.
Falcinellus resplendens, *Less. Traité,* p. 324.
Seleucides acanthylis, *Less. H. N. Ois. Parad.* pls. 36–38; *id. Syn.* p. 29.
Nematophora alba, *Gray, List Gen. B.* 1840, p. 12.
Seleucides alba, *Bp. Consp.* i. p. 412; *Wallace, Malay Archip.* ii. p. 253; *Elliot, Monogr. Parad.* pl. xxii.; *Salvad. Ann. Mus. Civic. Genov.* vii. p. 785; *Beccari, t. c.* p. 713; *Scl. P. Z. S.* 1876, p. 252.
Seleucides resplendens, *Rosenb. J. f. O.* 1864, p. 123.
Seleucides ignota, *Salvad. Ann. Mus. Civic. Genov.* ix. p. 191.

Adult male. General colour above velvety black, with a strong gloss of oil-green when viewed from the light, with coppery bronze reflections; scapulars and wing-coverts resembling the back; greater coverts and secondaries fiery purple, the primaries black with an external gloss of violet; tail fiery purple; head all round of a velvety texture, coppery purple above, oily green on the sides of the face and throat; fore neck and chest velvety black, forming a shield, somewhat shaded with oily green in the centre, the lateral plumes all tipped with bright metallic emerald-green, forming a fringe;

* Count Salvadori, in his recently published list of the Birds of Paradise, quotes this bird as *Seleucides ignota,* and founds this name on the *Paradisea ignota* of Forster (Zool. Ind. p. 31). Neither this name nor *P. alba* of Gmelin appear to me available, as they are both founded on names in the above-mentioned work of Forster. On examining the latter we find that he is there enumerating the species of Birds of Paradise, of which he says there are "about" six species known, and these are :—1. *P. major aruana*; 2. *P. minor papuana*; 3 & 4. *P.* species duæ nigræ; 5. *P. candida*; 6. *P. ignota,* &c. It is evident that Forster does not use these names in a binomial sense, but that by employing the name *ignota* he really meant that the species was *unknown* to him. The next name which will take precedence is *P. nigra* of Shaw.

rest of the under surface of body buffy yellow, the plumes of the flanks elongated and silky, and furnished with six thread-like shafts, produced to a great length and curved backwards on the body; under wing-coverts black; bill black. Total length 12 inches, culmen 2·7, wing 6·45, tail 3·15, tarsus 1·75, threads reaching 10·2 inches beyond the flank-feathers.

Adult female. General colour above bright chestnut-red; back of the neck and sides of the same black; the feathers of the mantle also mottled with black, the bases of the feathers being of this colour; crown of head and nape velvety black, with a purplish gloss when seen away from the light; wing-coverts and secondaries chestnut-red, like the back, the primaries black, chestnut on their outer webs; tail uniform chestnut; space round and behind the eye bare, as also a spot on the auricular region; ear-coverts black; sides of face and throat greyish white, faintly mottled with dusky bars of blackish; rest of under surface of body buffy brown, washed here and there with pale rufous, the whole transversely barred with somewhat irregular cross lines of blackish brown, broader on the fore neck and breast, and more faintly indicated on the abdomen, and especially on the long flank-feathers and under tail-coverts; under wing-coverts bright chestnut, with dusky blackish cross bars. Total length 12·5 inches, culmen 2·55, wing 6·5, tail 4·3, tarsus 1·7.

Young male. At first resembles the adult female. A specimen collected by Mr. Wallace is in perfect plumage as regards its head, mantle, and breast, the rest of the body being in the chestnut plumage of the female, the tail being still entirely chestnut. At the same time the beautiful purple colour is being put on the wings by a gradual change of feather and not by a moult; half the inner secondaries are chestnut, but more or less mottled with blackish, the purple colour appearing very plainly on the inner webs.

Hab. New Guinea.

a, b. ♂ ad. sk.	New Guinea.	A. R. Wallace, Esq. [C.].
c, d. ♀ ad. sk.	New Guinea.	A. R. Wallace, Esq. [C.].
e. ♂ juv. sk.	New Guinea.	A. R. Wallace, Esq. [C.].

3. DREPANORNIS.

Type.

Drepanophorus, *Sclater, Nature*, 1873, p. 192 (*nec Egerton*) D. albertisi.
Drepanornis, *Sclater, P. Z. S.* 1873, p. 560 . . . . D. albertisi.

Range. North-eastern New Guinea.

1. Drepanornis albertisi.

Drepanophorus albertisi, *Scl. Nat.* 1873, p. 192.
Drepanornis albertisi, *Scl. Nat.* 1873, p. 305, cum fig.; *id. P. Z. S.* 1873, p. 560, pl. xlvii.; *D'Albertis, t. c.* p. 558; *Elliot, Monogr. Parad.* pl. 21; *Gould, B. N. Guin.* part i.; *Salvad. Ann. Mus. Civic. Genov.* vii. p. 785; *Beccari, t. c.* p. 711; *Scl. Ibis*, 1876, p. 250; *Salvad. Ann. Mus. Civic. Genov.* ix. p. 190.
Epimachus wilhelminæ, *Meyer, Nat. Tijds.* 1873, p. 415; *J. f. O.* 1873, p. 404; *id. Ibis*, 1874, p. 303; *Finsch, t. c.* p. 54.
Epimachus veithi, *Rosenb. Reist. N. Guin.* pl. xviii.

Adult male. General colour above brown, the rump tawny chestnut, shading into fawn-colour on the upper tail-coverts, the tail being uniform fawn-colour; wing-coverts chestnut-brown, the rest of the wing paler and more tawny brown, the quills blackish brown on the inner webs, the innermost secondaries pale tawny brown like the back; crown of head rich brown with a purple gloss; sides of head bare, with a line of brown feathers drawn across the ear-coverts; lores and feathers in front of the eye and cheeks velvety brown glossed with purple; throat rich purplish brown; rest of the under surface of body chocolate-brown, with a lilac gloss on the breast and abdomen; across the lower breast a bar of dull green; from the sides of the chest springs a tuft of fine purple feathers, brown at the base and tip; from the sides of the breast another long tuft of plumes, chocolate-brown glossed with lilac, each feather tipped with a bar of amethystine purple; under wing-coverts brown, the innermost of the greater series whitish at the tip; quills ashy brown, edged with tawny on both webs; bill blackish; "iris violescent brown" (*Beccari*). Total length 13·3 inches, culmen 2·8, wing 5·95, tail 5·5, tarsus 1·4.

Female. Head chestnut brown; back and wings rufous brown; primaries and secondaries blackish brown on inner web, outer web brown; upper tail-coverts and tail light red; chin and throat blackish brown, each feather with a central streak of light brown; breast light brown, irregularly barred with dark brown; flanks and lower parts of body yellowish brown, indistinctly barred with dark brown, except in the centre of the abdomen, which is light reddish white; thighs reddish, banded with brown; under tail-coverts pale reddish; iris chestnut; bill long, curved, and slender, like that of the male, black; feet and tarsi lead-colour. (*Elliot.*)

Hab. Arfak Mountains, North-eastern New Guinea.

a. ♂ ad. sk.	Atam, Dec. 12, 1867.	Purchased.

4. EPIMACHUS.

	Type.
Epimachus, *Cuvier, Règne Anim.* i. p. 407 (1817)	E. speciosus.
Cinnamolegus, *Lesson, H. N. Ois. Parad. Syn.* p. 32 (1833)	E. speciosus.

Range. Confined to New Guinea.

Key to the Species.

a. Size very large; tail black, the centre feathers with steel-blue and purple reflections; under surface of body dark olive-green .... *speciosus*, p. 162.

b. Size much smaller; tail brilliant violet-purple; sides of breast, flanks, and under tail-coverts dark green; throat and upper part of breast deep maroon colour with purple reflections; across the lower breast a narrow band of reddish purple .. *elliotî*, p. 163.

1. Epimachus speciosus.

Le Grand Promérops de la Nouvelle Guinée, *Sonn. Voy. N. Guin.* p. 163, pl. 101 (♂).
Le Promérops brun de la Nouvelle Guinée, *id. t. c.* p. 164, pl. 100 (♀).
Grand Promérops à paremens frisés, *Buff. H. N. Ois.* vi. p. 472.
Promérops de la Nouvelle Guinée, *Buff. Pl. Enl.* vi. pls. 638, 639.
Upupa speciosa, *Bodd. Tabl. Pl. Enl.* p. 39 (*ex Pl. Enl.* 638).
Upupa striata, *Bodd. t. c.* p. 39 (♀, *ex Pl. Enl.* 639).
New-Guinea Brown Promerops, *Lath. Gen. Syn.* i. pt. 2, p. 694.
Grand Promerops, *Lath. t. c.* p. 695.
Upupa fusca, *Gm. S. N.* i. p. 468 (*ex Lath.*).
Upupa magna, *Gm. t. c.* p. 468 (*ex Lath.*).
Le Promérops rayé, *Audeb. et Vieill. Ois. Dor.* i. pl. 7.
Le Promérops à large parure, *Levaill. H. N. Promér. et Guép.* pls. 13, 15.
Promerops striata, *Shaw, Gen. Zool.* viii. p. 144.
Promerops superbus, *Shaw, t. c.* p. 145.
Falcinellus superbus, *Vieill. N. Dict. d'Hist. Nat.* xxviii. p. 166.
Falcinellus magnificus, *Vieill. t. c.* p. 167.
Epimachus magnus, *Cuvier, Règne Anim.* i. p. 407.
Epimachus superbus, *Steph. Gen. Zool.* xiv. p. 77; *Wagler, Syst. Av.* Epimachus, sp. 1; *Less. Traité,* p. 321, *Atlas,* pl. 73. fig. 1; *Rosenb. J. f. O.* 1864, p. 123.
Cinnamolegus papuensis, *Less. Ois. Parad. Syn.* p. 32; *id. H. N.* pls. 39, 40.
Epimachus speciosus, *Gray, Gen. B.* i. p. 94; *Schl. Mus. P.-B.* Coraces, p. 94; *Elliot, Monogr. Parad.* pl. xix.; *Salvad. Ann. Mus. Civic. Genov.* vii. p. 785, ix. p. 190.
Epimachus magnus, *Bp. Consp.* i. p. 411; *Wall. Ibis,* 1861, p. 287; *id. P. Z. S.* 1862, p. 160; *id. Malay Archip.* ii. p. 255; *Schl. J. f. O.* 1861, p. 386; *id. N. T. D.* i. p. 332.
Epimachus maximus, *Gray, P. Z. S.* 1861, p. 433; *id. Hand-l. B.* i. p. 105; *Beccari, Ann. Mus. Civic. Genov.* vii. p. 710; *Scl. Ibis,* 1876, p. 249.

Adult male. Above velvety black, with metallic feathers of coppery green on the head, middle of the back, and rump; lores and feathers on the side of the head metallic like the crown; entire under surface of body velvety black, with a purplish brown gloss on the sides of the body; on each side of the breast springs a tuft of sickle-shaped plumes in the shape of a fan velvety black, tipped with a broad band of steel-blue, before which is a narrow subterminal band of purplish blue; flank-feathers long and drooping, the outer ones broadly tipped with metallic bronzy green, before which is a double subterminal band of velvety black and purplish blue; wings velvety black, with a gloss of steel-blue; tail-feathers black, all but the three outermost feathers washed with steel-blue, the two centre ones entirely of this colour; bill and legs black: "iris dark brick-red" (*Beccari*). Total length 26 inches, culmen 2·85, wing 7·2, tail 16·7.

Female. Upper part of head brownish red; rest of upper parts olive-brown, becoming slightly rufous on the rump and upper tail-coverts; secondaries reddish brown, edged with rufous; primaries dark brown, edge of outer web rufous; cheeks, throat, and upper part of breast brownish black; underparts white, narrowly barred with black; tail light brown, with a rufous tinge; bill long and slender, much curved, and, with the feet and tarsi, jet-black.

Hab. New Guinea.

a. ♂ ad. st.	New Guinea.	Baron Laugier.
b. ♂ jun. sk.	New Guinea.	Purchased.

2. Epimachus ellioti.

Epimachus ellioti, *Ward, P. Z. S.* 1873, p. 743; *Elliot, Monogr. Parad.* pl. 20; *Beccari, Ann. Mus. Civic. Genov.* vii. p. 710; *Salvad. op. cit.* ix. p. 190.

Top of head rich amethyst; occiput and side of neck also amethyst-colour, changing in certain lights to a rich light greenish gloss; back, wings, upper tail-coverts, and tail brilliant violet-purple; the wings and the tail also marbled with a dark amethyst hue, like watered silk, changing according to the light; throat and upper portion of breast deep maroon-colour, with purple reflections, a narrow reddish purple band crosses the lower part of the breast; sides of the breast, flanks, and rest of underparts dark green, the flank-feathers much elongated and stretching beyond the wings; beneath the shoulder of the wing spring two rows of plumes which are greenish at the base, graduating into deep purple, and terminating in a brilliant metallic blue, very much narrower on the upper row than the lower one. The plumage of the entire bird is very velvety in texture, and, with the exception of the metallic parts, appears black in ordinary lights; bill black, rich orange-yellow at the gape. (*Elliot, l. c.*)

Hab. ? Waigiou.

Subfam. II. PARADISEINÆ.

Key to the Genera.

a. Tail very long and graduated, but the central plumes not elongated into wire-like shafts.
  a′. Lores and angle of mouth thickly feathered; tail longer than the body; head surrounded by an erect frill, round the throat a metallic frill of golden copper .... 5. ASTRAPIA, p. 165.
  b′. Lores with an erect orange-yellow wattle, angle of mouth with a pendent one of azure-blue; tail shorter than the body; no frills round the head or throat .... 6. PARADIGALLA, p. 165.
b. Centre tail-feathers produced into wire-like shafts.
  c′. With no dorsal shield.

a''. Flank-plumes produced to an extraordinary length and forming two enormous bunches of feathers reaching far beyond the tail; two centre tail-feathers thread-like or horny in structure .... 7. PARADISEA, p. 166.
b''. Flank-plumes not produced beyond the tail; on each side of the breast a large tuft of plumes spread like a fan; two centre tail-feathers thread-like and ending in a racket ...... 8. CICINNURUS, p. 171.
d'. With a dorsal shield.
c''. Head thickly plumed with close-set feathers.
a'''. With a fan-shaped tuft of feathers springing from each side of the breast .... 9. RHIPIDORNIS, p. 172.
b'''. Without fan-shaped tuft of pectoral feathers .... 10. DIPHYLLODES, p. 173.
d''. Head bare, with a few transverse tracts of feathers .... 11. SCHLEGELIA, p. 175.
c. Tail square or rounded, the two centre feathers not exceeding the others in any great degree.
e'. Head ornamented with six thread-like plumes furnished with a terminal racket, a large gular shield of metallic golden; no dorsal shield of feathers; plumes of the sides of the body enormously developed, sufficient to overshadow the whole of the back ...... 12. PAROTIA, p. 176.
f'. Head without the long thread-like plumes.
e''. With a large metallic pectoral shield.
c'''. With no dorsal shield of plumes, from each wing two projecting long white plumes .... 13. SEMIOPTERA, p. 178.
d'''. With an enormous dorsal shield of velvety black plumes spread out like a fan from the interscapulary region; no standard plumes from the wing .... 14. LOPHORHINA, p. 179.
f''. With no pectoral shield, tail slightly rounded.
e'''. With no dorsal shield of plumes.
a^4. Plumage metallic, with two elongated occipital tufts of plumes. 15. PHONYGAMA, p. 180.
b^4. Plumage metallic, the feathers of the crown and breast more or less crinkled............. 16. MANUCODIA, p. 181.
c^4. Plumage not metallic .... 17. LYCOCORAX, p. 185.
f'''. With an erectile shield of dorsal plumes; entire plumage yellow.. 18. XANTHOMELUS, p. 186.

5. ASTRAPIA.

	Type.
Astrapia, *Vieill. N. Dict.* iii. p. 36 (1816)	A. nigra.

Range. Confined to New Guinea.

1. Astrapia nigra.

Gorget Paradise-bird, *Lath. Gen. Syn.* i. p. 478, pl. 20.
Paradisea nigra, *Gm. S. N.* i. p. 401 *(ex Lath.)*.
Paradisea gularis, *Lath. Ind. Orn.* ii. p. 196; *Shaw, Gen. Zool.* vii. pp. 69, 70.
Le Hausse-col doré, *Aud. et Vieill. Ois. Dor.* ii. p. 22, pls. 8, 9.
La Pie de Paradis ou l'Incomparable, *Levaill. H. N. Ois. Parad.* i. pls. 20–22.
Astrapia gularis, *Vieill. N. Dict. d'Hist. Nat.* iii. p. 37; *Less. Traité*, p. 338; *id. Ois. Parad. Syn.* p. 18; *id. Hist. Nat.* p. 106, pls. 21–23; *Schl. J. f. O.* 1861, p. 386; *Beccari, Ann. Mus. Civ. Genov.* vii. p. 711; *Sclater, Ibis*, 1876, p. 249.
Astrapia nigra, *Steph. Gen. Zool.* xiv. p. 75; *Gray, Gen. B.* ii. p. 326; *Bp. Consp.* i. p. 414; *Gray, P. Z. S.* 1861, p. 436; *Wall. P. Z. S.* 1862, p. 154; *id. Malay Archip.* ii. p. 257; *Elliot, Monogr. Parad.* pl. ix.; *Salvad. Ann. Mus. Civ. Genov.* ix. p. 190.

Adult male. General colour above velvety black, with a purplish gloss; the wings black externally, glossed with purple; tail-feathers black, with wavy lines of dusky black under certain lights, the two centre feathers very long and glossed with rich purple; feathers of the head black, dense and velvety in texture, with a steel-blue gloss; from the nape a shield of golden-green feathers springs, and the feathers of the hind neck are very long and tipped with the same golden green; from each side of the nuchal shield spreads a ruff of velvety steel-black plumes; the feathers of the throat also steel-black, and encircled by a band of brilliant golden copper, which extends from behind the eye down the sides of the neck and encircles the throat; rest of under surface of body rich velvety grass-green, the lateral plumes of the breast tipped with burnished emerald-green; the sides of the body and under wing- and tail-coverts dusky black; bill and legs black; "iris almost black" (*Beccari*). Total length 28 inches, culmen 1·6, wing 8·8, tail 7, centre feathers 18.

Hab. New Guinea.

a. Ad. sk.	New Guinea.	Baron Laugier.

6. PARADIGALLA.

	Type.
Paradigalla, *Less. Ois. Parad.* 1835, p. 242	P. carunculata.

Range. Confined to New Guinea.

1. Paradigalla carunculata.

Astrapia carunculata, *Eyd. & Souleyet, Voy. Bonite*, p. 83, pl. 4.
Paradigalla carunculata, *Less. Ois. Parad.* p. 242; *id. Rev. Zool.* 1840, p. 1; *Bp. Consp.* i. p. 414; *Scl. P. Z. S.* 1857, p. 6; *Wall. P. Z. S.* 1862, p. 160; *id. Malay Arch.* ii. p. 257; *Elliot, Monogr. Parad.*

pl. xvii.; *Salvad. Ann. Mus. Civic. Genov.* vii. p. 784; *Beccari, t. c.* p. 711; *Scl. Ibis,* 1876, p. 250; *Salvad. Ann. Mus. Civic. Genov.* ix. p. 190.
Astrapia carunculata, *Gray, Gen. B.* ii. p. 326; *Schl. J. f. O.* 1861, p. 386; *Rosenb. J. f. O.* 1864, p. 131; *Gray, Hand-l. B.* ii. p. 17.

Adult male. General colour velvety black above and below, a little browner on the under surface; wings and tail black, the inner secondaries with a purplish gloss under certain lights; head glossed with metallic steel-green; forehead, lores, and base of lower mandible bare; over each nostril a small tuft of black feathers; on each side of the base of the bill an erect wattled skin; round the eye a ring of black plumes; space below and behind the eye bare; bill and legs black; "upper wattles, which are attached one on each side of the forehead, of a yellowish green colour; those at the base of the lower mandible blue, having a small patch of orange underneath" (*Beccari*). Total length 11·2 inches, culmen 0·55, wing 6·15, tail 4·85, tarsus 1·9.

Adult female. Similar to the male, but smaller.

Hab. New Guinea.

a. ♂ ad. sk. Arfak Mountains, New Guinea, 1875. Dr. O. Beccari [C.].

7. PARADISEA.

	Type.
Paradisea, *Linn. S. N.* i. p. 166 (1766)	P. apoda.
Uranornis, *Salvad. Ann. Mus. Civic. Genov.* ix. p. 191 ..	P. sanguinea.

Range. New Guinea, Aru Islands, Waigiou, Mysol, Jobie, Mysore.

Key to the Species.

a. Long flank-plumes yellow, white towards their tips; green on the throat produced to the maroon breast, and not separated by a yellow collar.
  a'. Size larger; mantle maroon-brown like the rest of the back; female maroon-brown .... ***apoda***, p. 166.
  b'. Size smaller; mantle dull straw-colour; female white on the breast...................... ***minor***, p. 168.
b. Long flank-plumes red; green of the throat separated from the maroon breast by a collar of straw-yellow.
  c'. Mantle maroon-brown; flank-plumes very long, as in *P. apoda*; two centre tail-feathers produced into a long wire-like plume turned round at the end .... ***raggiana***, p. 169.
  d'. Mantle straw-yellow; flank-plumes shorter, and brownish at tip; two centre tail-feathers produced into a broad horny plume .... ***sanguinea***, p. 170.

1. Paradisea apoda.

The Greater Bird of Paradise, *Edwards, Birds,* iii. pl. 110.
L'Oiseau de Paradis, *Briss. Orn.* ii. p. 130, pl. xiii.
Paradisea apoda, *Linn. S. N.* i. p. 165; *Wagler, Syst. Av.* Paradisea, sp. 1; *Gray, Gen. B.* ii. p. 323; *Bp. Consp.* i. p. 412; *Gray, P. Z. S.*

1861, p. 436, 1862, p. 436; *Wall. Ibis*, 1859, p. 111, 1861, p. 289; *Schl. J. f. O.* 1861, p. 384; *id. Mus. P.-B.* Coraces, p. 78; *Wall. Malay Arch.* ii. p. 238; *Rosenb. J. f. O.* 1864, p. 42; *Gray, Hand-l. B.* ii. p. 16; *Elliot, Monogr. Parad.* pl. 1; *Salvad. Ann. Mus. Civic. Genov.* ix. p. 191.

L'Oiseau de Paradis des Moluques, *Buff. Pl. Enl.* p. 190, pl. 254.

L'Émeraude, *Audeb. et Vieill. Ois. Dor.* ii. p. 9.

Le Grand oiseau de Paradis émeraude, *Levaill. H. N. Ois. Parad.* i. pls. 1–3.

Paradisea major, *Shaw, Gen. Zool.* vii. p. 480, pl. 58; *Less. Ois. Parad. Syn.* p. 6; *id. H. N.* p. 155, pl. 6.

Paradisea apoda, var. wallaciana, *Gray, P. Z. S.* 1858, p. 181.

Adult male. General colour above maroon-brown, including the entire back and wings; head and neck clothed in compressed thick-set feathers of a straw-yellow colour, velvety in texture; round the eye a narrow line of black; forehead, lores, cheeks, throat, and fore neck dark metallic green, all the plumes close-set and velvety, the forehead, lores, and chin having a velvety black appearance under certain lights, and forming as it were a mask; breast deep purplish brown, shading gradually into maroon-brown on the abdomen and under tail-coverts; under wing-coverts and inner lining of wings maroon-brown; tail maroon-brown like the back, the two centre feathers enormously elongated into two wire-like shafts about 30 inches long; from the flanks spring two enormous tufts of beautiful plumes, bright yellow for two thirds of their length, and then shading into chocolate-brown, the shafts produced at the tips of the feathers and white, the plumes towards their extremities very lax, the webs separate and very distinct; at the base of these side-tufts are several rigid plumes of bright yellow, some of which end in blood-red, giviug the appearance of being streaked with blood-red; bill lead-colour, inclining to greenish white at the tip; legs and feet flesh-colour. Total length 18 inches, culmen 1·6, wing 9·35, tail 7·2, tarsus 2·2.

Adult female. General colour all over maroon-brown, deepening to purplish chestnut on the head, neck, and chest; plumes of the head close-set and velvety, and the nape somewhat tinged with straw-yellow; flank-plumes lax and elongated, maroon-brown, like the breast and abdomen; two centre tail-feathers rather pointed. Total length 15·3 inches, culmen 1·45, wing 7·5, tail 6·4, tarsus 1·85.

Obs. A young male can always be told from an old female by its larger size. Total length 7·2 inches, culmen 1·6, wing 8·7, tail 7·1, tarsus 2·25. The young male here measured exactly resembles the foregoing old female in coloration; but the size is very much larger. They next get the yellow head and green throat and face, and the two centre tail-feathers commence to get elongated; the latter are lengthened to their full thread-like extent, generally, however, with a feathery tip, before the plumes on the flanks are at all developed. The green throat does not extend so far in the young male as in the old, and the yellow of the neck encroaches further down on each side, so as to form a half-collar.

Hab. Aru Islands.

a–c. ♂, ♀, ♂ juv. st.	Aru Islands.	A. R. Wallace, Esq. [C.].
d, e. ♂ and ♀ ad. sk.	Aru Islands.	A. R. Wallace, Esq. [C.].
f, g. ♂ juv. sk.	Aru Islands.	A. R. Wallace, Esq. [C.].
h, i. ♂ ad. sk.	Aru Islands.	W. Harvey, Esq. [P.].
k, l. ♂ ad. sk.	Aru Islands.	

2. Paradisea minor.

Paradisea minor papuana, *Forster, Ind. Zool.* p. 33.
Lesser Bird of Paradise, *Lath. Gen. Syn.* ii. p. 474.
La Petite Émeraude, *Audeb. & Vieill. Ois. Dor.* ii. p. 12, pls. 2, 11.
Le Petit Oiseau de Paradis émeraude, *Levaill. Ois. Parad.* i. pls. 4, 5.
Paradisea minor, *Shaw, Gen. Zool.* vii. p. 486; *Wagler, Syst. Av.* Paradisea, sp. 2; *Gray, Ill. Ind. Zool.* ii. pl.; *Less. Ois. Parad. Syn.* p. 3; *id. H. N. Ois. Parad.* pls. 1–5; *Gray, Hand-l. B.* ii. p. 16; *Elliot, Monogr. Parad.* pl. iii.; *Salvad. Ann. Mus. Civic. Genov.* vii. p. 781, ix. p. 191.
Paradisea papuana, *Bechst. Kurze Uebers.* p. 131; *Gray, Gen. B.* ii. p. 323; *Bp. Consp.* i. p. 413; *Gray, P. Z. S.* 1859, p. 157; *Wall. Ibis,* 1859, p. 111, 1861, p. 287; *Schl. J. f. O.* 1861, p. 385; *Scl. P. Z. S.* 1862, p. 123; *Wall. t. c.* p. 160; *Rosenb. J. f. O.* 1864, p. 129; *Schl. N. T. D.* iv. pp. 17, 49.
Paradisea bartletti, *Goodwin, P. Z. S.* 1860, p. 244.

Adult male. A narrow frontal band, lores, cheeks, and throat velvety green, slightly metallic, the forehead and chin as well as a spot at the base of the lower mandible appearing velvety black; crown of head and hind neck, as well as the sides of the latter, converging towards the lower throat, bright golden straw-colour, all the plumes velvety in texture; mantle and scapulars dull ochraceous straw-colour, all the feathers ruddy at the base; wing-coverts maroon-brown, the least ones washed with ochraceous straw-colour, the median and greater series tipped with golden straw-colour, forming a double bar across the wing-quills, which are maroon-brown, as also the lower back, rump, upper tail-coverts, and tail, the two centre tail-feathers produced into two long thread-like shafts about 18 inches in length; under surface of body from the throat downwards to the under tail-coverts maroon-brown; from the flanks are developed two large tufts of plumes, bright yellow for their basal half, white for the terminal half, at the base of the flank-tufts a few rigid blood-red plumes; under wing-coverts and inner lining of quills maroon-brown, like the breast; bill leaden grey; feet black. Total length 14·5 inches, culmen 1·35, wing 7·3, tail 6·2, tarsus 2.

Adult female. Wants the long flank-plumes and the metallic green about the forehead and throat which are seen in the male. General colour above, including the wings and tail, maroon-brown; head, sides of face, and throat maroon-brown, darker than the back, the hind neck and mantle dull ochraceous straw-colour, with which also the wing-coverts are washed; the whole of the under surface from the throat downwards silky white; thighs pale maroon-brown, the sides of the breast and flanks also slightly washed with maroon, the under wing-coverts white, those near the outer edge of the wing maroon-brown. Total length 13·5 inches, culmen 1·35, wing 6, tail 4·9, tarsus 1·6.

Young male. At first resembling the old female, and gaining the adult plumage only very gradually. Specimens in the Museum show that, by the time the yellow head and green throat are assumed, the middle tail-feathers commence to elongate and become narrower, the maroon-brown on the breast gradually extends over the whole of the under surface; but even when it has overspread the whole of the breast the long flank-plumes are not developed, although the two wire-like centre tail-feathers are seen to their full extent.

Hab. New Guinea and neighbouring islands.

a. ♂ ad. sk.	Sorong, New Guinea.	A. R. Wallace, Esq. [C.].
b, c. ♂ ♀ ad. sk.	Dorey, New Guinea.	A. R. Wallace, Esq. [C.].
d, e, f. ♂ juv. sk.	Dorey, New Guinea.	A. R. Wallace, Esq. [C.].
g. ♀ ad. sk.	Mysol.	A. R. Wallace, Esq. [C.].
h. ♂ vix ad. sk.		Zoological Society. (Type of *P. bartletti.*)

3. Paradisea raggiana.

Paradisea raggiana, *Scl. P. Z. S.* 1873, p. 559; *Elliot, Monogr. Parad.* pl. 2; *Salvad. & D'Albert. Ann. Mus. Civic. Genov.* vii. p. 829, ix. p. 191; *Gould, B. N. Guinea,* part iv.

Adult male. Head and neck all round straw-buff, the feathers close-set and of a velvety texture; a frontal band and cheeks, ear-coverts, and throat bright metallic green, the feathers velvety to the touch, this green patch leaving a narrow band of straw-buff plumes across the lower throat; forehead appearing black in the light; chin velvety greenish black, contrasting with the throat; entire back ruddy chestnut-brown, the wings slightly paler than the back and not so glossy, the shafts of the quills bright reddish, the least wing-coverts broadly edged with straw-buff; tail ruddy brown, with red shafts like the quills, the two centre feathers elongated into two long thread-like shafts 20 inches in length; fore neck and chest velvety purplish brown, forming a small shield; rest of under surface ruddy brown, rather paler on the abdomen and under tail-coverts, the flank-feathers elongated into two large tufts of plumes of a brilliant crimson, the ends of these plumes paler and whity brown, the barbules being lax near the end of these long plumes; under wing-coverts and inner lining of quills pale ruddy brown; "bill pearly grey; feet reddish brown; iris yellow" (*D'Albertis*). Total length 13·5 inches, culmen 1·5, wing 6·9, tail 5·4, tarsus 1·85.

Adult female. General colour above dull ruddy brown, including the wings and tail; hinder part of crown, hind neck, and sides of the latter dull straw-buff; fore part of crown-feathers round the eye, sides of face, throat, and breast deep purplish brown; rest of under surface very pale ruddy brown, including the thighs and under wing- and tail-coverts; bill pearly grey; feet plumbeous; iris yellow. Total length 12·5 inches, culmen 1·4, wing 6·15, tail 5, tarsus 1·5.

Young male. Similar to the adult female, but of a much richer brown. A male shot on Mount Epa on the 23rd of April begins to show indications of the approaching bright plumage on the head and throat.

Hab. South-eastern New Guinea.

a. ♂ ad. sk.	S.E. New Guinea, April 13, 1875.	Signor A. D'Albertis [C.].
b. ♀ ad. sk.	S.E. New Guinea, May 1, 1875.	Signor A. D'Albertis [C.].
d. ♂ juv. sk.	Mount Epa, N. G. April 24, 1875.	Signor A. D'Albertis [C.].

4. Paradisea sanguinea.

Le Paradis rouge, *Audeb. & Vieill. Ois. Dor.* ii. p. 14, pl. 3.
L'Oiseau de Paradis rouge, *Levaill. Ois. Parad.* i. pl. 6.
Paradisea sanguinea, *Shaw, Gen. Zool.* vii. p. 487, pl. 59; *Gray, Hand-l. B.* ii. p. 16; *Elliot, Monogr. Parad.* pl. iv.; *Salvad. Ann. Mus. Civic. Genov.* vii. p. 782.
Paradisea rubra, *Vieill. Gal. Ois.* i. p. 152, pl. 99; *Less. Voy. Coquille, Zool.* p. 660, pl. 27; *Wagler, Syst. Av.* Paradisea, sp. 3; *Less. Traité,* p. 336, pl. 3; *id. Ois. Parad. Syn.* p. 8; *id. Hist. Nat.* p. 160, pls. 7, 8; *Gray, Gen. B.* ii. p. 323, pl. 79; *Bp. Consp.* i. p. 413; *Wall. Ibis,* 1859, p. 111, 1861, p. 287; *id. P. Z. S.* 1862, p. 160; *Schl. J. f. O.* 1861, p. 385; *id. N. T. D.* ii. p. 249; *id. Mus. P.-B.* Coraces, p. 84.
Uranornis rubra, *Salvad. Ann. Mus. Civic. Genov.* ix. p. 191.

Adult male. Fore part of crown enclosing the eye, sides of face, and entire throat brilliant metallic green, the plumes of a velvety texture and elevated into a small tuft of raised feathers above the eye; lores, plumes in front of the eye, and chin velvety black, with an oily green lustre; hinder part of head, nape, hind neck, sides of neck, as well as the fore neck (forming a jugular band), and the entire mantle bright golden straw-colour, deepening into orange-brown on the sides of the mantle, scapulars, and middle of the back, the latter being deep ruddy brown shaded with orange; rump straw-yellow; upper tail-coverts and tail ruddy chestnut-brown, the two centre feathers consisting of two long shafts 21 inches in length, of a metallic horny structure; wing-coverts golden straw-colour like the back; the bastard wing, primary-coverts, and quills ruddy chestnut-brown, with reddish shafts to the feathers; breast deep purplish chestnut, the yellow on the fore neck running a little way down on each side of the upper breast, and forming a kind of incipient shield, as all the plumes are stiff in texture; abdomen and under tail-coverts rich chestnut-brown, but not so deeply coloured as the breast; the flank-plumes elongated into two immense tufts of bright crimson feathers hoary white at their tips, where the plumes become more lax; under wing-coverts deep chestnut-brown. Total length 13 inches, culmen 1·35, wing 6·65, tail 5·2, tarsus 1·7.

Adult female. Fore part of head, sides of face, and entire throat rich purplish brown; hinder part of head, neck all round, and mantle straw-yellow, deepening into orange, the wing-coverts also deep straw-yellow; remainder of the upper surface, including the back, wings, and tail, ruddy chestnut-brown, as also the under surface of the body from below the fore neck, which forms a band of straw-yellow somewhat descending on the sides of the neck. Total length 12·5 inches, culmen 1·3, wing 6·35, tail 5, tarsus 1·55.

Young male. At first exactly similar to the female. It is apparently a long time before the adult plumage is gained; and specimens are often seen with the metallic velvety plumes of the head completely assumed, as well as the yellow head and neck; the long middle tail-feathers are at first spatulated, the shaft hard, but not particularly horny. All the webs of these feathers gradually disappear, leaving a remarkable hard horny shaft, the latter being sometimes perfectly formed, and yet retaining at its tip the feathered spatula of the younger stage.

Hab. Islands of Waigiou, Ghemien, and Batanta.

a, b.	♂ ad. sk.	Waigiou.	A. R. Wallace, Esq. [C.].
c–i.	♂ imm. sk.	Waigiou.	A. R. Wallace, Esq. [C.].
k, l.	♀ ad. sk.	Waigiou.	A. R. Wallace, Esq. [C.].

8. CICINNURUS.

Type.
Cicinnurus, *Vieill. N. Dict. d'Hist. Nat.* xix. p. 247 (1818).. C. regius.

Range. New Guinea and Salwatti, Aru Islands, Mysol, Jobi.

1. Cicinnurus regius.

The supposed king of the Greater Birds of Paradise, *Edwards, Birds,* iii. pl. 111.

Le Petit Oiseau de Paradise, *Briss. Orn.* ii. p. 136, pl. 13.

Paradisea regia, *Linn. S. N.* i. p. 166 (*ex Edwards*); *Shaw, Gen. Zool.* vii. p. 497, pl. 67; *Less. Voy. Coquille,* i. p. 658, pl. 26; *Wagler, Syst. Av.* Paradisea, sp. 7; *Gray, Gen. B.* ii. p. 323; *Schl. Hand-l. Dierk.* i. p. 332, pl. iv. fig. 46; *id. J. f. O.* 1861, p. 385; *id. Mus. P.-B.* Coraces, p. 88; *id. N. T. D.* iv. pp. 17, 49; *Gray, P. Z. S.* 1858, p. 181, 1861, p. 436; *id. Hand-l. B.* ii. p. 16; *Wall. Ibis,* 1859, p. 111; *id. P. Z. S.* 1862, p. 160; *id. Malay Arch.* ii. p. 132; *Wagler, Zool. Gart.* 1873, p. 10.

Le Roi des Oiseaux de Paradis, *Sonn. Voy. N. Guin.* i. p. 156, pl. 95.

Le Manucode, *Buff. H. N. Ois.* iii. p. 163, pl. 13; *id. Pl. Enl.* p. 192, pl. 496; *Audeb. & Vieill. Ois. Dor.* ii. p. 16, pl. 5; *Levaill. H. N. Ois. Parad.* i. pls. 7, 8.

Cicinnurus spinturnix, *Less. Ois. Parad. Syn.* p. 14; *id. H. N. Ois. Parad.* p. 182, pls. 16–18.

Cicinnurus regius, *Vieill. N. Dict. d'Hist. Nat.* xix. p. 247; *id. Gal. Ois.* i. p. 146, pl. xcvi.; *Less. Traité,* p. 338; *Bp. Consp.* i. p. 413; *Wall. Ibis,* 1861, p. 287; *Rosenb. J. f. O.* 1864, p. 130; *Elliot, Monogr. Parad.* pl. xvi.; *Salvad. Ann. Mus. Civic. Genov.* vii. p. 783; *id. & D'Albert. t. c.* p. 832; *Salvad. op. cit.* ix. p. 192; *Gould, B. N. Guin.* part iii.

Adult male. General colour above brilliant metallic crimson, slightly orange under certain lights, and more particularly so in the frontal plumes; above the eye a spot of green velvety feathers, appearing black in most lights; sides of head and sides of neck crimson like the back; entire throat and fore neck rich purplish crimson, the lower feathers tipped with orange-buff, forming a band across the fore neck, followed by a small pectoral band of rich metallic green, while from each side of the breast springs a tuft of long plumes ashy brown in colour, tipped with metallic green, each feather with a subterminal line of buff and a second narrower one of reddish brown; remainder of under surface, including the under wing- and tail-coverts, pure white; scapulars and wings crimson, like the back; the quills orange-brown, externally washed with crimson, the innermost secondaries being entirely of the latter colour; tail ashy brown, the feathers margined with crimson or orange-red, the whole tail hidden with the long tail-coverts; two centre feathers elongated into a thread-like wire, with a curved disk of metallic green at the tip. Total length 6·6 inches, culmen 0·8, wing 3·75, tail 1·6, middle tail-feathers 6·5, tarsus 1·1.

Adult female. Dissimilar to the male. General colour above uniform brown, with a slight olive or golden shade, more particularly on the head; least and median wing-coverts uniform with the back; greater coverts and quills darker brown, externally washed with reddish orange; tail brown, with a slight golden gloss; loral plumes, feathers round the eye, sides of face, and throat dusky, varied with minute straw-coloured mesial streaks; rest of under surface pale fulvous, washed with golden buff on the fore neck, all transversely barred with dusky brown; under wing-coverts and axillars rufous obscurely barred with brown, the edge of the wing golden orange. Total length 7·2 inches, wing 3·9, tail 2·5, tarsus 1·05.

Young male. Similar to the adult female at first, but with a dull crimson shade on the wing. When the crimson plumage of the adult is nearly complete the two central feathers of the tail are put on, the shafts being feathered the whole way, and the ends of the feathers having the web entire and curved round. From the excellent series in the Museum it would appear that the curve is gradually continued upon the feather itself, and that when the twist is complete the metallic green colour is assumed without a change of feather. One young male before me has one racket brown, the other metallic green.

Hab. The same as that of the genus.

a, b. Ad. sk.	[New Guinea.]	Native skins. Purchased.
c–f. ♂ and ♀ ad., ♂ juv. st.	Aru Islands.	A. R. Wallace, Esq. [C.].
g. ♀ ad. sk.	Aru Islands.	A. R. Wallace, Esq. [C.].
h, i. ♂ juv. sk.	Aru Islands.	A. R. Wallace, Esq. [C.].
k.–q. ♂ imm. sk.	Aru Islands.	Mr. Cockerell [C.].
r, s. ♂ ♀ ad. sk.	Mysol.	A. R. Wallace, Esq. [C.].
t. ♂ ad. sk.	New Guinea.	A. R. Wallace, Esq. [C.].

9. RHIPIDORNIS.

Rhipidornis, *Salvad. Ann. Mus. Civic. Genov.* ix. p. 192.

Range. Confined to Waigiou.

1. Rhipidornis respublica.

Lophorhina respublica, *Bp. C. R.* xxx. p. 131.
Diphyllodes respublica, *Bp. C. R.* xxx. p. 291; *id. Consp.* i. p. 413.
Diphyllodes gulielmi iii., *Meyer, P. Z. S.* 1875, p. 31 (*ex Musschenbr. MS.*); *id. Mitth. zool. Mus. Dresden,* 1875, p. 4, pl. 1; *Gould, B. N. Guin.* part ii.
Paradisea gulielmi iii., *Van Musschenbr. Zool. Gart.* 1875, p. 29.
Rhipidornis gulielmi tertii, *Salvad. Ann. Mus. Civic. Genov.* ix. p. 192.

Adult male. General colour fiery orange, the nuchal shield reddish orange, as also the sides of the crown; top of head olivaceous brown; mantle and lower part of back red; wings fiery orange; tail-feathers dark brown, the feathers externally orange-brown, the two centre feathers produced and ending in a narrow recurved plume of bright green; throat and sides of neck purplish brown; breast-shield rich green; the shield separated from the throat by a narrow line of brilliant metallic emerald-green, the entire shield fringed on the outer edge with emerald-green; on each side of the breast spring two tufts in the shape of a fan, composed of purple plumes tipped with emerald-green; rest of under surface purple, the vent and under tail-coverts white.

Obs. The above description is taken from Mr. Gould's plate in the 'Birds of New Guinea.'

Adult female. Whole upperside olive-brown; chin, throat, breast, belly, abdomen, and under wing-coverts light yellow with fine brown stripes; each feather has several such light and dark markings, the lines getting smaller towards the upper part of the body; upperside of the wings brownish, the secondaries and tertiaries margined with yellow on the outer webs; underside silver-grey, the basal portions of the inner webs cream-coloured; upper part of shafts brown, lower part whitish; underside of tail changing into grey, outer web lighter; lower parts of shafts white, upper part brown. (*Meyer, in Gould's 'Birds of New Guinea.'*)

Hab. Mountains of Eastern Waigiou.

Obs. Respecting the synonymy of this species and *Schlegelia wilsoni*, *vide infrà*, p. 176.

10. DIPHYLLODES.

	Type.
Diphyllodes, *Less. Ois. Parad. Syn.* p. 16 (1835)	D. magnifica.
Cricocercus, *Gloger, Handb. Naturg.* p. 345 (1842)......	D. magnifica.

Range. New Guinea and Salwatti, Waigiou, Mysol, Jobi.

Key to the Species.

a. Secondaries and margins of primaries ochraceous.. *magnifica,* p. 174.
b. Secondaries and margins of primaries golden yellow *chrysoptera,* p. 175.

1. Diphyllodes magnifica.

Oiseau de Paradis de la Nouvelle Guinée dit le Magnifique, *Montb. Pl. Enl.* 631; *Sonn. Voy. N. Guin.* p. 163, pl. 98; *Levaill. H. N. Ois. Parad.* i. pls. 9, 10.
Magnificent Paradise-bird, *Lath. Gen. Syn.* ii. p. 477, pl. 19.
Paradisea magnifica, *Penn. in Forst. Ind. Zool.* p. 40; *Gm. S. N.* i. p. 401; *Audeb. et Vieill. Ois. Dor.* ii. p. 15, pl. 4; *Shaw, Gen. Zool.* vii. p. 492, pl. 62; *Wagler, Syst. Av.* Paradisea, sp. 4; *Less. Traité*, p. 338; *Wall. P. Z. S.* 1862, p. 160.
Paradisea speciosa, *Bodd. Tabl. Pl. Enl.* p. 38; *Gray, Gen. B.* ii. p. 323; *Schl. Mus. P.-B.* Coraces, p. 86; *id. N. T. D.* iv. p. 171; *Gray, Hand-l. B.* ii. p. 16.
Paradisea cirrhata, *Lath. Ind. Orn.* i. p. 195.
Diphyllodes seleucides, *Less. Ois. Parad. Syn.* p. 16, pls. 19, 20.
Diphyllodes magnifica, *Gray, List Gen. B.* 1841, p. 53; *Bp. Consp.* i. p. 413; *Wall. Ibis*, 1861, p. 287; *Rosenb. J. f. O.* 1864, p. 130; *Salvad. Ann. Mus. Civic. Genov.* ix. p. 192.
Diphyllodes speciosa, *Wall. Malay Arch.* ii. p. 247; *Elliot, Monogr. Parad.* pl. 12; *Scl. P. Z. S.* 1873, p. 697; *Wagler, J. f. O.* 1873, p. 11, pl. 1. map 5; *Gould, B. N. Guin.* pt. ii.; *Beccari, Ann. Mus. Civic. Genov.* vii. p. 712; *Salvad. t. c.* p. 783; *Sclater, Ibis*, 1876, p. 251.

Adult male. Crown of head covered with close-set velvety brown feathers, deepening into dull orange on each side of the crown; forehead slightly washed with metallic green; a spot at the base of the bill and another above the eye velvety black, the latter slightly glossed in front with metallic green; from the nape a little fan-shaped tuft of light-brown feathers, each with a subterminal spot of darker brown; from the hind neck a broad fan-shaped shield of light yellow, flanked on each side by two long tufts of plumes brown in colour, with a subterminal mark of darker brown on each, the inner ones inclining to orange; feathers of the back metallic chestnut; lower back and rump orange, the upper tail-coverts and tail dark brown, the two centre feathers produced and curved back on themselves, metallic steel-green; wing-coverts dull ochraceous, the primary-coverts and quills dark brown, externally washed with ochraceous straw-colour, the whole of the inner secondaries bright ochraceous straw-colour; cheeks and throat metallic olive-green; the under surface of the body from the throat, occupying the whole of the breast in the form of a shield, rich green, the external feathers tipped with metallic steel-blue, forming a fringe; down the centre of the breast a broad band of metallic emerald-green, formed by the tips to the feathers being of this colour; rest of under surface of body purplish black; under wing-coverts deep ochraceous, as also the lining of the quills. Total length 6·8 inches, culmen 0·95, wing 4·55, tail 1·95, middle tail-feathers 9·5, tarsus 1·35.

Adult female. General colour dark olive-brown, the crown and sides of head rather inclining to ochraceous brown; wings and tail dark brown, the former washed externally with olivaceous; tail clearer brown; cheeks and under surface of body fulvescent, the throat and cheeks whiter, the whole barred across very distinctly with black. Total length 7·8 inches, culmen 0·95, wing 4·45, tail 2·6, tarsus 1·35,

Hab. New Guinea and Mysol.

a, b. ♂ ♀ ad. sk.	New Guinea.	Purchased.
c, d. ♂ ad. sk.	New Guinea.	Native skins.

2. Diphyllodes chrysoptera.

Diphyllodes speciosa, var. chrysoptera, *Elliot, Monogr. Parad.* pl. 13.
Diphyllodes chrysoptera, *Gould, B. N. Guin.* part ii.; *Beccari, Ann. Mus. Civic.* vii. p. 710; *Salvad. t. c.* p. 971, ix. p. 192.

Similar to *D. magnifica*, but having the secondaries and edgings to the primaries golden.

Hab. Island of Jobi.

11. SCHLEGELIA.

Schlegelia, *Bernst. N. T. D.* iii. p. 4, pl. 7 (1864)........ Type. S. calva.

Range. Islands of Waigiou and Batanta.

1. Schlegelia wilsoni.

Paradisea wilsoni, *Cass. Journ. Acad. N. Sci. Philad.* ii. p. 133, pl. 15; *Gray, P. Z. S.* 1861, p. 436; *Sclater, P. Z. S.* 1865, p. 465; *Schl. N. T. D.* iii. p. 249; *id. Mus. P.-B.* Coraces, p. 87; *Gray, Hand-l. B.* ii. p. 16; *Wagner, J. f. O.* 1873, p. 11, Cart. vii.
Diphyllodes respublica (nec *Bp.*) *Sclater, P. Z. S.* 1857, p. 6; *Rosenb. J. f. O.* 1864, p. 130; *Elliot, Monogr. Parad.* pl. 14; *Gould, B. N. Guin.* part iii.
Diphyllodes wilsoni, *Wall. P. Z. S.* 1862, p. 160; *Newt. Ibis,* 1865, p. 343; *Wall. Malay Arch.* ii. p. 248; *Beccari, Ann. Mus. Civ. Genov.* vii. p. 713; *Sclater, Ibis,* 1876, p. 251.
Paradisea calva, *Schl. N. T. D.* ii. p. 1.
Schlegelia calva, *Bernst. N. T. D.* iii. p. 4, pl. 7.
Schlegelia respublica, *Salvad. Ann. Mus. Civic. Genov.* ix. p. 192.

Adult male. Head bare, of a bright blue colour in life; nasal plumes, lores and forehead, cheeks, ear-coverts, chin and upper throat covered with velvety feathers of an oily brown lustre; a narrow line of the same feathers drawn from the centre of the frontal patch down the back of the head to the hind neck, and transversely crossed above the occiput by a narrow line of velvety plumes which run down the side of the hinder skull along the sides of the neck to the hind neck; behind the ear-coverts a small bare patch; sides of neck and a band above the mantle and running down the sides of the latter metallic oily green; mantle bright yellow; remainder of the back bright metallic crimson, the lower plumes tipped with black, forming a terminal fringe; on each side of the back a border of greenish black, formed by the outer webs of the crimson plumes, which are somewhat curved inwards; rump brown, washed with reddish; upper tail-coverts olivaceous brown; tail brown, the two centre feathers elongate and narrow, after crossing each other twice, ending in a ram's-horn curve; scapulars, as well as the lesser and median wing-coverts, brown, all more or less broadly edged with reddish ochre, shaded, under certain lights, with metallic oil-green; greater coverts and quills brown, narrowly edged with metallic crimson like the back, these margins becoming broader on the innermost both of the coverts and of the secondaries, which are nearly entirely fiery crimson; lower throat and entire breast forming a shield of metallic grass-green, all the jugular feathers having a mesial spot of brilliant metallic blue, which is represented by a similar spot of brilliant green on the breast-plumes, which become developed into long square-tipped plumes on the lower breast, forming a kind of pectoral ruff, the lateral plumes of which are burnished copper, ending in metallic greenish blue of extreme brilliancy, with a narrow subterminal burnished line of brighter copper; remainder of under surface purplish brown, including the under wing- and tail-coverts; quills brown below, edged with dull orange-buff on the inner web. Total length 7·2 inches, culmen 0·95, wing 3·85, tail 2, elongated feathers extending 2·15 inches beyond, tarsus 1·15.

Adult female. Dissimilar in plumage to the male, but the head bare as in that sex, the feathers on the crown being velvety brown; mantle brown, remainder of the back olive-brown, somewhat shaded with golden, and on the upper tail-coverts with faint red; wings brown, the greater coverts narrowly edged with dull crimson, as also are the quills, the secondaries having a decided golden gloss; tail brown, also slightly shaded with golden; cheeks and under surface of body light yellowish brown, transversely barred with narrow cross lines of blackish brown, a little broader on the flanks and sides of the body; hinder part of cheeks and chin hoary white, minutely speckled with brown cross lines; under wing-coverts and axillaries reddish, barred with dusky brown; quills dark brown below, edged with reddish buff along the inner web. Total length 7 inches, culmen 0·9, wing 3·8, tail 2·4, tarsus 1·1.

Obs. Dr. Sclater (P. Z. S. 1857, p. 6), after examining Mr. Cassin's type of *S. wilsoni* in Philadelphia, says that it is not only the same as Bonaparte's *Lophorhina respublica*, but that it also served as the type before going to Philadelphia. In these identifications he is followed by Mr. Elliot and by other ornithologists. Bonaparte's description is, indeed, short, consisting of the words "fasciculo e plumis elongatis nuchæ rubris." He himself corrects the mistake of placing it in the genus *Lophorhina*, and in relegating it to *Diphyllodes* he repeats the above characters: these he apparently considered sufficient on account of the very distinct nature of the bird; and I contend that *Rhipidornis respublica* (*vide suprà*) is the only species which answers to this diagnosis. Consequently this name takes precedence over that of *gulielmi tertii*.

Hab. Islands of Waigiou and Batanta.

a, b. ♂ ♀ ad. sk.	Waigiou.	Dr. Bernstein [C.].

12. PAROTIA.

Type.
Parotia, *Vieill. N. Dict. d'Hist. Nat.* xxxi. p. 160 (1819) P. sefilata.
Otostylus, *Gloger, Handb. Naturg.* p. 344 (1842) ...... P. sefilata.

Range. Confined to New Guinea.

1. Parotia sefilata.

Manucode à six filets, *Briss. Orn.* iii. p. 17.
Le Sifilet de la Nouvelle Guinée, *Montb. Pl. Enl.* iii. p. 198, pl. 633.
Paradisea sefilata, *Penn. in Forst. Ind. Zool.* p. 40.
L'Oiseau de Paradis à gorge dorée, *Sonn. Voy. Nouv. Guin.* p. 158, pl. 97.
Gold-breasted Bird of Paradise, *Lath. Gen. Syn.* ii. p. 481; *id. Gen. Hist.* iii. p. 194, pl. xlvii.
Paradisea sexpennis, *Bodd. Tabl. Pl. Enl.* p. 38; *Salvad, Ann. Mus. Civic. Genov.* ix. p. 190.
Paradisea penicillata, *Scop. Del. Faun. et Flor. Insubr.* ii. p. 88 (*ex Sonn.*).
Paradisea aurea, *Gm. S. N.* i. p. 402 (*ex Lath.*).
Paradisea sexsetacea, *Lath. Ind. Orn.* i. p. 196; *Shaw, Gen. Zool.* vii. p. 496, pl. 66; *Bonn. et Vieill. Enc. Méth.* iii. p. 909, pl. cxliv. fig. 1; *Wagler, Syst. Av.* Paradisea, p. 6; *Less. Traité*, p. 337.
Le Sifilet, *Audeb. et Vieill. Ois. Dor.* ii. p. 18, pl. 6; *Levaill. Ois. Parad.* i. pls. 12, 13.
Parotia sexsetacea, *Vieill. N. Dict. d'Hist. Nat.* xxxi. p. 160; *id. Gal. Ois.* i. p. 148, pl. xcvii.; *Less. Traité*, p. 337; *id. Ois. Parad. Syn.* p. 10; *id. Hist. Nat. Ois. Parad.* p. 172, pls. 10, 11, 11 bis, 12.
Paradisea aurea, *Bp. Consp.* i. p. 414.
Parotia aurea, *Gray, List Gen. B.* 1840, p. 38; *Wall. Ibis*, 1861, p. 287.
Paradisea sexpennis, *Gray, P. Z. S.* 1861, p. 436; *Schl. J. f. O.* 1861, p. 385; *id. Mus. P.-B.* Coraces, p. 92; *id. N. T. D.* iv. p. 50; *Gray, Hand-l. B.* ii. p. 16.
Parotia sexpennis, *Wall. Malay Arch.* ii. p. 250; *Elliot, Monogr. Parad.* pl. 10; *Salvad. Ann. Mus. Civic. Genov.* vii. p.782; *Gould, B. N. Guin.* part 1; *Beccari, Ann. Mus. Genov.* vii. p. 712; *Scl. Ibis*, 1876, p. 250.

Adult male. General colour above rich purplish velvety black, including the wings and tail; plumes of head and neck also close-set and velvety, purplish black like the back; above the ear-coverts a tuft of elongated silky hair-like plumes, from among the anterior ones of which spring three thread-like shafts on either side of the head, each ornamented with an ovate racket of velvety black; frontal plumes purplish, the feathers tipped with shining white, which forms a band across the forehead, these stiff plumes being capable of depression forwards nearly to the tip of the bill; across the upper crown a band of brilliant metallic plumes, the centre ones bright green, with a double subterminal bar of purple and blue, the outermost ones somewhat more shining with metallic purple and lilac; throat velvety purplish black, the lower throat and fore neck forming a brilliant metallic shield, composed of golden-copper feathers, shining with steel-green, blue, and purple, and relieved by a median spot of velvety black; rest of under surface of body glossy black, on each side of the breast the plumes elongated into two lax velvety shields; bill and legs black; "iris light blue, with a circle of pale yellowish green colour" (*D'Albertis*). Total length 12·5 inches, culmen 1·3, wing 6·1, tail 5·25, tarsus 2·1.

Hab. The same as that of the genus.

a. Ad. st.	New Guinea.	Purchased.
b. Ad. sk.	New Guinea.	Purchased.

13. SEMIOPTERA.

Type.
Semioptera, *Gray, P. Z. S.* 1859, p. 130 . . . . S. wallacii.

Range. Confined to the islands of Batchian and Halmahéra or Gilolo.

1. Semioptera wallacii.

Paradisea wallacii, *Gray, P. Z. S.* 1859, p. 130; *Schl. Mus. P.-B.* Coraces, p. 385.
Semioptera wallacii, *Wall. Ibis*, 1859, p. 210, note, 1860, p. 61; *Scl. Ibis*, 1860, p. 26, pl. 2; *Gray, P. Z. S.* 1860, p. 355; *Gould, B. Austr. Suppl.* pl. 3; *Rosenb. J. f. O.* 1864, p. 125; *Wall. Malay Arch.* ii. p. 17; *Elliot, Monogr. Parad.* pl. 18; *Salvad. Ann. Mus. Civic. Genov.* vii. p. 785, ix. p. 190.
Epimachus wallacii, *Gray, Hand-l. B.* i. p. 105.

Adult male. General colour above brown, shaded with purple under certain lights, the shafts whity brown; scapulars dark brown; wings rather paler than the back, especially on the outer coverts, the quills whitish at tip, the shafts also whitish; from the middle of the least wing-coverts two long whitish plumes, ending in an elongate spatula; tail rather lighter brown than the back, with white shafts; nasal plumes pale fulvous brown, slightly erect and recurved; crown pale metallic greyish lilac; lores, sides of face, sides of neck, and hinder part of the latter pale brown, with a purplish lilac gloss, only visible under a change of light; cheeks and throat light brown, mottled with metallic green edgings to the feathers, becoming more uniform on the lower throat, and gradually merging in a triangular shield of green, embracing the lower throat and fore neck, and produced into two long tufts of brilliant metallic green feathers, reaching down each side of the breast to the level of the thighs; chest-feathers light brown, fringed with metallic green; breast brown, washed with metallic green; sides of body, abdomen, thighs, and under wing- and tail-coverts light brown, paler on the thighs; "bill horny olive; feet orange, claws horny; iris deep olive" (*Wallace*). Total length 11·5 inches, culmen 1·65, wing 6·15, tail 3·4, tarsus 1·7.

Adult female. General colour above and below brown, slightly glossed with purple, the quills whitish externally and at tip; nasal plumes and frontal feathers sandy buff; crown of head and nape glossed with dull purplish; sides of face and throat lighter brown than the back; no green gloss on any part of the plumage; pectoral shield and standard plumes of wing absent. Total length 11·6 inches, culmen 1·45, wing 5·55, tail 4, tarsus 1·6.

Obs. Birds from Gilolo are decidedly larger than those from Batchian. The males have longer pennants and pectoral shields; and the females are darker, with a purplish instead of lilac gloss on the head.

Hab. Batchian and Gilolo.

a. ♂ ad. sk.	Batchian.	A. R. Wallace, Esq. [C.]. (Type of species.)
b. ♀ ad. sk.	Batchian.	A. R. Wallace, Esq. [C.].
c. ♂ juv. sk.	Batchian.	A. R. Wallace, Esq. [C.].
d, e. ♂ ad. sk.	E. Gilolo.	A. R. Wallace, Esq. [C.].
f, g. ♀ ad. sk.	E. Gilolo.	A. R. Wallace, Esq. [C.].
h–k. ♂ ♀ ad. st.	Batchian.	A. R. Wallace, Esq. [C.].

14. LOPHORHINA.

Type.

Lophorina, *Vieill. N. Dict. d'Hist. Nat.* xviii. p. 184 (1817). L. superba.

Range. Confined to New Guinea.

1. Lophorhina superba.

L'Oiseau de Paradis de la Nouvelle Guinée, dit le Superbe, *Briss. Orn.* iii. p. 169; *Buff. Pl. Enl.* iii. p. 197, pl. 632.

Paradisea superba, *Penn. in Forster Ind. Zool.* p. 40; *Scop. Del. Faun. et Flor. Insubr.* ii. p. 88; *Gm. S. N.* i. p. 402; *Shaw, Gen. Zool.* vii. p. 494, pls. 63–65; *id. & Nodd. Nat. Misc.* xxiv. pl. 1021; *Wagler, Syst. Av.* Paradisea, sp. 5; *Wall. Ibis,* 1859, p. 111; *Schl. J. f. O.* 1861, p. 385.

Oiseau de Paradis à gorge violette, *Sonn. Voy. Nouv. Guin.* p. 157, pl. 96.

Superb Bird of Paradise, *Lath. Gen. Syn.* p. 479.

Paradisea atra, *Bodd. Tabl. Pl. Enl.* p. 38.

Le Superbe, *Audeb. et Vieill. Ois. Dor.* ii. pl. vii.; *Levaill. H. N. Ois. Parad.* i. pls. 14, 15.

Paradisea furcata, *Bechst. Kurze Uebers.* p. 132.

Lophorhina superba, *Vieill. N. Dict. d'Hist. Nat.* xviii. p. 184; *id. Gal. Ois.* p. 149, pl. xcviii.; *Less. Traité,* p. 337; *id. Ois. Parad. Syn.* p. 12; *id. H. N. Ois. Parad.* pls. 13, 14; *Bp. Consp.* i. p. 414; *Wall. Ibis,* 1861, p. 287, 1862, p. 100; *Salvad. Ann. Mus. Civic. Genov.* ix. p. 190.

Epimachus ater, *Schl. Mus. P.-B.* Coraces, p. 96, note.

Lophorhina atra, *Wall. Malay Arch.* ii. p. 249; *Elliot, Monogr. Parad.* pl. 11; *Salvad. Ann. Mus. Civic. Genov.* vii. p. 783; *Beccari, t. c.* p. 712; *Scl. Ibis,* 1876, p. 251.

Adult male. General colour above velvety black, somewhat glossed with bronzy purple; mantle produced into an elevated shield, composed of velvety black plumes, glossed under certain lights with bronze; wing-coverts velvety black, rather more distinctly glossed with purple than the back; quills and tail-feathers deep black, glossed with steel-blue; lores and nasal plumes forming an elevated crest of purplish black feathers; crown of head, nape, and hind neck spangled with metallic steel-coloured feathers, each of which has a subterminal bar of purple; sides of face, sides of neck, and entire throat deep coppery bronze; on the fore neck and breast a pectoral shield of bright metallic green plumes, most of which have a narrow edging of copper; remainder of under surface purplish black. Total length 9 inches, culmen 1·15, wing 4·55, tail 3·6.

Adult female. Above deep chocolate-brown, the feathers of the top and sides of the head blackish brown; over the eye a few white-spotted plumes; wing-coverts and quills blackish brown, externally reddish; tail brown, externally dull rufous brown; throat white, all the feathers being black tipped with white; rest of the under surface buffy white, inclining to rufous on the flanks and under tail-coverts, the whole under surface barred across with dull brown; under wing-coverts rufous, barred across with brown. Total length 8·8 inches, culmen 1·05, wing 5·1, tail 4, tarsus 1·3.

Hab. New Guinea.

a. ♂ ad. st.	New Guinea.	
b. ♀ ad. sk.	Atam, Dec. 10, 1875.	Purchased.

15. PHONYGAMA.

	Type.
Phonygama, *Less. Voy. Coquille*, i. p. 636 (1826)......	P. keraudreni.
Chalybeus, *Cuvier, Règn. Anim.* i. p. 354 (1829)......	P. keraudreni.

Range. New Guinea and North Australia.

Key to the Species.

a. Wings externally purple.
  a'. General colour above and below burnished steel-*blue*, verging somewhat towards a purple tinge .... *keraudreni*, p. 180.
  b'. General colour above and below burnished *green* .... *jamesii*, p. 181.
b. Wings externally green, like the rest of the upper and under surface........................ *gouldi*, p. 181.

1. Phonygama keraudreni.

Phonygama keraudreni, *Less. Voy. Coquille, Ois.* i. p. 636; *id. Man. d'Orn.* i. p. 141; *Gray, Gen. B.* ii. p. 303; *Bp. Consp.* i. p. 368; *Rosenb. J. f. O.* 1864, p. 123.
Barita keraudreni, *Less. Voy. Coquille, Ois. Atlas*, pl. 13.
Chalybeus cornutus, *Cuv. Règn. Anim.* i. p. 354 (1829); *Gould in Voy. Rattlesnake*, ii. p. 357.
Phonygama lessonia, *Swains. Classif. B.* ii. p. 264.
Manucodia keraudreni, *Gray, P. Z. S.* 1859, p. 158, 1861, p. 436; *id. Hand-l. B.* ii. p. 17; *Elliot, Monogr. Parad.* pl. viii.; *Salvad. Ann. Mus. Civic. Genov.* vii. p. 781; *id. op. cit.* ix. p. 189; *Scl. Ibis*, 1876, p. 252.
Chalybeus keraudreni, *Schl. Mus. P.-B.* Coraces, p. 120.

Adult male. General colour burnished steel-blue, the feathers of the crown velvety in texture, as well as those of the sides of the face and chin; from the occiput two long tufts of steel-blue feathers extend; the plumes of the hind neck and throat also lanceolate; under surface of body burnished steel-blue, washed with green on the abdomen; wings and tail purple, the edges of the quills steel-blue, the latter black on their inner web; bill and legs black. Total length 13 inches, culmen 1·3, wing 6·35, tail 5·55, tarsus 1·35.

Hab. New Guinea.

a. ♂ ad. sk.	Dorey, N. G.	A. R. Wallace, Esq. [C.].
b, c, d. ♂ ♀ ♂ juv. et ad. st.	Dorey, N. G.	A. R. Wallace, Esq. [C.].
e. ♂ ad. sk.	New Guinea.	A. R. Wallace, Esq. [C.].

2. Phonygama jamesii.

Manucodia keraudreni, *Salvad. Ann. Mus. Civic. Genov.* ix. p. 40 (*nec Less.*).

Adult. General colour above burnished green, with a slight shade of purplish blue here and there on the back; under surface of body burnished green like the upper, with a subterminal lustre of purplish blue on some of the feathers; head and neck all round of a burnished oily green, the plumes of the crown close-set and velvety, those of the neck and throat narrowly lanceolate; from the occiput two long tufts of green feathers; wings and tail purple, the wing-coverts burnished green like the back, the quills and tail black on their inner webs, except the innermost secondaries, which are entirely purple; bill and legs black. Total length 12·3 inches, culmen 1·3, wing 6·2, tail 4·9, tarsus 1·45.

Hab. South-eastern New Guinea.

a. Ad. sk.	Aleya, S.E. New Guinea, April 1876.	Dr. James [C.]. (Type of species.)

3. Phonygama gouldi.

Manucodia keraudreni, *Gould, B. Austr. Suppl.* pl. (*nec Less.*).

Manucodia gouldi, *Gray, P. Z. S.* 1859, p. 158, note; *Gould, Handb. B. Austr.* i. p. 236; *Gray, Hand-l. B.* ii. p. 17; *Masters, Pr. Linn. Soc. N. S. W.* i. p. 50.

Adult. General colour steel-green, of a somewhat oily-green cast on the rump and upper tail-coverts; wings coloured like the back, the coverts and the outer webs of the quills with a slight shade of purplish blue, this colour being also faintly indicated on the interscapulary region; wings black on the inner webs, excepting the innermost secondaries, which are bluish green; tail black, glossed with deep purplish blue, the feathers greenish on their outer edges under certain lights; head green, as also the two long tufts projecting from each side of the occiput; sides of face and throat and underparts green, the abdomen with an oily-green shade, the feathers of the throat pointed and lanceolate, these parts somewhat shaded with steel-blue under certain lights; bill and legs black. Total length 11·5 inches, culmen 1·25, wing 6·15, tail 5·15, tarsus 1·55.

Hab. North-eastern Australia.

a. Ad. sk.	Port Albany.	Purchased.
b. ♂ ad. st.	Cape York.	J. Macgillivray, Esq. [C.]. (Type of species.)

16. MANUCODIA.

Type.
Manucodia, *Bodd. Tabl. Pl. Enl.* p. 39 (1783) .... M. chalybata.

Range. N. Australia and adjacent islands of the Papuan region.

Key to the Species.

a. Head covered with velvety plumes, but not having any elongated occipital tufts.
  a'. Plumes of the throat and fore neck oily green, crinkled and recurved.
    a''. Plumes of the head velvety, but not curled; bill moderately long, 1·65 inch from gape; centre tail-feathers ordinary .... *chalybata*, p. 182.
    b''. Plumes of the head strongly curled; bill very long, 2·2 inches from gape; two centre tail-feathers recurved....................... *comrii*, p. 183.
  b'. Plumes of the throat and fore neck steel-black, the latter not crinkled, but apparently with an edging of velvety black.................... { *atra*, p. 183. *jobiensis*, p. 184. }

1. Manucodia chalybata.

Le Chalybé, *Buff. Pl. Enl.* iii. pl. 634; *Audeb. et Vieill. Ois. Dor.* ii. p. 24, pl. 10; *Levaill. Ois. de Parad.* p. 64, pl. 23.
Paradisea chalybata, *Penn. Faunula Indica in Forst. Zool. Ind.* p. 40.
Blue-green Paradise-bird, *Lath. Gen. Syn.* ii. p. 482.
Manucodia chalybea, *Bodd. Tabl. Pl. Enl.* p. 39; *Elliot, Monogr. Parad.* pl. vi.; *Salvad. Ann. Mus. Civic. Genov.* vii. p. 781.
Paradisea viridis, *Gm. S. N.* i. p. 402.
Paradisea chalybea, *Lath. Ind. Orn.* i. p. 197; *Shaw, Gen. Zool.* vii. p. 504, pl. 71.
Cracticus chalybeus, *Vieill. N. Dict. d'Hist. Nat.* v. p. 355.
Phonygama chalybeus, *Less. Traité*, p. 344.
Phonygama viridis, *Gray, Gen. B.* ii. p. 303; *Bp. Consp.* i. p. 368; *Rosenb. J. f. O.* 1864, p. 122.
Chalybea viridis, *Schl. Mus. P.-B.* Coraces, p. 122; *id. N. T. D.* p. 49.
Manucodia chalybeata, *Salvad. Ann. Mus. Civic. Genov.* vii. p. 968; *id. op. cit.* ix. p. 189.

Adult male. Head purple, the feathers compressed and close-set, the nape slightly washed with steel-greenish, as also the hinder neck and mantle; back rich purple, the feathers of the interscapulary region rather recurved; wings and tail rich purple, the inner webs of the feathers blackish, the outer wing-coverts somewhat shaded with steel-black; sides of the face and sides of the neck deep green, the feathers compressed and velvety like those of the crown; those of the chin, throat, and fore neck extending onto the sides of the neck, crinkled and curled and of an oily-green colour; the rest of the under surface deep purple, the feathers being tipped with this colour, less broadly on the vent and under tail-coverts, a few of the abdominal plumes with a slight greenish reflection; under wing-coverts black, the outer edge of the wing washed with green; bill and legs black. Total length 14·5 inches, culmen 1·65, wing 6·85, tail 5·9, tarsus 1·55.

Obs. The British-Museum series shows that immature birds are much duller than the adult, and greener on the head and on the underparts; the feathers of the throat are oily green, but are not crinkled as in the adult. This peculiar conformation of the throat-plumes is evidently only acquired by age.

Hab. North-western New Guinea.

a, b. ♂ ♀ ad. st.	Dorey, New Guinea.	A. R. Wallace, Esq.
c. ♂ ad. sk.	Dorey, New Guinea.	A. R. Wallace, Esq.
d. Ad. sk.	New Guinea.	Purchased.
e, f, g. Juv. sk.	New Guinea.	Purchased.

2. Manucodia comrii.

Manucodia comrii, *Sclater, P. Z. S.* 1876, p. 459, pl. xlii.; *Salvad. Ann. Mus. Civic. Genov.* ix. p. 189.

Adult (type of species). General colour above metallic steel-green, somewhat shaded with purplish round the eyes, the entire head adorned with a large crest of curled and crinkled feathers, the dorsal feathers also all slightly curled; wings and tail metallic steel-green, with somewhat of a purplish gloss, the two centre tail-feathers recurved at their ends; feathers of the throat, chest, and sides of the neck steel-green, all of them crinkled and recurved, giving a velvety texture to the whole of these parts; remainder of the under surface rich purple, all the feathers recurved, and most of them showing a subterminal gloss of violet or steel-green; thighs and under wing- and tail-coverts entirely steel-green. Total length 17 inches, culmen 2·4, wing 9·7, tail 7·4, tarsus 2. (*Mus. Lord Tweeddale.*)

Hab. Huon Gulf, S.E. New Guinea.

Obs. Dr. Sclater writes (*l. c.*):—"This is by far the finest and largest of the peculiar genus now generally termed *Manucodia* (formerly *Phonygama*) yet discovered. It is immediately distinguishable from *M. chalybeia* and *M. atra* by its much larger size and longer bill, which is deeply sulcated at the nostrils. The characteristic curling of the feathers is extended to a greater degree than in *M. chalybeia*, and pervades the whole of the head and neck. The feathers of the abdomen are black at the base, broadly margined with purple."

3. Manucodia atra.

Phonygama atra, *Lesson, Voy. Coquille,* i. p. 638; *id. Traité,* p. 344; *Gray, Gen. B.* ii. p. 303.

Manucodia atra, *Gray, P. Z. S.* 1859, p. 158, 1861, p. 436; *id. Hand-l. B.* ii. p. 17; *Elliot, Monogr. Parad.* pl. vii.; *Salvad. Ann. Mus. Civic. Genov.* vii. p. 781; *id. & D'Albert. t. c.* p. 829; *Salvad. op. cit.* ix. pp. 40, 189.

Adult male. Head all round covered with compressed velvety plumes, steel-green without any purple reflections; the neck all round greenish like the head, the plumes slightly recurved and glistening with metallic endings; general colour of upper surface steel-black, shaded, according to the light, with metallic reflections of greenish or purple; wings and tail purple, with steel-black shades under certain lights, the quills externally glossed with greenish; under surface of body glossy steel-black, with purplish or greenish reflections, the under wing- and tail-coverts uniform with the breast; bill black; feet blackish; iris red. Total length 16·5 inches, culmen 1·6, wing 7·2, tail 6·9, tarsus 1·75.

Obs. From the following measurements it will be seen that the size varies a good deal; but there is no specific difference in coloration between any of these specimens:—

	Total length.	Culmen.	Wing.	Tail.	Tarsus.
	in.	in.	in.	in.	in.
1. ♂. Dorey (*Wallace*)..	16·5	1·6	7·2	6·9	1·5
2. Ad. Dorey (*Wallace*)..	15·0	1·45	7·25	6·2	1·55
3. Ad. New Guinea......	14·0	1·45	7·0	5·9	1·6
4. ♂ ad. Aru Islands (*Wall.*)	16·5	1·6	7·1	6·5	1·7
5. ♀ ad. Aru Islands (*Wall.*)	16·5	1·45	7·6	6·8	1·75
6. ♂ ad. Mysol (*Wallace*) ..	13·7	1·5	7·1	6·1	1·6
7. ♂ ad. Mysol (*Wallace*) ..	13·7	1·5	7·15	6·4	1·55
8. ♂ ad. Waigiou (*Wallace*)	16·0	1·5	7·15	6·65	1·6

Hab. New Guinea and Salwatti, Aru Islands, Waigiou, Ghemien, Mysol.

a. Ad. sk.	New Guinea.	Purchased.
b. Jun. sk.	New Guinea.	Purchased.
c. ♂ ad. sk.	Dorey, New Guinea.	A. R. Wallace, Esq. [C.].
d, e. ♂ ♀ ad. sk.	Aru Islands.	A. R. Wallace, Esq. [C.].
f, g. ♂ ad. sk.	Mysol.	A. R. Wallace, Esq. [C.].
h. ♂ ad. sk.	Waigiou.	A. R. Wallace, Esq. [C.].

4. Manucodia jobiensis.

Manucodia jobiensis, *Salvad. Ann. Mus. Civic. Genov.* vii. p. 969; *id. op. cit.* ix. p. 189.

Bright green, the wings and tail above bluish violet; the plumes of the head short, angulate, in the form of scales, green, under certain lights blackish velvet margined with green; feathers of the neck also rather short, the hinder ones blackish green, of velvety appearance, margined with bright green; feathers of the fore neck and upper breast having the webs near the edge separated and waved, bright green, under the light marked by a very beautiful transverse band of velvet-black; remainder of the breast, abdomen, and under tail-coverts black, margined with bright green; interscapulary region bright steel-green, each plume furnished with a velvety band of blackish and green; rest of the back steel-green; quills and tail-feathers black below; bill and feet black. Total length about 14·5 inches, culmen 1·55, wing 7·5, tail 6·1, tarsus 1·6. (*Salvadori, l. c.*)

Hab. Island of Jobi.

Obs. This species is certainly very closely allied to *M. chalybata*, but differs in its green head and in other characters pointed out by Count Salvadori (*l. c.*). I quite agree with him that the differences between *M. chalybata* and *M. atra* are not shown at all in the plates in Mr. Elliot's work; and this makes it difficult to compare allied species with them. Count Salvadori has such an immense series of *M. atra* wherewith to compare his *M. jobiensis*, that he has doubtless not separated the species without good grounds for so doing. From the description it seems to be intermediate between *M. atra* and *M. chalybata*.

17. LYCOCORAX.

		Type.
Lycocorax, *Bp. C. R.* xxxvii. p. 829 (1853)	....	L. pyrrhopterus.
Manucodia (pt.) *Gray, Hand-l.* ii. p. 17 (1870)	....	

Range. Molucca Islands: Batchian, Gilolo, Morty Island, Raou, Obi.

Key to the Species.

a. Primaries ashy brown below, rather paler, but not white at base of inner web.	
a'. General colour black, slightly washed with green; secondaries brown	*pyrrhopterus*, p. 185.
b'. General colour dark rifle-green; secondaries black, washed with green	*obiensis*, p. 185.
b. Primaries white for basal half of inner web ..	*morotensis*, p. 186.

1. Lycocorax pyrrhopterus.

Corvus pyrrhopterus, *Bp. Consp.* i. p. 384 (*ex Temm. MS. in Mus. Lugd.*).
Pica pyrrhoptera, *Schl. Bijdr. tot de Dierk.* fol. pt. viii. p. 1, pl. i. (1858).
Lycocorax pyrrhopterus, *Bp. C. R.* xxxvii. p. 829; *Gray, P. Z. S.* 1860, p. 355; *Schl. N. T. D.* iii. p. 191; *id. Mus. P.-B.* Coraces, p. 131; *Finsch, Neu-Guinea*, p. 172.
Manucodia pyrrhopterus, *Gray, Hand-l. B.* ii. p. 17.

Adult female. General colour above and below black, with a slight wash of dull oily green; tail black, with an almost imperceptible wash of green on the outer web of some of the feathers; least wing-coverts resembling the back, the rest of the wing brown, becoming paler and more reddish on the outer primaries, the lower surface of the quills rather ashy brown; bill and feet black. Total length 15·5 inches, culmen 1·75, wing 7·1, tail 6·1, tarsus 1·75.

Hab. Batchian, Gilolo.

a ♀ ad. st.	Batchian.	A. R. Wallace, Esq. [C.].
b. ♀ ad. sk.	Batchian.	A. R. Wallace, Esq. [C.].
c. Ad. sk.	Gilolo.	A. R. Wallace, Esq. [C.].
d. ♀ ad. sk.	Gilolo.	A. R. Wallace, Esq. [C.].

2. Lycocorax obiensis.

Lycocorax obiensis, *Bernst. J. f. O.* 1864, p. 410; *Schl. N. T. D.* iii. p. 192; *id. Mus. P.-B.* Coraces, p. 132.
Manucodia obiensis, *Gray, Hand-l. B.* ii. p. 17.

Adult. General colour above and below of a dull rifle-green, somewhat glistening; tail black, the feathers slightly washed with green on the outer web; quills blackish brown, the least wing-coverts edged with dull green like the scapulars, the rest of the coverts and secondaries slightly washed with green on the outer web, the primaries much paler brown; bill and feet black. Total length 13·5 inches, culmen 1·95, wing 7·75, tail 6·75, tarsus 1·9.

Hab. Moluccas: Obi group.

a. Ad. sk.	Obi Islands.	Dr. Bernstein [C.].

3. Lycocorax morotensis.

Lycocorax morotensis, *Bernst. J.f. O.* 1864, p. 408; *Schl. Ibis,* 1863, p. 120; *id. N. T. D.* iii. p. 192; *id. Mus. P.-B.* Coraces, p. 132.
Manucodia mortiensis, *Gray, Hand-l. B.* ii. p. 17 (*lapsu*).

Adult. General colour black above and below, the wings brown, the primaries lighter than the secondaries and having the base of the inner web conspicuously white; tail black, shaded with dull green on the outer web; bill and feet black. Total length 17 inches, culmen 2, wing 8·4, tail 7, tarsus 2.

Hab. Morty Island (Morotai).

a, b, c. Ad. sk.	Morty Island.	A. R. Wallace, Esq. [C.].

18. XANTHOMELUS.

	Type.
Xanthomelus, *Bp. Ann. Sci. Nat. Zool.* (4), i. p. 122 (1854)	X. aureus.

Range. Confined to New Guinea.

1. Xanthomelus aureus.

Golden Bird of Paradise, *Edwards, Birds,* iii. p. 112.
Oriolus aureus, *Linn. S. N.* i. p. 163; *Gray, Hand-l. B.* i. p. 293.
Paradisea aurea, *Lath. Ind. Orn.* ii. p. 195.
Le Paradis orangé, *Audeb. et Vieill. Ois. Dor.* ii. p. 26, pls. 11, 12.
Le Loriot de Paradis, *Levaill. Ois. Parad.* i. pls. 18, 19.
Paradisea aurantia, *Shaw, Gen. Zool.* vii. p. 499, pl. 68.
Lophorhina aurantia, *Steph. Gen. Zool.* xiv. p. 76.
Sericulus aurantiacus, *Less. Traité,* p. 339; *id. Ois. Parad. Syn.* p. 20; *id. H. N.* p. 201, pls. 25 bis, 25 ter.
Sericulus aureus, *Gray, Gen. B.* i. p. 232; *Bp. Consp.* i. p. 349; *Gray, P. Z. S.* 1861, p. 435; *Wall. Ibis,* 1861, p. 287; *id. P. Z. S.* 1862, p. 160; *id. Malay Arch.* ii. p. 257; *Schl. Mus. P.-B.* Coraces, p. 98; *Scl. Ibis,* 1876, p. 248.
Xanthomelus aureus, *Bp. C. R.* xxxviii. p. 538; *Elliot, Ibis,* 1872, p. 112; *id. Monogr. Parad.* pl. 15; *Salvad. Ann. Mus. Civic.* vii. p. 783; *id. op. cit.* ix. p. 192.
Sericulus xanthogaster, *Schl. N. T. D.* iv. p. 50; *Salvad. Ibis,* 1876, p. 267; *id. Atti R. Accad. Torino,* xi. p. 688.
Chlamydodera xanthogastra, *Elliot, Ibis,* 1872, p. 113; *id. Monogr. Parad.* pl. xxxiii.; *Scl. P. Z. S.* 1873, p. 697.

Adult male. General colour above fiery orange-red, the head crested, the back with a large dorsal shield; scapulars as well as the lower back, rump, and upper tail-coverts orange-yellow mixed with black, the feathers of the rump olive-brown at base; wings deep orange, the feathers black at the base; quills orange, shading off into olive-brown towards the tips, the inner secondaries olive-brown, with more or less orange-yellow towards the base; tail black; a narrow line running from the base of the bill, as well as the lores and the feathers round the eye, sides of face, and throat, black; rest of under surface of body orange-yellow; sides of the head and neck, as well as the long plumes on the side of the latter, fiery orange-red like the dorsal shield; bill black; "iris clear straw-yellow" (*Beccari*). Total length about 9 inches, culmen 0·95, wing 5·1, tail 3·65.

Female. General plumage uniform olive-colour; throat blackish brown, spotted with olive; bill and feet brown. (*Elliot.*)

Young. Upper part of head and neck light brown, darkest in the centre of the feathers; mantle covering the upper part of the back dark brown, with the shafts yellow; rest of upper parts dark rufous-brown, the centres of the feathers of the back bright yellow; wings rufous brown; tail lighter brown than the back, the shafts rufous; cheeks and throat light rufous brown; entire underparts bright golden yellow; bill, feet, and tarsi black. (*Elliot.*)

Hab. New Guinea.

a, b. Ad. sk. New Guinea.

Family ORIOLIDÆ.

Key to the Genera.

a. Lores feathered, but generally with a little bare space behind the eye .... 1. ORIOLUS, p. 188.
b. Lores bare, as also all the region around and behind the eye .... 2. SPHECOTHERES, p. 223.

I cannot determine any valid characters which shall serve for the discrimination of the genera *Psarophulus*, *Analcipus*, or *Mimeta*. They appear only to form sections of the large genus *Oriolus*, distinguished by the colour of their plumage, which is not, in my mind, a sufficient generic character.

1. ORIOLUS*.

	Type.
Oriolus, *Linn. S. N.* i. p. 160 (1766)	O. galbula.
Mimetes, *King, Survey Austr.* ii. p. 419 (1827) ..	O. viridis.
Mimeta, *Vigors, Tr. Linn. Soc.* xv. p. 326 (nom. emend.)..................................	O. viridis.
Analcipus, *Swains. Faun. Bor.-Amer.* p. 482 (1831)	O. cruentus.
Ptilocarpus, *Müll. Tijdschr. Nat. Gesch. Phys.* ii. p. 331 (1835)	O. cruentus.
Psarophulus, *Jard. & Selby, Ill. Orn.* iv. pl. 26 (c. 1837)	O. trailli.
Erythrolanius, *Less. R. Z.* 1840, p. 274	O. cruentus.

---

* The following species are unknown to me.

1. ORIOLUS PHILIPPENSIS.

Oriolus philippensis, *J. E. Gray, Zool. Misc.* p. 3; *Bp. Consp.* i. p. 346; *G. R. Gray, Hand-l. B.* i. p. 292. no. 4321; *Walden, Tr. Z. S.* ix. p. 186.

Head, upper part of neck, back, epigastrium, and tail coal-black; lower part of neck, chest, belly, and rump bright golden yellow, paler on the back of the neck; bill yellow; wings and feet wanting. Length 7½ inches.

Hab. Philippine Islands (*Capt. Hay*).

2. ORIOLUS AFFINIS.

Oriolus affinis, *Gould, B. Austr.* folio, i. Intr. p. liii.; *Gray, Hand-l. B.* i. p. 293.
Mimeta affinis, *Bp. Consp.* i. p. 346; *Gould, Handb. B. Austr.* i. p. 465.

"This species differs from *Mimeta viridis* in having a smaller body, a shorter wing, a much larger bill, and in the white spots at the tips of the lateral tail-feathers being much smaller in extent. Although I have not at this moment any specimens wherewith to institute a comparison, I have but little doubt that this bird is quite distinct from its southern representative *M. viridis*."—*Gould.*

Hab. Port Essington.

Obs. Mr. Gould has no specimens of this bird, which seems to me to be a very doubtful species. All the Orioles in the British Museum from the locality quoted are, in my opinion, the young of *O. flavicinctus*.

Galbulus, *Bp. C. R.* xxxviii. p. 535 (1854) ...... O. galbula.
Broderipus, *Bp. t. c.* p. 536 (1854) .... O. broderipi.
Baruffius, *Bp. t. c.* p. 538 (1854) .... O. brachyrhynchus.
Xanthonotus, *Bp. t. c.* p. 538 (1854) .... O. xanthonotus.
Euchlorites, *Heine, J. f. O.* vii. p. 402 (1859).... O. broderipi.

Range. Europe, excepting the extreme north; the whole of Africa; Turkestan; India; China; Indo-Chinese subregion; Malayan subregion and entire archipelago; Australia.

Key to the Species.

a. Head entirely golden yellow.
  a'. Wings black, without broad yellow edgings to the feathers.
    a''. With a black loral spot, not continued round the eye; outer tail-feather black at base .... *galbula*, p. 191.
    b''. The black loral streak continued past the eye above the ear-coverts; outer tail-feather entirely yellow .... *kundoo*, p. 194.
  b'. Wings black, all the coverts and the secondaries broadly margined with golden yellow.
    c''. Outer tail-feather black at base .... *auratus*, p. 195.
    d''. Outer tail-feather entirely golden yellow *notatus*, p. 196.
b. Head golden yellow, with a horseshoe band of black encircling the nape, leaving a frontal patch of yellow sometimes occupying nearly the entire crown, sometimes confined to a small patch on the anterior part of the forehead.
  c'. A distinct yellow alar speculum.
    e''. Outer tail-feather black at the base.
      a'''. Inner secondaries entirely yellow on the outer web .... { *diffusus*, p. 197. *tenuirostris*, p. 198. }
      b'''. Inner secondaries yellow at tip, forming a more or less distinct spot, but not occupying the entire outer web.
        a⁴. Yellow of crown much wider than the black horseshoe.
          a⁵. Alar speculum large, as also the yellow spot at the end of the inner secondaries .... *maculatus*, p. 199.
          b⁵. Alar speculum small, as also the tips of the inner secondaries .... *andamanensis*, p. 200.
        b⁴. Yellow of crown confined to a frontal spot much less than the black coronal patch in width .... *broderipi*, p. 201.
    f''. Outer tail-feather entirely yellow.
      c'''. Larger; bill 1·4 inch; inner secondaries black on the outer web; two middle tail-feathers black .... *macrurus*, p. 202.
      d'''. Smaller; bill 1·15 inch; inner secondaries externally yellow along the outer web; two centre tail-feathers olive-yellow .... *celebensis*, p. 203.

d'. No yellow alar speculum.
  g''. Innermost secondaries with a distinct yellow spot at the tip of each .... *chinensis*, p. 203.
  h''. Innermost secondaries entirely black, like the rest of the quills, or externally greenish yellow.
    e'''. Size moderate; back brilliant golden yellow .... *frontalis*, p. 204. / *suluensis*, p. 205.
    f'''. Size very large; back yellowish green *formosus*, p. 205.

c. Head olivaceous or greenish yellow, sometimes brown with black streaks.
  e'. Belly uniform brown or yellow.
    i''. Primary-coverts broadly tipped with yellow, so as to form a broad and distinct alar speculum .... *flavicinctus*, p. 206.
      g'''. Alar speculum and yellow tips to tail-feathers large. *a. flavicinctus*, p. 206.
      h'''. Alar speculum and yellow tips to tail-feathers much smaller. *β. viridissimus*, p. 207.
    k''. Primary-coverts never having more than a slight edging or lighter tip to the feathers, not forming an alar speculum.
      i'''. Ear-coverts ashy brown; cheeks and throat slaty grey; sides of neck grey washed with olive .... *viridifuscus*, p. 208.
      k'''. Ear-coverts dark brown, washed with olivaceous; sides of neck silvery grey, in strong contrast to the ear-coverts; throat grey, washed strongly with olivaceous yellow, which pervades the whole of the under surface.......... *forsteni*, p. 209.
      l'''. Ear-coverts blackish, as also the cheeks, contrasting with the sides of the neck; under surface of body pale brown.
        c⁴. Ear-coverts streaked, as also the entire throat; breast light brown, narrowly streaked with black .... *striatus*, p. 210.
        d⁴. Ear-coverts uniform; throat whitish, mottled at the sides with black, but not streaked down the centre; rest of under surface of body entirely uniform...................... *bouruensis*, p. 211.
      m'''. Ear-coverts uniform dark brown, like the rest of the plumage, which is uniform brown above and below........ *phæochromus*, p. 211.
  f'. Belly white, broadly streaked with black.
    l''. Under tail-coverts white, as also the spot at the tip of the tail-feathers .... *viridis*, p. 212.
    m''. Under tail-coverts bright yellow, as also the spot at the tip of the tail-feathers .. *steerii*, p. 213.

d. Head black all round, as well as the fore neck.
  g'. Centre of breast white, streaked with black; size small; wing not exceeding 4·5 inches: back bright yellow .... *xanthonotus*, p. 213.
  h'. Centre of breast bright yellow, uniform; size larger; wing 4·7 to 5·5 inches.

n″. Primary-coverts tipped with yellow, forming an alar speculum of that colour .... *melanocephalus*, p. 215.
o″. Primary-coverts tipped with white, forming an alar speculum of that colour.
n‴. Outer tail-feather entirely yellow .... *monachus*, p. 216.
o‴. Outer tail-feather black at base.
e⁴. Bill very stout .... *crassirostris*, p. 217.
f⁴. Bill slender.
c⁵. Larger; above golden yellow, inclining to olive on the lower back; wing-coverts edged with yellow; alar speculum large, about one third of the length of the primary-coverts...... .... *larvatus*, p. 217.
d⁵. Smaller; above olivaceous yellow; wing-coverts edged with grey; alar speculum small, formed by a narrow white tip to the primary-coverts .... *brachyrhynchus*, p. 218.
p″. Primary-coverts entirely black, not showing any speculum at all .... *nigripennis*, p. 220.
′. Centre of breast bright crimson.
q″. Back black, of the same colour as the head and lower parts, leaving only the centre of the breast crimson .... *cruentus*, p. 221.
r″. Back red, like the breast and lower parts generally.
p‴. Colour of back and lower parts deep crimson; black of throat occupying also the jugulum and fore neck...... *ardens*, p. 221.
q‴. Colour of back and lower parts deep maroon, intensifying here and there to blackish; black of throat not extending beyond the jugulum .... *traillii*, p. 222.

1. Oriolus galbula.

The Golden Oriole.
Le Loriot, *Briss. Orn.* ii. p. 320; *Montb. Pl. Enl.* 26.
Oriolus galbula, *Linn. S. N.* i. p. 160; *Naum. Vög. Deutschl.* ii. Taf. 61; *Wagler, Syst. Av.* Oriolus, sp. 3; *Werner, Atlas*, Omnivores, pl. 15; *Roux, Orn. Prov.* pls. 125, 126, 127; *Brehm, Vög. Deutschl.* p. 155; *Macgill. Br. B.* ii. p. 69; *Nordm. in Démid. Voy. Russie Mér.* iii. p. 195; *Keys. & Blas. Wirb. Eur.* p. 50; *Gould, B. Eur.* ii. pl. 71; *Yarr. Br. B.* i. p. 212; *Gray, Gen. B.* i. p. 232; *Bp. Consp.* i. p. 347; *Kjærb. Orn. Dan.* pl. xiii. fig. 5; *Schl. u. Susem. Vög. Eur.* Taf. 16; *Schl. Vog. Nederl.* pls. 113, 114; *Horsf. & Moore, Cat. B. Mus. E.I. Co.* i. p. 271; *Hartl. Orn. Westafr.* p. 80; *Jaub. & Barth.-Lap. Rich. Orn.* p. 199 (1859); *Schl. Dier. Nederl. Vog.* pl. 14. figs. 3, 4, 4 *b*; *Linderm. Vög. Griechenl.* p. 82; *Hartl. Faun. Madag.* p. 43; *Fritsch, Vög. Eur.* Taf. 14. figs. 7, 8; *Sundev. Sv. Fogl.* pl. lxvii. fig. 2; *Gould, B. Gt. Br.* ii. pl. 31; *Filippi, Viagg. Pers.* p. 350; *Layard, B. S. Afr.* p. 135; *Loche, Expl. Sci. Algér. Ois.* ii. p. 54; *Degl. et Gerbe, Orn. Eur.* i. p. 392 (1867); *Schl. Mus. P.-B.* Coraces, p. 99; *Bettoni, Ucc. Lomb.* ii. Taf. 75; *Gray, Hand-l. B.* i. p. 291; *Shelley, B. Egypt*, p. 156; *Doderl. Avif. Sicil.* p. 104; *Salvad. Faun. Ital. Ucc.* p. 168; *Sharpe*,

Ibis, 1870, p. 215; *id. Cat. Afr. B.* p. 53; *Heugl. Orn. N.O.-Afr.* i. p. 400; *Newt. ed. Yarr. Br. B.* i. p. 233; *Gurney in Anderss. B. Dam. Ld.* p. 124; *Hume, S. F.* 1873, p. 182; *Irby, B. Gibr.* p. 125; *Dresser, B. Eur.* part xliii.
Coracias oriolus, *Scop. Ann.* i. p. 41.
Coracias galbula, *Bechst. Gemeinn. Naturg. V. D.* i. p. 1292.
Oriolus galbula, var. virescens, *Ehr. Symb. Physic.* fol. *x.*
Oriolus aureus, *Brehm, Vög. Deutschl.* p. 156.
Oriolus garrulus, *Brehm, t. c.* p. 157.

Nestling. Above dull yellowish green, the feathers with margins of very pale yellow; wings greyish black, the primaries edged and tipped with white, the secondaries externally washed with dull olive-yellow and tipped with brighter pale yellow, as are also the wing-coverts; primary-coverts tipped with yellowish white; tail-feathers olive-greenish, tipped with bright yellow, extending gradually further up the inner web on the outer feathers, most of the rectrices with a more or less perceptible subterminal shade of black; under surface of body greyish white, purer on the abdomen, the breast streaked down the centre of the feathers with greyish brown, these streaks obsolete on the throat; thighs and under tail-coverts lemon-yellow, as also the under wing-coverts; bill dark horn-brown, paler on the under mandible.

Young. Golden yellow above, tinged with olive-green; quills and wing-coverts brownish black, tinged with olive-green; tail-feathers marked as in the adult, but the feathers olive-green, blackish towards the tip; throat and breast grey, with black stripes; flanks bright yellow, with fainter stripes; loral spot indistinct, dusky; bill brownish red; feet black. Total length 9·5 inches, of bill from front 1, from gape 1·2, wing 5·9, tail 3·3, tarsus 8, middle toe 7, hind toe 4.

Adult male. Above rich golden yellow; wing-coverts black, the cubital coverts with a narrow tip of yellow; primary-coverts black, broadly edged with yellow, forming a conspicuous spot; quills black, lighter underneath, the secondaries edged towards the tip with yellowish white; tail black, the middle feathers entirely of this colour, with a spot of yellow at the tip, all the other feathers black at the base, yellow towards the tip, the outer feathers having more yellow than black; a spot between the base of the bill and the eye black; entire under surface golden yellow; bill dull reddish; feet leaden grey; iris blood-red. Total length 9 inches, of bill from front 1, from gape 2·2, wing 5·9, tail 3·1, tarsus 8, middle toe ·7, hind toe ·4.

Obs. The bird described above as a young bird is the same as that characterized by me in my paper on African Orioles as the "Adult female;" this identification I now believe to be an error. Mr. Blyth, indeed, has insisted that in the adult birds the plumage is exactly alike, the female being only slightly tinged with green on the upper surface. It follows, then, that the birds with greyish white breasts are young, and that streaks on the breast are a sign of immaturity. The specimens now before me bear out Mr. Blyth's surmise; for although I have not yet seen a fully golden bird determined to be a female by dissection, a great number of the striped-breasted birds are rapidly becoming yellow below; some even have nearly completed the full golden plumage, but have remains of narrow streaks still visible. The change of plumage appears to take place gradually by a change of the colour of the feather, and not by a direct moult.

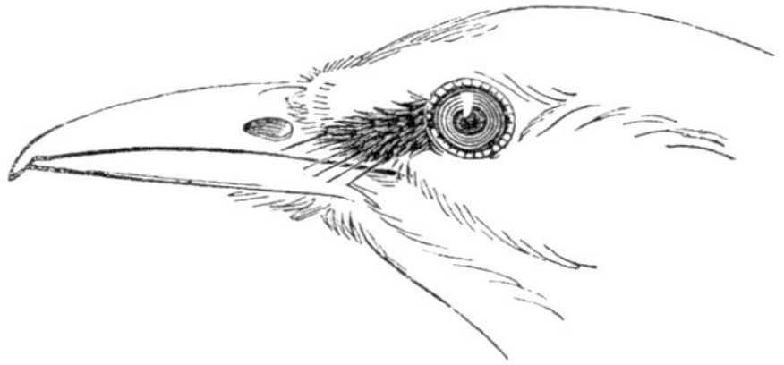

Head of *Oriolus galbula*.

Hab. The whole of Europe: Persia and S.W. Asia, visiting South Africa in winter.

a. ♂ ad. st.	Europe.	
b. Pull. st.	Europe.	
c. Pull. sk.	Westphalia, July 1868.	Baron A. von Hügel [P.].
d. Ad. sk.	Turkey (*Robson*).	R. B. Sharpe, Esq. [P.].
e. Juv. st.	Athens.	C. W. L. Merlin, Esq. [P.].
f. ♂ ad. sk.	Athens.	C. W. L. Merlin, Esq. [P.].
g. ♂ ad. sk.	Algeria, June 6, 1850.	Purchased.
h. ♂ jun. sk.	Algeria, May 28, 1850.	Purchased.
i. ♀ imm. sk.	Banias, Mount Hermon.	Canon Tristram [C.].
k. Ad. sk.	Plain of Zangaban, Zagros, May 14, 1851.	Kennett Loftus, Esq. [P.].
l. ♂ ad. sk.	Eklid, Persia, 7500 feet.	Major St.-John [C.].
m. ♂ juv. sk.	Eklid, Persia, 6700 feet.	Major St.-John [C.].
n. ♀ ad. sk.	Shiraz, Persia.	Major St.-John [C.].
o. Ad. sk.	Persia.	Charles Darwin, Esq. [P.].
p. ♀ juv. sk.	Mesopotamia.	Commander Jones [C.]. (*Cf.* Horsf. & Moore, Cat. p. 271, No. 417, *b.*)
q. Ad. sk.	Bogos Land.	Mr. Esler [C.].
r, s. Imm. sk.	White Nile (*Brun-Rollet*).	R. B. Sharpe, Esq.
t. Ad. sk.	Mombas, E. Africa (*Wakefield*).	R. B. Sharpe, Esq. [P.].
u, v. Imm. sk.	Near Rustenburg, S.E. Africa.	F. A. Barratt, Esq. [C.].
w. ♀ imm. sk.	Knysna, Jan. 31, 1866 (*Andersson*).	R. B. Sharpe, Esq.
x. Imm. sk.	Ondonga, Ovampo Land, Oct. 30, 1866 (*Andersson*).	R. B. Sharpe, Esq.
y. Juv. sk.	South Africa.	Sir A. Smith [C.].
z. Skeleton.		

2. Oriolus kundoo.

Oriolus kundoo, *Sykes, P. Z. S.* 1832, p. 87; *Gray, Gen. B.* i. p. 232; *Blyth, J. A. S. B.* xv. p. 49; *id. Cat. B. Mus. A. S. B.* p. 216; *Fraser, Zool. Typ.* pl. 38; *Bp. Consp.* i. p. 348; *Horsf. & Moore, Cat. B. Mus. E.I. Co.* i. p. 272; *Jerd. B. Ind.* ii. p. 107; *Schl. Mus. P.-B.* Coraces, p. 101; *Gray, Hand-l. B.* i. p. 291; *Hume, Lahore to Yark.* p. 167; *Henders. & Hume, t. c.* p. 200, pl. xi.; *Hume, Nests and Eggs Ind. B.* p. 299.

Oriolus galbula, *Sykes, P. Z. S.* 1832, p. 17; *Hodgs. in Gray's Zool. Misc.* p. 83.

Oriolus aureus, *Jerd. Madr. Journ.* x. p. 261; *Blyth, J. A. S. B.* xi. p. 797.

Oriolus galbuloides, *Gould, P. Z. S.* 1841, p. 6.

Oriolus melanoris, *Hodgs. in Gray's Zool. Misc.* p. 83. no. 263.

Head of *Oriolus kundoo*.

Nestling. Above light yellowish green, with edgings of paler yellow to the feathers; wings olive-brown washed with yellowish green, with light-yellow tips; primaries externally edged with greyish white; tail olive-green, tipped with yellow, the external feathers yellow on the inner web; rest of under surface of body dull white, slightly washed with yellow here and there, the feathers mesially streaked with brown.

Full-grown young bird. Similar in plumage to the nestling before described, but rather brighter yellow above and wanting the yellow margins to the feathers of the upper surface, excepting on the wing-coverts, where a few still remain; quills brown, the primaries externally edged with greyish white, the secondaries washed with olive-yellow, more distinct on the inner ones, which resemble the wing-coverts; tail dull yellowish olive, tipped with bright yellow, both increasing in extent at the tip and occupying more of the base towards the outer tail-feathers, some of which show a slight subterminal blackish shade; under surface of body dull white, with mesial blackish streaks to all the feathers, more obscure on the throat; the sides of the body, thighs, and under wing- and tail-coverts bright yellow. Total length 9·5 inches, culmen 1·15, wing 5·35, tail 3·6, tarsus 0·85.

Adult male. Bright golden yellow above and below, including the thighs and under wing- and tail-coverts; lores, feathers round the eye, and a narrow eyebrow black; most of the scapulars externally black; wings black, the edge of the wing golden yellow, as also the tips of some of the bastard-wing-feathers; primary-coverts golden yellow, with black bases; quills black, the primaries edged and tipped with yellow, these yellow tips being very large on the secondaries, especially on the outer web, where they form a large spot; tail golden yellow, the feathers with black bases to the outer webs, increasing in extent towards the centre of the tail, the two middle feathers being black with narrow yellow tips, the two next feathers black with the terminal third yellow; "bill deep lake-red; legs plumbeous; iris rich blood-red" (*Jerdon*). Total length 9·4 inches, culmen 1·2, wing 5·7, tail 3·8, tarsus 0·95.

Hab. India; Turkestan.

a. Ad. st.	Nepal.	B. H. Hodgson, Esq. [P.].
b, c, d. Ad. sk.	Nepal.	B. H. Hodgson, Esq. [P.].
e, f. Juv. sk.	Nepal.	B. H. Hodgson, Esq. [P.].
g. Pull. sk.	Nepal.	B. H. Hodgson, Esq. [P.].
h, i. Ad. sk.	Behar.	B. H. Hodgson, Esq. [P.].
k, l. Ad. sk.	N.W. Himalayas.	Capt. Stackhouse Pinwill [P.].
m. Imm. sk.	N.W. Himalayas.	Capt. Stackhouse Pinwill [P.].
n. Ad. sk.	Ladakh.	Purchased.
o, p. ♂ ad. sk.	Cashmere.	E. M. Langworthy, Esq. [C.].
q. Ad. sk.	Kamptee, C. India.	Dr. Hinde [P.].
r, s. Imm. sk.	Kamptee, C. India.	Dr. Hinde [P.].
t. Juv. sk.	Kamptee, C. India.	Dr. Hinde [P.].
u. Ad. sk.	Deccan.	Dr. F. Day [C.].

3. Oriolus auratus.

Le Loriodor, *Levaill. Ois. d'Afr.* pl. 260; *Sundev. Crit. om Levaill.* p. 53.

Oriolus auratus, *Vieill. N. Dict. d'Hist. Nat.* xviii. p. 194; *id. et Oud. Gal. Ois.* pl. 83; *Swains. B. W. Afr.* ii. p. 33; *Bp. Consp.* i. p. 348; *Cab. Mus. Hein.* Th. i. p. 209; *Hartl. J. f. O.* 1854, p. 24; *Antin. Cat. Met. Ucc. Affr. C–N*, p. 44; *Schl. Mus. P.-B.* Coraces, p. 101; *Heugl. Orn. N.O.-Afr.* i. p. 401; *Gray, Hand-l. B.* i. p. 291; *Sharpe, Ibis*, 1870, p. 216; *id. Cat. Afr. B.* p. 53.

Le Loriot bicolor, *Temm. Cat. Syst. Cab. d'Orn.* pp. 46, 202.

Oriolus bicolor, *Licht. Verz. Doubl.* p. 20; *Hartl. Orn. W.-Afr.* p. 80; *id. J. f. O.* 1861, p. 163; *Bocage, Jorn. Lisb.* i. p. 137.

Oriolus chryseos, *Heugl. Syst. Uebers.* p. 31. no. 294; *id. Naum.* 1857, p. 433.

Oriolus icterus, *Paul v. Würt. MS.*; *Heugl. J. f. O.* 1867, p. 299.

Tail of *Oriolus auratus*.

Adult male. Above very rich golden yellow, a little lighter on the rump; least wing-coverts entirely yellow; lesser wing-coverts black at the base and very broadly edged with yellow, so that the black scarcely shows, the feather nearest the edge of the wing entirely black; primary-coverts black, with a few of the smaller feathers tipped with yellow; quills black, light grey underneath, the innermost secondaries broadly, the outer ones more narrowly edged with yellow; middle tail-feathers black, with a yellow tip, the next two on each side black, with a little broader edging of yellow, the next feathers for the most part yellow till the last, which are black only at the base; a line of black feathers from the base of the bill to the extremity of the ear-covert, encircling the eye; entire under surface of the body rich golden yellow; bill brownish red; feet black. Total length 9 inches, of bill from front 1·1, from gape 1·3, wing 5·5, tail 3·3, tarsus ·75, middle toe ·7, hind toe ·4.

Young. Similar to the male, but with the upper surface tinged with olive-green, the rump alone being bright yellow; the quills more dusky black, and the tail-feathers tinged with olive where they are black in the adult male; throat and breast whitish; sides of the body, flanks, and abdomen yellow, the whole striped with longitudinal lines. Total length 9 inches, bill from front 1·2, from gape 1·35, wing 5·5, tail 3·4, tarsus ·8, middle toe ·7, hind toe ·4.

Hab. W. Africa, from Senegambia to Angola; N.E. Africa.

a. ♂ ad. sk.	Casamance.	R. B. Sharpe, Esq.
b, c. Imm. sk.	River Gambia.	R. B. Sharpe, Esq.
d. ♀ ad. sk.	[W. Africa.]	R. B. Sharpe, Esq.
e. Ad. sk.	River Gambia.	Lieut.-Col. Strachan [P.].

4. Oriolus notatus.

Oriolus auratus, *Gurney, P. Z. S.* 1864, p. 2 (*nec V.*); *Anderss. t. c.* p. 6; *Layard, B. S. Afr.* p. 135.

Oriolus notatus, *Peters, J. f. O.* 1868, p. 132; *Cab. in Von der Decken's Reis. Vög.* iii. p. 33; *Gray, Hand-l. B.* i. p. 391; *Sharpe, Ibis,* 1870, p. 218, pl. vii. fig. 2; *id. Cat. Afr. B.* p. 53; *Gurney in Anderss. B. Dam. Ld.* p. 124.

Oriolus anderssoni, *Bocage, Jorn. Lisb.* ii. p. 342.

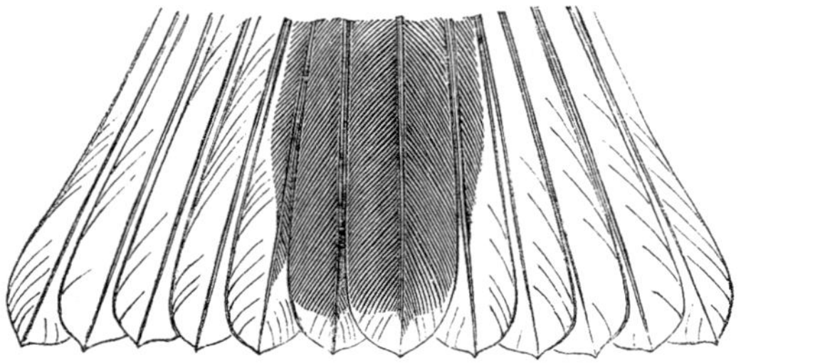

Tail of *Oriolus notatus.*

Adult male. Above rich golden yellow, a little paler on the scapularies; least wing-coverts yellow, with a black shaft down the centre of the feather; cubital coverts black at the base, with a yellow edging, gradually getting broader towards the tip; quills black, greyish on the underside; secondaries margined with yellow, and the outer primaries with white; middle tail-feathers black, with a yellow tip, the two next to the middle tail-feathers black at the base and for the greater part of the inner web, the four outer tail-feathers yellow; a black streak from the base of the bill, passing through the eye and produced to the extremity of the ear-covert; entire under surface rich golden yellow; bill brownish red; feet black. Total length 8 inches, of bill from front 1·1, from gape 1·2, wing 5·4, tail 3·1, tarsus 0·8, middle toe 0·7, hind toe 0·35.

Hab. Eastern Africa, from Mombas to the Zambesi, and South-western Africa.

a. Ad. sk.	Tete, Zambesi.	Dr. Kirk [C.].
b. ♂ ad. sk.	Ondonga, Ovampo Land, Jan. 22, 1867 (*Andersson*).	R. B. Sharpe, Esq.
c. Ad. sk.	Maconjo (*Anchieta*).	R. B. Sharpe, Esq.
d. Juv. sk.	Benguela.	J. J. Monteiro, Esq. [C.]*.

5. Oriolus diffusus †.

Oriolus sinensis, *Swains. An. in Menag.* p. 342 (sub *O. coronatus*); *Eyton, P. Z. S.* 1839, p. 104.

Oriolus chinensis (*nec Linn.*), *Jerd. Madr. Journ.* x. p. 262; *Blyth, J. A. S. B.* xi. p. 797; *David, N. Arch. Mus. Bull.* p. 8; *Swinh. P. Z. S.* 1871, p. 374; *Hume, Str. F.* 1874, p. 477, 1875, p. 132; *Prjevalsk. Orn. Misc.* ii. p. 271.

Oriolus indicus, *Jerd. Ill. Ind. Orn.* pl. 15; *Blyth, J. A. S. B.* xv. p. 47; *id. Cat. B. Mus. A. S. B.* p. 216; *Bp. Consp.* i. p. 348; *Horsf. & Moore, Cat. B. Mus. E.I. Co.* i. p. 270; *Jerd. B. Ind.* ii. p. 109; *Schl. Mus. P.-B.* Coraces, p. 102; *Gray, Hand-l. B.* i. p. 292; *Holdsw. P. Z. S.* 1872, p. 453; *Blyth & Wald. B. Burm.* p. 139.

Euchlorites chinensis, *Heine, J. f. O.* 1859, p. 402.

Euchlorites indicus, *Heine, t. c.* p. 402.

Adult male. General colour above and below bright yellow; fore part of head bright yellow; lores, feathers round the eye, and the hinder part of the head and neck, black, forming a distinct horse-shoe; wing-coverts bright yellow, the inner webs of the greater series black; primary-coverts black, broadly tipped with yellow, forming a large speculum; quills black, tipped with greyish white,

---

* This is the specimen referred to by Dr. Hartlaub (P. Z. S. 1865, p. 93) as *O. bicolor*; but from its smaller size, light coloration, and nearly obsolete basal markings to the tail-feathers, it is most probably an immature *O. notatus*; from its immaturity, however, it is rather difficult to refer the specimen accurately.

† This Oriole is the bird generally called *O. indicus* by authors; but Brisson's description is totally inadmissible, unless, indeed, a Golden Oriole with blue in its coloration remains to be discovered, which is hardly likely. The Chinese bird (*O. chinensis*, auct.) is not separable from the Indian; and the name *chinensis* being undoubtedly referable to the Oriole of the Philippines, I have proposed the name of *diffusus* for the Indian bird, as it is the most widely spread of all the black-naped species.

the primaries externally margined with that colour, the secondaries with yellow, the innermost almost entirely yellow, leaving only the half of the inner web black; tail black, the feathers broadly tipped with bright yellow, increasing towards the outer feathers of the tail, the yellow extending further up the inner than the outer web; "bill pinky red; feet plumbeous; iris rich blood-red" (*Jerdon*). Total length 9·6 inches, culmen 1·25, wing 5·85, tail 3·85, tarsus 1.

Young. Greenish yellow, the feathers tipped with bright yellow; quills brownish, the primaries externally edged with grey, the primary-coverts and secondaries externally washed with greenish yellow, inclining to brighter yellow on the outer margin; tail greenish, tipped with yellow, more broadly on the outer feathers, which have a subterminal shade of blackish of greater or less extent; under surface of body white, the feathers mesially streaked with black; the sides of the body, thighs, and under wing- and tail-coverts bright yellow.

Obs. The bird described is an adult from China, as at present the Museum does not possess a specimen from India. Lord Tweeddale has kindly lent me a Malabar skin, which measures as follows:—Total length 9·8 inches, culmen 1·25, wing 6·2, tail 4·1, tarsus 0·9. There are some slight differences between Indian and Chinese specimens, but not sufficient to found species upon. (*Cf.* also Hume, S. F. 1875, p. 132.)

Head of *Oriolus diffusus*.

Hab. Indian peninsula and China: Tenasserim, as far south as Penang.

a, b. Ad. sk.	Tenasserim.	J. C. D. Packman, Esq. [P.].
c. Juv. sk.	Tenasserim.	J. C. D. Packman, Esq. [P.].
d, e. Ad. sk.	Penang.	Dr. Cantor [C.].
f, g. Ad. sk.	China.	Capt. Blakiston [C.].
h. Juv. sk.	China.	Capt. Blakiston [C.].
i. Var. sk.	N. China.	Capt. Blakiston [C.].
k. Ad. sk.	N. China.	Mr. Fortune [C.].
l. Ad. sk., *m*. Juv. sk.	S.W. Formosa.	R. Swinhoe, Esq. [C.].

6. **Oriolus tenuirostris.**

Oriolus tenuirostris, *Blyth, J. A. S. B.* xv. p. 48; *id. Cat. B. Mus. A. S. B.* p. 216; *Gray, Hand-l. B.* i. p. 292. no. 4308; *Hume, Str. F.* 1874, p. 477, 1875, p. 131; *Blyth & Wald. B. Burm.* p. 140. no. 484.

Adult. General colour pale greenish yellow, the head brighter yellow, the lores and feathers round the eye black, encircling the nape in a circular band; wing-coverts coloured like the back, but edged with brighter yellow; primary-coverts black, with the terminal half yellow, forming a large alar speculum; quills black, broadly tipped with yellow, the secondaries externally margined with bright yellow, the primaries with greyish; the innermost secondaries olive-yellow, broadly tipped with yellow, the inner web only blackish; middle tail-feathers black, tipped with yellow, all the other feathers bright yellow with black bases, the latter decreasing in size towards the outermost; "bill pale pink, inside of mouth fleshy; legs plumbeous; iris crimson; eyelids grey" (*Oates*). Total length 9·5 inches, culmen 1·3, wing 5·35, tail 3·5, tarsus 1.

Young. Differs from the adult, as do the young birds of the allied species. Much duller in plumage than the adult, and whitish below, with black shaft-streaks.

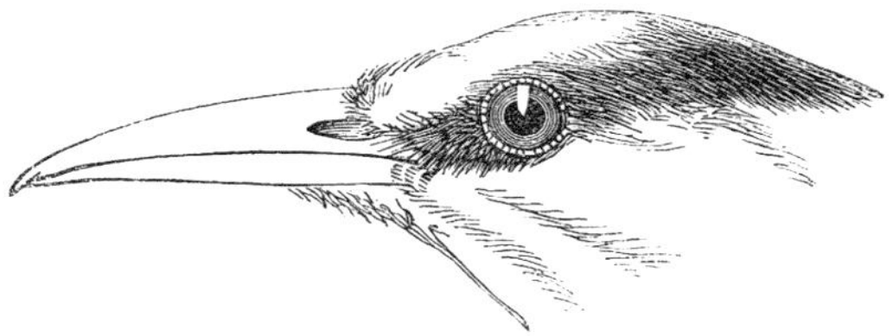

Head of *Oriolus tenuirostris.*

Hab. Burmah and Pegu.

a, b. Ad. sk.	Moulmein.	W. T. Blanford, Esq. [P.].
c. Juv. sk.	Thayet-myo, Pegu.	W. T. Blanford, Esq. [P.].

7. Oriolus maculatus *.

Oriolus maculatus, *Vieill. N. Dict.* xviii. p. 194.
Oriolus galbula, *Horsf. Tr. Linn. Soc.* xiii. p. 152 (*nec Linn.*).
Oriolus chinensis, *Raffl. t. c.* p. 303 (*nec Linn.*).
Oriolus coronatus, *Swains. An. in Menag.* p. 342; *Gray, Gen. B.* i. p. 232; *Blyth, J. A. S. B.* xv. p. 48; *id. Cat. B. Mus. A. S. B. App.* p. xxiii.
Oriolus hippocrepis, *Horsf. & Moore, Cat. B. Mus. E.I. Co.* i. p. 271; *Gray, Hand-l. B.* i. p. 292; *Hume, Str. F.* 1873, p. 457.
Oriolus horsfieldi, *Bp. Consp.* i. p. 348.

Adult male. General colour above and below bright yellow; crown of head bright yellow, the lores and a band passing above and below the eye and encircling the nape black; least wing-coverts

* Vieillot states that his *O. maculatus* is from Java; and his description suits the bird from that island. As the name of *O. kundoo* is admitted, although founded on the young bird, so I consider Vieillot's name must be restored. That of *O. hippocrepis*, Wagler, Syst. Av. Oriolus, sp. 5, cannot be identified, the description suiting several species equally well.

coloured like the back, the greater series bright yellow, not quite so intense as the back, the inner webs of the feathers black; primary-coverts black, broadly tipped with yellow, forming an alar speculum; quills black, the primaries externally edged and tipped with greyish white, the secondaries margined with yellow, increasing on the innermost to a large yellow spot occupying the end of the outer web and the tip of the inner one also; two centre tail-feathers black, tipped with bright yellow, the rest of the feathers bright yellow with black bases, the black increasing towards the centre of the tail. Total length 10·5 inches, culmen 1·2, wing 5·55, tail 4, tarsus 0·95.

Young. Much duller in colour than the adult and shaded with greenish, the wing-coverts tipped with bright yellow; no distinct black coronal circlet; but the lores dusky greenish, and an indication of a darker colour round the nape; wings brown, the primary-coverts narrowly tipped with yellow; the secondaries externally olivaceous, inclining to brighter yellow on the outer webs, and at the tips; four centre tail-feathers olivaceous, slightly tipped with yellow, the remainder blackish, inclining to olivaceous more or less near the base, with a large yellow triangular spot at the tip of each feather, increasing in size towards the outermost; general colour underneath bright yellow, more or less white on the breast, the feathers of which are mesially streaked with black.

Hab. Java, Sumatra, Bali, Bangka, Borneo.

a. ♂ ad. sk.	W. Java, 1861.	A. R. Wallace, Esq. [C.].
b. Ad. sk.	W. Java, 1863.	A. R. Wallace, Esq. [C.].
c. Ad. sk.	E. Java, 1861.	A. R. Wallace, Esq. [C.].
d. Ad. st.	Java.	Purchased.
e. Ad. sk.	Bali, 1856.	A. R. Wallace, Esq. [C.].
f. ♂ juv. sk.	Sumatra, 1861.	A. R. Wallace, Esq. [C.].
g. Ad. sk.	Borneo.	Hugh Low, Esq. [C.].
h. Skeleton.		Purchased.

8. **Oriolus andamanensis.**

Oriolus horsfieldi, *Blyth in Mouatt's Andaman Is.* p. 352 (*nec Swains.*).

Oriolus andamanensis, *Tytler & Beavan, Ibis,* 1867, p. 326; *Gray, Hand-l. B.* i. p. 292. no. 4311; *Ball, Str. F.* 1873, p. 72; *Hume, Str. F.* 1874, p. 227.

Broderipus andamanensis, *Walden, Ibis,* 1873, p. 305.

Adult male. General colour above and below brilliant yellow; crown of head yellow, like the back, from which it is separated by a horseshoe band of black, drawn from the lores, enclosing the eye and passing round the nape; wing-coverts yellow like the back, the greater series black on the inner web; rest of the wings black, with only a faint terminal edging of yellow to the quills, forming a little larger spot on the inner secondaries; primary-coverts black, tipped with yellow, but not very broadly; centre tail-feathers black, tipped with yellow and slightly washed with olivaceous near the base, all the other feathers black at base, these black bases decreasing in extent towards the outer tail-feathers; "bill carneous, or delicate pink; legs and feet plumbeous, sometimes tinged with green, the soles dirty grey; iris carmine, the eyelids brownish red" (*Hume*). Total length 9·5 inches, culmen 1·2, wing 5·3, tail 4·1, tarsus 0·95.

Adult female. Has the yellow everywhere less bright, and nearly the whole mantle suffused with a dusky greenish shade; the portions of the wings and tail, but especially of the former, which are black or almost black in the male, are in the female much more decidedly brown; the whitish edgings of the third to the fifth primaries are much larger, longer, and more conspicuous than in the male, and in some specimens extend to the second and the sixth also. The outer webs of the secondaries and tertiaries, and some of the later primaries, are more or less broadly tinged with dull olive-yellow, the amount varying greatly in different specimens. The basal portions of the central tail-feathers similarly but more strongly suffused.

Young. Duller than the adult, olive-yellow with a greenish tinge, many of the feathers with pale yellow tips, especially those of the wing-coverts; wings dark brown, externally washed with olivaceous and tipped narrowly with light yellow, the primaries externally edged with grey; tail olive-green, with yellow tips to the feathers, becoming broader on the external ones, most of them having a slight subterminal shade of blackish; sides of face greenish yellow; under surface of body white, slightly washed with yellow, all the feathers mesially streaked with black, the throat less perceptibly; sides of body, thighs, and under wing- and tail-coverts bright yellow; "bill almost blackish, in some dingy pink" (*Hume*).

Hab. Andaman Islands.

a. ♂ ad. sk.	South Andaman, Feb. 10, 1873 (*Wardlaw Ramsay*).	R. B. Sharpe, Esq. [P.].
b. ♂ ad. *c.* juv. sk.	Port Blair, S. Andamans, July 7, 1873 (*Wimberley*).	R. B. Sharpe, Esq. [P.].

9. **Oriolus broderipi.**

Oriolus broderipii, *Bp. P. Z. S.* 1850, p. 279, pl. xviii.; *id. Consp.* i. p. 348; *Wall. P. Z. S.* 1863, p. 485; *Schl. Mus. P.-B.* Coraces, p. 106; *id. Dirent.* p. 179, cum fig.; *Finsch, Neu-Guinea*, p. 173; *Gray, Hand-l. B.* i. p. 291. no. 4303.
Broderipus refulgens, *Bp. C. R.* xxxviii. p. 538 (1854).
Euchlorites broderipi, *Heine, J. f. O.* 1859, p. 402.

Adult male. General colour above and below brilliant golden yellow, deepening into orange; forehead golden; crown of head and nape, as well as the lores and feathers round the eye, black; least wing-coverts orange, like the back, the greater series bright yellow, the inner webs of this series black; bastard wing, primary-coverts, and quills jet-black, the primary-coverts tipped with yellow, forming a speculum, the primaries externally edged with grey, the secondaries narrowly margined with grey near the tips, the innermost more broadly tipped with yellow on the outer web; tail black, the two centre feathers broadly tipped with yellow, the rest orange-yellow at the tip, the black bases markedly decreasing towards the outer feathers of the tail; "bill pink; feet black; iris red" (*Wallace, MS.*). Total length 12 inches, culmen 1·5, wing 6·2, tail 4·7, tarsus 1·15.

Young. Much paler and more yellow, the black on the head forming a ring round the nape, leaving the rest of the head yellow, whereas in the adult the yellow is confined to a small frontal patch, the rest of the crown being black; wings brownish, the primaries externally margined with greyish, the secondaries washed with olive and externally edged with yellow near the tip; two centre tail-feathers olive-greenish, tipped with yellow, the rest of the feathers olive-green at base, bright yellow at the tip, with a more or less subterminal shade of black; entire under surface of body bright yellow, including the under wing- and tail-coverts; the lower surface of the quills dark brown, buffy white at the base of the inner web.

Hab. Sumbawa, Lombock, Flores.

a, b. ♂ ad. sk.	Lombock.	A. R. Wallace, Esq. [C.].
c. ♂ juv. sk.	Lombock.	A. R. Wallace, Esq. [C.].
d. Vix ad. sk.	Lombock.	A. R. Wallace, Esq. [C.].
e, f. Ad. sk.	Flores.	A. R. Wallace, Esq. [C.].
g. Juv. sk.	Flores.	A. R. Wallace, Esq. [C.].

10. **Oriolus macrurus.**

Oriolus macrourus, *Blyth, J. A. S. B.* xv. pp. 46, 370; *id. Cat. B. Mus. A. S. B.* p. 216; *Horsf. & Moore, Cat. B. Mus. E.I. Co.* i. p. 270; *Gray, Hand-l. B.* i. p. 292. no. 4306; *Ball, Str. F.* 1873, p. 71; *Hume, Str. F.* 1874, p. 228, 1875, p. 132.
Euchlorites macrurus, *Heine, J. f. O.* 1859, p. 403.

Adult. Above and below bright yellow; forehead yellow, the lores, feathers round the eye, and remainder of the head and nape black; wings black, the wing-coverts yellow, like the back, the greater series black on the inner web, the primaries externally edged with grey, the secondaries with a very narrow yellowish edging near the tip, almost imperceptible; primary-coverts tipped with yellow, forming an alar speculum; two centre tail-feathers black with a narrow yellow tip, the rest of the tail yellow, the feathers with black bases, the extent of black greater on the inner web; the outer tail-feather almost entirely yellow, with a very little black near the base, this colour evidently disappearing altogether, the centre feathers slightly olivaceous near the base; "bill carneous; legs and feet plumbeous, claws horny; iris carmine" (*Hume*). Total length 11 inches, culmen 1·4, wing 5·9, tail 4·7, tarsus 1·1.

The young bird has the bill dull fleshy colour; iris dull red-brown. (*Hume.*)

Hab. Nicobar Islands.

a. Ad. sk.	Nicobar Islands.	Secretary of State for India [P.]. (*Cf.* Horsf. & Moore, Cat. i. p. 270. no. 413.)

11. Oriolus celebensis.

Broderipus coronatus, *Walden, Tr. Z. S.* viii. p. 60 (1874, *nec Sw.*).
Broderipus celebensis, *Walden, t. c.* p. 112; *Salvad. Ann. Mus. Civ. Genov.* vii. p. 609.
Oriolus coronatus, var. celebensis, *Brügg. Abhandl. Brem.* v. p. 62.

Adult male. General colour above and below bright yellow; head yellow, with a broad black band drawn from the lores above and below the eye and encircling the nape; wing-coverts yellow, like the back, the greater series with the inner webs black; the remainder of wing black, the primary-coverts slightly tipped with yellow; primaries externally edged with greyish white, the secondaries with yellow, more broadly developed on the latter as they approach the innermost, which have the outer web entirely yellowish; tail-feathers black, yellow at the tips, the black bases increasing gradually in extent towards the middle of the tail, the two centre feathers olive-greenish, narrowly tipped with yellow, before which is a subterminal spot of black; under surface of body bright yellow, including the under wing- and tail-coverts. Total length 10·8 inches, culmen 1·15, wing 5·95, tail 4·3, tarsus 0·95.

Obs. Lord Tweeddale has examined the specimens collected by Mr. Wallace; and he considers that they are not in full and complete plumage, but in the intermediate stage which he correctly describes as found in the Javan Oriole. I believe, however, that the Celebes specimens are all adult, and that the black on the centre feathers, instead of increasing, *decreases*, in fact, with age, and disappears entirely in a very old bird. The black base to the outer feather also disappears with age, leaving it entirely yellow in the fully adult bird. The greenish centre tail-feathers are found in some of the African Orioles; and the blackish subterminal colouring vanishes as the birds grow older. *Cf.* Brüggem. *l. c.* pp. 62, 63.

I may add that the alar speculum varies in size, and is nearly obsolete in some individuals.

Hab. Celebes.

a, b. Ad. sk.	Celebes.	J. Gould, Esq.	
c, d. ♂ ad. sk.	Macassar, 1857.	A. R. Wallace, Esq.	[C.].
e. ♀ ad. sk.	Macassar, 1857.	A. R. Wallace, Esq.	[C.].
f. Ad. sk.	Menado.	A. R. Wallace, Esq.	[C.].

12. Oriolus chinensis.

Le Loriot de la Cochinchine, *Briss. Orn.* ii. p. 326, pl. 33. fig. 1.
Oriolus chinensis, *Linn. S. N.* i. p. 160 (*ex Briss.*); *Gray, Gen. B.* i. p. 232; *Blyth, J. A. S. B.* xv. p. 46; *id. Cat. B. Mus. A. S. B.* p. 215; *Horsf. & Moore, Cat. B. Mus. E.I. Co.* i. p. 270; *Gray, Hand-l. B.* i. p. 292.
Le Couliavan de la Cochinchine, *Montb. Pl. Enl.* 570.
Oriolus acrorhynchus, *Vigors, P. Z. S.* 1831, p. 97; *Gray & Mitch. Gen. B.* i. p. 232, pl. 58; *Bp. Consp.* i. p. 348; *Schl. Mus. P.-B. Coraces,* p. 104 (pt.); *Walden & Layard, Ibis,* 1872, p. 101; *Hume, S. F.* 1875, p. 132.
Oriolus cochinchinensis, *Bp. Consp.* i. p. 348.
Broderipus acrorhynchus, *Walden, Tr. Z. S.* ix. p. 185.

Adult. Above bright golden yellow; lores, feathers round the eye, hinder crown and nape black, forming a broad horseshoe; sides of face, neck, and entire under surface of body bright yellow, like the back; wing-coverts bright golden yellow; bastard wing, primary-coverts, and quills black, with a spot of yellow at the tip of the outer web of the inner secondaries, increasing on the innermost, where it occupies nearly the entire outer web; tail black, broadly tipped with bright yellow, this occupying the terminal third of the outermost rectrix and gradually decreasing towards the centre of the tail, the two centre feathers being merely tipped with yellow; "bill pink rose-coloured; feet and claws blue-grey" (*Meyer*). Total length 11·5 inches, culmen 1·45, wing 6·2, tail 4·35, tarsus 1·1.

Young male. Differs from the adult in being greener on the back, the quills and tail-feathers brown instead of black; two centre tail-feathers olive greenish, the rest greenish at the base, the tips broadly yellow, with a subterminal blackish shade; head dull yellow, with a horseshoe-mark on the hinder part of the crown dusky black streaked with yellow; sides of face and under surface of body bright yellow, with a few narrow streaks of black on the chest. Total length 10·2 inches, culmen 1·35, wing 5·6, tail 4·3, tarsus 1.

Hab. Philippine Islands.

a. Ad. st.	Cataguan, Luzon.	Hugh Cuming, Esq. [C.].
b. Ad. sk.	Cataguan, Luzon.	Hugh Cuming, Esq. [C.].
c. Ad. sk.	[Philippine Islands.]	Admiral Sir E. Belcher [P.].
d. ♀ ad. sk.	San Bernardino, Negros, Jan. 1875.	Prof. J. B. Steere [C.].
e. ♂ juv. sk.	Dumalon, Mindanao.	Prof. J. B. Steere [C.].

13. Oriolus frontalis.

Orielus frontalis, *Wall. P. Z. S.* 1862, p. 340, pl. xl.; *Finsch, Neu-Guinea*, p. 173; *Gray, Hand-l. B.* i. p. 292.
Broderipus frontatus, *Walden, Tr. Z. S.* viii. p. 61.

Adult male. Above and below brilliant golden orange; head black, with a frontal band of yellow; wing-coverts coloured like the back; bastard wing, primary-coverts, and quills black, with no yellow tips to any of the feathers; two centre tail-feathers black, with a narrow yellow tip, the rest of the feathers black, yellow at the tip, this colour gradually increasing in extent towards the outer feathers, where it occupies more than the terminal third. Total length 11·8 inches, culmen 1·45, wing 6·8, tail 4·95, tarsus 1·1.

Young. Much paler yellow than the adults, and having faint streaks of yellow on the head, the frontal plumes being narrowly tipped with black; wings brown externally, shaded with olive, the primaries with grey; least wing-coverts bright yellow like the back, the greater series brown, externally with a broad margin of yellow; four centre tail-feathers dull olive-yellow, the second pair slightly blackish towards the tips, the rest of the feathers olive-green at base, and tipped with yellow principally on the inner web, the blackish subterminal shade greatly increasing towards the outermost feathers; entire under surface of body bright yellow, including the under wing- and tail-coverts.

Hab. Sula Islands.

a. ♂ ad. sk.	Sula Islands, 1861.	A. R. Wallace, Esq. [C.]. (Type of species.)
b, c. ♂ ad. sk.	Sula Islands, 1861.	A. R. Wallace, Esq. [C.].
d, e. ♀ ad. sk.	Sula Islands, 1861.	A. R. Wallace, Esq. [C.].
f, g. Juv. sk.	Sula Islands, 1861.	A. R. Wallace, Esq. [C.].

Subsp. *a.* **Oriolus suluensis.**

Like *O. frontatus*, but stouter; primary-coverts and quills black, without any yellow tips to the feathers; terminal third of outer tail-feathers yellow. Total length 11·5 inches, culmen 1·45, wing 6·65, tail 4·8, tarsus 1·05.

Hab. Sooloo archipelago.

a, b, c. Ad. sk.	Si Butu, Sooloo Islands.	Hugh Low, Esq. [P.].
d. Juv. sk.	Si Butu, Sooloo Islands.	Hugh Low, Esq. [P.].

14. **Oriolus formosus.**

Oriolus acrorhynchus, *Schl. Mus. P.-B.* Coraces, p. 105.
Oriolus formosus, *Cab. J. f. O.* 1872, p. 392; *Brüggem. Abhandl. Brem.* v. p. 61.
Broderipus formosus, *Rowley, Orn. Misc.* ii. p. lvi.

Adult male. General colour above olive-greenish, the scapulars and wing-coverts being of the same colour, most of the coverts dull blackish at the base and on the inner webs of the greater series; bastard wing and primary-coverts black, the latter slightly shaded with grey at the tips; quills blackish, the primaries externally washed with grey, inclining to white on the margins of the longest; the secondaries olive-green on the outer web, blackish on the inner; a narrow frontlet of bright golden yellow; crown of head and nape black, as also the feathers in front of and below the eye and those above the ear-coverts; cheeks, ear-coverts, sides of neck, and a collar round the hinder neck bright golden yellow; rump and upper tail-coverts bright yellow; tail-feathers bright yellow, with black bases, the latter diminishing towards the outermost, the two centre feathers olive-green, shaded with black near the shaft and towards the end of the feathers, the tips bright yellow; entire under surface of body bright golden yellow, including the thighs and under wing- and tail-coverts; bill flesh-colour. Total length 12·1 inches, culmen 1·5, wing 6·8, tail 5·2, tarsus 1·15.

Adult female. Similar to the male, but a little more green on the back. Wing 6·4 inches, tail 5·2.

Young female. Similar to the adult female, but with all the colours much duller; the frontal patch of yellow more extended, the black crown less so, and slightly streaked with yellow; greater wing-coverts somewhat tipped with yellow; tail olive-greenish, the yellow as in the adult bird, but less bright and only subterminally shaded with black; underneath bright yellow, lighter on the throat; the chest narrowly streaked with black, the sides of the breast shaded with greenish; bill horn-brown. Total length 10·5 inches, culmen 1·55, wing 6·25, tail 5.

Hab. Sanghir Islands.

a. ♂ ad. sk. Sanghir Islands. Darmstadt Museum [E.].

15. Oriolus flavicinctus.

Mimetes flavocinctus, *King, Survey Intertr. Coasts Austr.* ii. p. 419.
Mimeta flavocincta, *Vig. & Horsf. Tr. Linn. Soc.* xv. p. 327; *Bp. Consp.* i. p. 346; *Gould, Handb. B. Austr.* i. p. 466; *Finsch, New Guinea*, p. 173; *Ramsay, P. Z. S.* 1868, p. 385.
Oriolus flavocinctus, *Gray, Gen. B.* i. p. 232; *Gould, B. Austr.* iv. pl. 14; *Schl. M. P.-B.* Coraces, p. 110; *Diggles, Orn. Austr.* part xvi.; *Gray, Hand-l. B.* i. p. 293.

Adult male. Above bright olivaceous yellow, the hinder neck, lower rump, and upper tail-coverts uniform, the rest of the upper surface mottled with arrow-shaped centres of black to all the feathers, the head less distinctly streaked with black; scapulars and least wing-coverts resembling the back; rest of the wing-coverts and quills blackish, washed externally with olivaceous, and rather broadly tipped with pale yellow, inclining more or less to white; edge of the wing bright yellow, with black bases to the feathers; primary-coverts black, very broadly tipped with yellow, and forming a speculum; primaries edged with greyish white towards the tips; tail blackish, all the feathers washed with olivaceous green and tipped with yellow, more largely on the inner web, the olive colour increasing more towards the centre feathers, which have very small yellow tips, and are washed with olive-greenish for the entire extent of both their webs; sides of face olive-yellow, the ear-coverts greener and more dingy; under surface of body olive-yellow, brighter on the abdomen and lower tail-coverts; thighs yellow, with black bases to the feathers; under wing-coverts and axillaries greyish brown, edged and tipped with bright yellow, the outermost yellow, with black bases, giving a mottled appearance; "bill dull red; feet lead-colour; iris reddish orange." Total length 11·5 inches, culmen 1·3, wing 5·8, tail 4·4, tarsus 1·05.

Young. Brown, strongly washed with grey, and slightly tinged with olivaceous here and there, all the feathers indistinctly streaked with brown down the centre, the head thickly streaked with black; lores dull greyish, as also the ear-coverts, the latter streaked with brown and slightly washed with olive; over the eye a few lighter-brown feathers forming a faintly indicated eyebrow; under surface of body white, broadly streaked with brown, the throat much more dingy, tinged with sandy buff, and very thickly streaked with dark brown; under wing-coverts and axillaries light tawny buff, the outermost coverts whitish, with very distinct dark brown bases, giving a mottled appearance; wings above brown, the feathers more or less washed with greyish externally, the coverts distinctly edged with tawny rufous, the primaries externally edged with greyish white, the primary-coverts slightly tipped with whitish; tail dull brown, tipped on the inner web with reddish white, all the feathers washed with grey.

Hab. Northern Australia.

a. Ad. st.	Port Essington.		Capt. Chambers [P.].
b. Juv. st.	Port Essington.		Capt. Chambers [P.].
c. Juv. sk.	Port Essington.		Capt. Chambers [P.].
d. ♀ juv. sk.	Port Essington (*J. Macgillivray*).		Capt. Stanley [P.].
e. ♂ ad. sk.	Port Essington (*J. Macgillivray*).		Capt. Stanley [P.].
f. Juv. sk.	Port Essington.		J. B. Jukes, Esq. [P.].
g. Ad. sk.	Cape York.		Capt. Stanley [P.].

Subsp. *α*. **Oriolus viridissimus.**

Mimeta muelleri, *Bp. Consp.* i. p. 346*; *Finsch, Neu Guinea,* p. 173.
Oriolus muelleri, *Gray, P. Z. S.* 1858, p. 175; *id. Hand-l. B.* i. p. 293.
Mimeta viridissima, *Heine, J. f. O.* 1859, p. 403 (*ex Temm. MS. in Mus. Lugd.*).

Adult male. Above dull greenish yellow, nearly uniform on the hinder neck; the head, mantle, scapulars, and back centred with black in spear-shaped spots; least wing-coverts resembling the back, the median and greater series black, broadly margined with greenish yellow, the latter tipped with very pale yellow, forming a band across the wing; primary-coverts black, narrowly margined with greenish yellow, and broadly tipped with pale yellow, almost white; quills blackish, the primaries externally edged with whitish, the secondaries with greenish yellow, and all tipped with yellowish white, more plainly on the secondaries; tail-feathers blackish on the inner web, washed with olive-green on the outer one, and tipped with a spot of very pale yellow, larger on the inner than on the outer web, the two centre tail-feathers greenish, inclining to black down the centre near the shaft; lores greyish; sides of face and neck nearly uniform yellowish green, the latter with narrow black mesial streaks to the feathers; throat dull yellowish green, with tiny shaft-streaks of black; rest of the under surface of the body rather lighter and brighter yellow, especially on the abdomen and under tail-coverts, most of the plumes narrowly streaked with black down the shafts; under wing-coverts and axillaries yellowish green like the breast, those near the edge of the wing black, tipped with yellow, giving a mottled appearance. Total length 10·6 inches, culmen 1·35, wing 5·95, tail 4·6, tarsus 1·05.

Young. Much more mottled above than the adult, the feathers being blackish, with broad margins of yellowish green, the head lighter than the back, and very thickly and broadly streaked with black; feathers encircling the eye and a broad eyebrow bright yellow, narrowly streaked with black; ear-coverts dull olive-green with black streaks; cheeks and under surface of body bright yellow,

* Bonaparte's description is totally irrecognizable, being merely "Minor: speculo alari albido." Heine's name must be adopted.

washed with green, especially on the lower throat and sides of the body, all the feathers broadly and distinctly streaked with black, these streaks disappearing on the lower abdomen and under tail-coverts, the black streaks so broad on the malar line as to form an indistinct moustache; wings brown, primaries and outer wing-coverts blacker, the latter all tipped with pale yellow or whitish, as also are the quills and the primary-coverts; primaries externally margined with ashy whitish, the secondaries with very pale yellow; tail as in adult.

Hab. Aru Islands.

a, b. ♂ ad. sk.	Aru Islands, 1857.	A. R. Wallace, Esq. [C.].
c, d. Juv. sk.	Aru Islands, 1857.	A. R. Wallace, Esq. [C.].

16. **Oriolus viridifuscus.** (Plate XI.)

Oriolus virescens, *Gray, Gen. B.* i. p. 232 (descr. nullâ, *ex Temm. MS.*); *Schl. Mus. P.-B.* Coraces, p. 111; *Gray, Hand-l. B.* i. p. 293.

Mimeta virescens, *Bp. Consp.* i. p. 346*.

Mimeta viridifusca, *Heine, J. f. O.* 1859, p. 405; *Wall. P. Z. S.* 1863, p. 485.

Oriolus variegatus, *Gray, Hand-l. B.* i. p. 293.

Adult male. General colour above greenish grey, washed with olivaceous yellow on the head and back, the lower back and rump rather paler brown, slightly washed with grey, the upper tail-coverts decidedly greyer; wings brown, the feathers all washed externally with light greyish, the least series washed with olive-yellow like the back, the primary-coverts and secondaries tipped with white, the primaries externally margined with pale rufous; tail-feathers brown, edged with paler and slightly olivaceous brown, tipped with dull whitish, more broadly on the inner web; lores and sides of face ashy grey, the ear-coverts darker brown; throat and breast grey, the latter washed with olivaceous, the lower breast, abdomen, and under tail-coverts creamy brown, slightly tinged with fulvous; under wing-coverts light brown like the belly, those along the edge of the wing dark brown, tipped with greyish. Total length 10 inches, culmen 1·3, wing 5·55, tail 4·6, tarsus 1·05.

Female. Above ashy brown, with whity-brown margins to the feathers, causing a streaked appearance to the upper back; lower back, rump, and upper tail-coverts pale ashy brown, inclining to greyish, especially on the latter; head olivaceous brown, narrowly streaked with black, the hind neck pale fulvous brown, with obsolete streaks of darker brown; over the eye a narrow line of whitish buff, broader on the sides of the head above the ear-coverts, and forming an eyebrow; lores greyish; ear-coverts blackish brown,

* This species must bear the name of *viridifuscus*, inasmuch as Bonaparte's description consists, as in the previous species, of four words. Mr. G. R. Gray reinstates the name of "*variegatus*, Cuv.;" but this appears never to have been published, to judge from Dr. Pucheran's remarks (Arch. Mus. d'Hist. N. vii. p. 339), where he observes that the specimens, though stated to be from Timor, do not appear to differ from *M. viridis* of Australia.

slightly streaked with paler brown; cheeks and malar region blackish, the feathers margined with buffy white, forming a distinct moustachial band running down the sides of the neck; throat white; remainder of under surface pale brown, the under wing-coverts uniform with the breast, the outermost dark brown, with broad fulvous tips; under tail-coverts pale brown, tipped with white; wings above brown, rather broadly edged with rusty red, the primary-coverts and the secondaries tipped with white; tail olivaceous brown, tipped with pale rufous, larger on the inner web. Total length 12·5 inches, culmen 1·25, wing 5·5, tail 5·1, tarsus 1·05.

Young. Similar to the old female, but distinguished by the very broad rufous edgings to the coverts and feathers of the wing; underneath white, washed with pale brown on the sides of the body, the breast streaked with black, the abdomen with pale brown.

Obs. I cannot be quite certain whether I am right in describing the brown birds as being of the same species as the greenish one. In this I am following Schlegel, who has in the Leiden Museum a specimen "dans la livrée de passage," and who ought therefore to be able to judge of the changes of plumage in the species. Gray admits two species; and the specimens collected by Mr. Wallace rather bear out this idea, as the brown birds are both from Semao, and the full-plumaged greenish bird from East Timor. The brown birds, however, are both ascertained females by dissection; so that they may represent the adult hen of the greenish bird, the young one differing as described above.

Hab. Timor group of islands.

a. Juv. sk.	Timor.	Massena Collection.
b. Ad. st.	Timor.	Leiden Museum [E.].
c. Juv. st.	Timor.	Leiden Museum [E.].
d, e. ♀ ad. sk.	Semao Island.	A. R. Wallace, Esq. [C.].
f. ♂ ad. sk.	E. Timor.	A. R. Wallace, Esq. [C.].

17. Oriolus forsteni.

Mimeta forsteni, *Bp. Consp.* i. p. 346; *Heine, J. f. O.* 1859, p. 404; *Finsch, Neu-Guinea*, p. 173.
Oriolus forsteni, *Schl. Mus. P.-B.* Coraces, p. 112; *Gray, Hand-l. B.* i. p. 293.

Adult. Above brown washed with olive, the feathers being obscurely edged with this colour, the hind neck decidedly browner; head dull olive-yellow, all the feathers mesially streaked with blackish; lores dull grey; round the eye a bare space, larger behind it; ear-coverts dark brown; sides of neck grey; under surface of body dull yellow, with brighter yellow margins to the feathers of the breast, the throat much duller and more greenish, with indications of dusky shaft-streaks; lower breast and abdomen ashy grey, with yellowish margins to the feathers, especially distinct on the under tail-coverts; under wing-coverts resembling the breast; wings above brown, edged with olive-yellow like the back, the greater coverts and quills brighter, and edged on the outer web with dull golden yellow; tail brownish, externally washed with olive-yellow, the inner webs slightly tipped with whitish. Total length 12·4 inches, culmen 1·4, wing 5·85, tail 5·3, tarsus 1·2.

Hab. Ceram.

a. Ad. sk.	Ceram, 1859.	A. R. Wallace, Esq. [C.].

18. **Oriolus striatus.**

Oriolus striatus, *Quoy et Gaim. Voy. de l'Astrol. Zool.* i. p. 195, pl. ix. fig. 2; *Gray, Gen. B.* i. p. 232; *Schl. M. P.-B.* Coraces, p. 113; *Gray, Hand-l. B.* i. p. 293.

Mimeta melanotis, *Bp. Consp.* i. p. 346 (*nec Müll. MS.*).

Mimeta striata, *Scl. Journ. Linn. Soc.* ii. p. 159; *Finsch, Neu-Guinea,* p. 173; *Meyer, SB. Akad. Wien,* lxx. p. 218; *Salvad. Ann. Mus. Civ. Genov.* vii. pp. 779, 828.

Adult male. General colour above brown, streaked with black down the centre of the feathers of the hind neck and mantle; the dorsal plumes washed with greyish; head dull greyish, all the feathers broadly streaked with black, these streaks narrower on the nape and hinder neck, which are pale greyish; lores and an indistinct eyebrow dull greyish, narrowly streaked with black; sides of face and throat light greyish, thickly streaked with black, the ear-coverts darker than the sides of the face, the black streaks much thicker on the malar line and sides of the neck, the latter appearing to have a large triangular black patch; rest of under surface of body light ashy brown, streaked with darker brown down the centres of the feathers, these streaks becoming narrower on the abdomen and disappearing on the under tail-coverts, which are tipped with whitish; under wing-coverts and axillaries light reddish brown, the outermost dark brown, margined and tipped with light fulvous brown; wings above dark brown, all the feathers edged with paler brown, some of the least wing-coverts with greyish; tail-feathers dark brown, narrowly edged with paler brown and tipped with light fulvous on the inner web; "eyes dark brown" (*Wallace*). Total length 12 inches, culmen 1·35, wing 5·75, tail 4·95, tarsus 1.

Female. Similar to the male. Total length 12·3 inches, culmen 1·35, wing 5·55, tail 5·2, tarsus 1·05.

Young. Similar to the adult, but more dingy, especially on the head, the breast more broadly and distinctly streaked with black, and having the wing-coverts and quills edged with rusty rufous.

Hab. New Guinea, Salwatti, Mysol, and Waigiou.

a. Juv. sk.	New Guinea.	Zoological Society.
b. ♂ juv. sk.	New Guinea.	A. R. Wallace, Esq. [C.].
c, d. ♂ ad. sk.	Dorey, N. G.	A. R. Wallace, Esq. [C.].
e. Ad. sk.	Port Moresby, N. G.	Rev. S. M'Farlane [C.].
f. ♂ ad. sk.	Waigiou.	A. R. Wallace, Esq. [C.].
g. ♂ ad. sk.	Mysol.	A. R. Wallace, Esq. [C.].

19. Oriolus bouruensis.

Philedon bouruensis, *Quoy et Gaim. Voy. de l'Astr. Zool.* i. p. 192, pl. 8. fig. 2.
Tropidorhynchus bouruensis, *Bp. Consp.* i. p. 390.
Mimeta bouruensis, *Wall. P. Z. S.* 1863, p. 26; *Finsch, Neu-Guinea*, p. 173.
Oriolus buruensis, *Schl. Mus. P.-B.* Coraces, p. 113.

Adult male. General colour above earthy brown, without any shade of olive-green or yellow; head lighter than the back, the feathers centred with darker brown, giving a striped appearance, especially on the forehead, where the plumes are edged with buffy whitish; over the eye a narrow line of whity-brown feathers, getting broader on the sides of the head, whence it encircles the nape, separating the crown from the hinder neck, which is darker brown than the rest of the back; lores, feathers round the eye, and ear-coverts very dark brown, contrasting with the sides of the neck, which are whity brown, here and there spotted with darker brown; cheeks and throat dull whitish, streaked with black, more thickly on the malar line and sides of the throat; rest of the under surface of the body ashy brown, the fore neck whity brown, like the throat; under wing- and tail-coverts slightly rufescent, especially the former; wings above dark brown, the coverts and the primaries narrowly edged with pale rufous, the secondaries almost imperceptibly tipped with whitish; tail-feathers dark brown, the margins rather lighter, tipped with very pale rufous, the terminal spots larger on the inner web. Total length 12·6 inches, culmen 1·5, wing 5·75, tail 5·1, tarsus 1·1.

Young. Similar to the adult, but with broader rufous margins to the wing-coverts; the head more obscurely brown, and not so distinctly striped as in the adult.

Hab. Bouru.

a. ♂ ad. sk.	Bouru.	A. R. Wallace, Esq. [C.].
b, c. Juv. sk.	Bouru.	A. R. Wallace, Esq. [C.].

20. Oriolus phæochromus.

Oriolus phæochromus, *Gray, P. Z. S.* 1860, p. 351; *Schl. M. P.-B.* Coraces, p. 114; *Gray, Hand-l. B.* i. p. 293.
Mimeta phænochroma, *Finsch, Neu-Guinea*, p. 173.
Mimeta phæochroma, *Salvad. Ann. Mus. Civic. Genov.* vii. p. 31.

Adult (type of species). General colour above and below brown, with more or less concealed whitish shafts to the feathers of the mantle; wings and tail rather lighter than the back, with pale fulvous-brown shafts, the primaries rufous at the base of the inner web; on the throat and fore neck a decided shade of grey; bill black; legs blackish in skin. Total length 10·2 inches, culmen 1·15, wing 5·4, tail 4·2, tarsus 0·95.

Hab. Island of Halmahera.

a. Ad. sk.	E. Halmahera.	A. R. Wallace, Esq. [C.]. (Type of species.)
b. Ad. sk.	Gilolo.	A. R. Wallace, Esq. [C.].

21. Oriolus viridis.

Green Grakle, *Lath. Gen. Syn. Suppl.* ii. p. 129.
Gracula viridis, *Lath. Ind. Orn. Suppl.* p. xxviii.
Striated Roller, *Lath. Gen. Syn. Suppl.* ii. p. 122.
Coracias sagittata, *Lath. Ind. Orn. Suppl.* p. xxvi.
Oriolus variegatus, *Vieill. N. Dict. d'Hist. Nat.* xviii. p. 196 (*nec Cuv.*).
Oriolus viridis, *Vieill. N. Dict.* xviii. p. 197; *Wagl. Syst. Av.* Oriolus, sp. 6; *Gould, B. Austr.* iv. pl. 13; *Gray, Gen. B.* i. p. 232; *Schl. M. P.-B.* Coraces, p. 110; *Diggles, Orn. Austr.* part xvi.; *Gray, Hand-l. B.* i. p. 293.
Philemon sagittatus, *Vieill. N. Dict.* xxviii. p. 430.
Mimetes viridis, *King, Survey Intertrop. Coasts Austr.* ii. p. 419.
Mimeta viridis, *Vig. & Horsf. Tr. Linn. Soc.* xv. p. 326; *Jard. & Selby, Ill. Orn.* ii. pl. 61; *Bp. Consp.* i. p. 346; *Cab. Mus. Hein.* Th. i. p. 211; *Gould, Handb. B. Austr.* i. p. 462.
Mimeta meruloides, *Vig. & Horsf. Tr. Linn. Soc.* xv. p. 327; *Gray, Gen. B.* i. p. 232.

Adult. Dull olive-yellowish above, the feathers with indistinct shaft-streaks of blackish, and somewhat of a greyish shade on the edges of many of them; lores dull greyish; sides of face dull olivaceous; wings brown, all the feathers externally washed with grey, the lesser wing-coverts olive-yellowish, duller than the back, the greater series, the primary-coverts, and secondaries narrowly tipped with white, the primaries externally edged with this colour also; tail brown, washed with grey, especially the two centre feathers, all the others tipped largely with white on the inner web, more minutely on the outer one; under surface of body white, all the feathers broadly streaked with black, these streaks being less pronounced on the centre of the abdomen and absent on the under tail-coverts; throat dull olivaceous, the fore neck greyish, with narrow black mesial streaks; breast and sides of body washed with olive-yellow; bill dull fleshy red; feet lead-colour; iris scarlet. Total length 10 inches, culmen 1·2, wing 5·95, tail 4·5, tarsus 1.

Obs. "The sexes, when fully adult, differ so little in colour that they can scarcely be distinguished; the male, however, is of a more uniform tint about the head, neck, and throat, and has the yellowish olive of the upper surface of a deeper tint than the female." (*Gould.*)

Young. Browner than the adult, slightly washed with olivaceous above, the feathers mesially streaked with blackish, more closely and numerously on the crown; over the eye an indistinct buffy white eyebrow, streaked with dark brown; sides of face and ear-coverts uniform dull brownish; fore part of cheeks and under surface of body white, broadly streaked with blackish brown, these streaks much narrower on the abdomen and absent on the under tail-coverts; wings brown, the primaries externally washed with greyish, the wing-coverts and secondaries rather broadly margined with pale rufous; tail greyish brown, tipped with white, much more largely on the inner web; under wing-coverts sandy buff, the outermost with black bases, giving a mottled appearance to the edge of the wing; bill blackish brown.

Hab. Australia.

a.	Ad. sk.	Australia.	
b.	Ad. sk.	Australia.	Capt. King [P.].
c.	Juv. sk.	Australia.	Linnean Society [P.]. (Type of *M. meruloides.*)
d.	♀ ad. sk.	Claremont Isles, N.E. Australia, Sept. 4, 1848 (*J. Macgillivray*).	Capt. Stanley [P.].
e,f.	Ad. sk.	Queensland.	Purchased.

22. **Oriolus steerii.** (Plate X.)

Adult. General colour olive-yellow, darker on the head and rather brighter towards the lower back and rump; wing-coverts olive-yellow like the back, the greater series dark grey, externally olive-yellow; quills blackish, externally washed with grey, the secondaries with olive-yellow, the innermost being entirely of this colour; two centre tail-feathers olive-yellow, with an indistinct spot of black near the tip; rest of the tail-feathers jet-black, yellow towards the base of the inner web, and having a large spot of yellow near the tip of the inner web, these spots smaller towards the centre of the tail, the outer webs also olivaceous at their tips; ear-coverts dull yellowish olive; lores, cheeks, throat, and entire breast uniform bluish grey; rest of under surface of body white, very broadly streaked with black; under tail-coverts bright yellow, with dusky basal streaks; under wing-coverts grey like the breast, the outermost with olivaceous edgings; bill red. Total length 9 inches, culmen 1·05, wing 4·75, tail 3·35, tarsus 0·9.

Hab. Philippine Islands: Negros and Basilan.

a.	Ad. sk.	Negros.	Dr. J. B. Steere [C.]. (One of the typical specimens.)

23. **Oriolus xanthonotus.**

Oriolus xanthonotus, *Horsf. Tr. Linn. Soc.* xiii. p. 153; *Temm. Pl. Col.* 214; *Horsf. Zool. Research. in Java, Birds,* pl. 12; *Wagl. Syst. Av.* Oriolus, sp. 9; *Gray, Gen. B.* i. p. 232; *Blyth, J. A. S. B.* xv. p. 49; *id. Cat. B. Mus. A. S. B.* p. 346; *Horsf. & Moore, Cat. B. Mus. E.I. Co.* i. p. 268; *Pelz. Reis. Novara, Vög.* pp. 75, 151; *Schl. Mus. P.-B.* Coraces, p. 109; *Gray, Hand-l. B.* i. p. 292; *Salvad. Ucc. Born.* p. 277.
Oriolus leucogaster, *Temm. Pl. Col.* 214 (texte, *ex Reinw. MS.*).
Oriolus castanopterus, *Blyth, J. A. S. B.* xi. p. 796; *Gray, Gen. B.* i. p. 232.
Xanthonotus leucogaster, *Bp. C. R.* xxxviii. p. 538.

Adult male. Above from the nape to the upper tail-coverts bright yellow; scapulars blackish, externally yellow; wings entirely black excepting the least wing-coverts, which are bright yellow like the back, some few of the secondaries exhibiting narrow yellowish tips, and some of the primaries with a narrow edging of whitish on the outer web; tail black, tipped with yellow, this latter colour principally confined to the inner web and increasing in size towards the outer feathers of the tail; head and neck all round, as well as the fore neck, black; rest of under surface white, with a yellow tinge on the chest near the black gorget; all the breast and abdomen with distinct mesial streaks of black; under tail-coverts bright yellow; thighs greyish black, with yellow tips to the feathers; under wing-coverts black, with distinct yellow edgings to the feathers near the bend of the wing. Total length 7·5 inches, culmen 0·9, wing 4·35, tail 3·1, tarsus 0·75.

Obs. Remains of immaturity are seen in some specimens in the shape of black edgings to the feathers of the back, the upper tail-coverts also being centred with black.

Adult female. Above dull olive-yellow, greener on the head; the rump and upper tail-coverts much brighter yellow; wing-coverts dull olive-yellow like the back, the median coverts slightly, the greater ones broadly edged with chestnut near the tips, these rufous edgings faintly indicated on the primary-coverts and on the inner secondaries; quills greyish black, the primaries edged with grey, inclining to white near the tip, the secondaries externally olive-yellow, a little brighter than the back; tail olive-green, all the feathers with a large spot of bright yellow at the tip of the inner web, increasing in extent towards the outer feathers, most of which have a more or less distinct subterminal shade of blackish; round the eye a ring of bright yellow feathers; sides of face greyish brown, washed with green; rest of under surface of body dull white, purer on the abdomen, all the feathers mesially streaked with black, these streaks browner and less distinct on the throat and fore neck, and narrowing considerably on the abdomen; under tail-coverts bright yellow; thighs greyish black, with yellow tips to the feathers; under wing-coverts greyish, washed with olive-yellow, especially bright near the edge of the wing, the lower series greyish, with whitish tips, the lower surface of the quills greyish brown, the inner webs light yellow for the greater part of their extent. Total length 7 inches, culmen 0·85, wing 4·1, tail 2·95, tarsus 0·8.

Hab. Malacca, Java, Sumatra, Borneo.

a. ♂ ad. sk.	Java (*Horsfield*).	Secretary of State for India [P.]. (Type of species.)
b. ♂ ad. sk.	Sumatra, 1861.	A. R. Wallace, Esq. [C.].
c, d. ♂ ♀ ad. sk.	Malacca, 1854.	A. R. Wallace, Esq. [C.].
e. [♀] ad. sk.	Malacca, 1854.	Dr. Cantor [C.].
f, g. ♂ ♀ ad. sk.	Malacca, 1854.	W. Harvey, Esq. [P.].
h. ♂ ad. sk.	Sarawak, Borneo.	A. R. Wallace, Esq. [C.].
i, k. ♂ ad. sk.	Borneo.	
l, m. ♂ ♀ ad. sk.	Borneo.	Hugh Cuming, Esq. [C.].
n. ♂ ad. sk.	Bintulu.	A. Everett, Esq. [C.].
o. ♀ ad. sk.	Labuan.	Hugh Low, Esq. [C.].
p. ♂ ad. st.	Indian archipelago.	F. Nightingale, Esq. [P.].
q. ♀ ad. st.	Indian archipelago.	F. Nightingale, Esq. [P.].

24. Oriolus melanocephalus.

The Black-headed Indian Icterus, *Edwards, Birds*, p. 77, pl. 77.
Le Loriot de Bengale, *Briss. Orn.* ii. p. 329.
Oriolus melanocephalus, *Linn. S. N.* i. p. 160; *Wagler, Syst. Av.* Oriolus, sp. 8; *Jerd. Madr. Journ.* x. p. 262; *Gray, Gen. B.* i. p. 232; *Blyth, J. A. S. B.* xv. p. 45; *id. Cat. B. Mus. A. S. B.* p. 215; *Bp. Consp.* i. p. 346; *Horsf. & Moore, Cat. B. Mus. E.I. Co.* i. p. 269; *Jerd. B. Ind.* ii. p. 110; *Schl. Mus. P.-B.* Coraces, p. 106; *Gray, Hand-l. B.* i. p. 292; *Ball, S. F.* 1873, p. 71, 1874, p. 412; *Hume, S. F.* 1874, p. 477, 1875, p. 133; *id. Nests and Eggs Ind. B.* p. 301; *Blyth & Walden, B. Burm.* p. 139.
Loriot de la Chine, *Montb. Pl. Enl.* 79.
Loriot rieur, *Levaill. Ois. d'Afr.* vi. pl. 263.
Oriolus maderaspatanus, *Frankl. P. Z. S.* 1831, p. 118.
Oriolus maccoshii, *Tickell, J. A. S. B.* ii. p. 577.
Oriolus hodgsonii, *Swains. An. in Menag.* p. 290; *Blyth, J. A. S. B.* xi. p. 400; *Hodgs. in Gray's Zool. Misc.* p. 83; *Gray, Gen. B.* i. p. 232.
Oriolus strigipectus, *Hodgs. in Gray's Zool. Misc.* p. 83.

Adult male. General colour, both above and below, bright golden yellow, intensifying more or less to orange; head all round, throat, and fore neck black; scapulars and wing-coverts bright yellow like the back; bastard wing and primary-coverts black, the latter yellow for their terminal half, forming a large alar speculum; quills black, yellow at the base, and having a longitudinal spot of yellow at the tip of the outer web, increasing in extent towards the secondaries, the innermost of which have the entire outer web yellow; tail bright yellow, the two middle feathers with a broad subterminal bar of black, the next feathers on either side with a patch of black on the inner web; bill pale lake-red; legs plumbeous; iris rich red. Total length 10 inches, culmen 1·15, wing 5·55, tail 4, tarsus 1·05.

Young. Not so bright yellow as the adult, the head being blackish, with olive margins to many of the feathers; frontal plumes yellow, streaked with black; sides of face dusky blackish; throat white, longitudinally streaked with black; rest of under surface bright yellow; wing-coverts black, with broad yellow margins; quills brownish black, the primaries externally edged with whitish, the secondaries with a broad longitudinal spot of yellow near the tip of the outer web; tail bright yellow, the two centre feathers dark olive-green, the rest with a blackish subterminal shade, decreasing in extent towards the outermost, on which the black is confined to a longitudinal patch along the outer web.

Hab. Peninsula of India and the Himalayas, extending through the Burmese countries to Pegu and Tenasserim.

a. Ad. st.	India.	
b, c, d. Ad. sk.	Nepal.	B. H. Hodgson, Esq. [P.].
e, f. Juv. sk.	Nepal.	B. H. Hodgson, Esq. [P.].
g, h. Ad. sk.	Behar.	B. H. Hodgson, Esq. [P.].
i. Juv. sk.	Behar.	B. H. Hodgson, Esq. [P.].
k. Juv. sk.	Kumaon (*Strachey*).	Secretary of State for India [P.].
l. Juv. sk.	Bootan.	Secretary of State for India [P.].

m, n. ♂ ♀ ad. sk.	N.W. Himalayas.	Capt. Stackhouse Pinwill [P.].
o. Juv. sk.	N.W. Himalayas.	Capt. Stackhouse Pinwill [P.].
p. Ad. sk.	Kamptee, C. India.	Dr. Hinde [P.].
q. Juv. sk.	Kamptee, C. India.	Dr. Hinde [P.].
r, s, t, u. Ad. sk.	Madras.	Rev. H. H. Baber [P.].
v. Ad. sk.	Travancore.	Capt. Biddulph [C.].
w. Juv. sk.	Burmah (*Day*).	Secretary of State for India [P.].
x. Juv. sk.	Bhamo, Upper Burmah, Jan. 28, 1868.	Dr. Anderson [C.].
y. Ad. sk.	Pegu.	W. T. Blanford, Esq. [P.].
z. Juv. sk.	Pegu.	W. T. Blanford, Esq. [P.].
a'. Ad. sk.	Tenasserim.	J. C. D. Packman, Esq. [P.].

Subsp. *a.* **Oriolus ceylonensis.**

Oriolus ceylonensis, *Bp. Consp.* i. p. 347; *Jerd. B. Ind.* ii. p. 111; *Schl. Mus. P.-B.* Coraces, p. 107; *Gray, Hand-l. B.* i. p. 292; *Holdsw. P. Z. S.* 1872, p. 453; *Hume, Str. F.* 1873, p. 439.
Oriolus melanocephalus, *Hume, Str. F.* 1874, p. 230.

Adult. Similar to *O. melanocephalus*, but smaller, and having a very much smaller alar speculum. Total length 8·5 inches, culmen 1·1, wing 4·7, tail 3·4, tarsus 1.

Obs. Dr. Jerdon (*l. c.*) separates the Black-headed Oriole of Southern India as *O. ceylonensis*, affirming that it is the same as the Ceylonese bird. Although rather smaller than Himalayan examples, the alar speculum is larger than in the bird of Ceylon; and I look upon it as an intermediate form between the continental and the insular birds. *O. ceylonensis* is not at all a good species, and perhaps in a large series would be found to be inseparable from *O. melanocephalus*. The Andaman bird seems to be the same as the Ceylonese one, with a small speculum; but I have only a single specimen to judge from, and that one not fully adult. The Madras specimens have the wing measuring from 5·05 to 5·25 inches.

Hab. Ceylon and the Andaman Islands.

a. Ad. sk., *b, c.* Juv. sk.	Ceylon.	E. L. Layard, Esq. [C.].
d, e. Ad. sk., *f.* Juv. sk.	Newara Ellia.	Mr. E. Boate [C.].
g. ♀ juv. sk.	Port Blair, S. Andamans, July 4, 1873 (*Wimberley*).	R. B. Sharpe, Esq. [P.].

25. **Oriolus monachus.**

Moloxita ou la Religieuse d'Abyssinie, *Montb. Hist. Nat. Ois.* iii. p. 405.
Nun Thrush, *Lath. Gen. Syn.* ii. p. 77.
Turdus monacha, *Gm. S. N.* i. p. 824.
Nun Oriole, *Lath. Gen. Hist. B.* iii. p. 142.
Oriolus monacha, *Steph. Gen. Zool.* xiii. p. 193; *Sharpe, Ibis*, 1870, p. 220; *Blanf. Geol. & Zool. Abyss.* p. 392.
Oriolus monachus, *Cab. Mus. Hein.* Th. i. p. 210; *Heugl. Orn. N.O.-Afr.* i. p. 402; *Gray, Hand-l. B.* i. p. 292; *Sharpe, Cat. Afr. B.* p. 53.

Oriolus moloxita, *Rüpp. N. W. Vög.* p. 29, t. 12. fig. 1; *id. Syst. Uebers.* p. 65; *Gray, Gen. B.* i. p. 232; *Bp. Consp.* i. p. 347; *Schl. Mus. P.-B.* Coraces, p. 108; *Gray, Hand-l. B.* i. p. 292.

Head black, as also the entire throat, extending a little way down the breast; upper surface of the body olive-yellow, brighter on the nape, sides of the neck, and rump; wing-coverts olive-yellow; primary-coverts black, white at the tip, forming a very distinct white alar speculum; quills black, very light grey on the underside, the inner secondaries olive-yellow on the outer web, the outer ones very light blue-grey; the primaries white on the basal portion of the outer web, and becoming light brown towards the apical portion; tail pure yellow, except the middle feathers, which are olive-green, edged and tipped with yellow; under surface of the body bright golden yellow; bill brownish red; feet bluish grey. Total length 9 inches, bill from front 0·9, from gape 1·1, wing 5·4, tail 3·1, tarsus 0·8, middle toe 0·8, hind toe 0·35.

Hab. North-eastern Africa.

a.	Ad. sk.	Abyssinia.	
b.	Ad. st.	Abyssinia.	Dr. Rüppell [C.].
c.	Ad. sk.	Abyssinia.	R. B. Sharpe, Esq.
d.	♀ ad. sk.	Mishek, south of Antalo, May 10, 1868.	W. T. Blanford, Esq. [C.].

26. Oriolus crassirostris.

Oriolus crassirostris, *Hartl. Orn. Westafr.* p. 266; *Sharpe, Ibis,* 1870, p. 221.

Similar to *O. larvatus*, but larger and with a much shorter bill. For differences between these two species see Dr. O. Finsch's remarks in my paper on African Orioles (*l. c.*), where a life-sized figure of the heads of these two species is given.

Hab. Island of S. Thomas, West Africa.

27. Oriolus larvatus.

Le Coudougnan, *Le Vaill. Ois. d'Afr.* pl. 261.
Oriolus coudougnan, *Temm. Rec. d'Ois.* livr. 54.
Le Loriot à masque noir, *Temm. Cat. Syst. Cab. d'Orn.* pp. 46, 203.
Oriolus larvatus, *Licht. Verz. Doubl.* p. 20; *Bp. Consp.* i. p. 347; *Cab. Mus. Hein.* Th. i. p. 210; *Grill, Zool. Anteckn.* p. 34; *Mont. P. Z. S.* 1865, p. 93; *Schl. Mus. P.-B.* Coraces, p. 107; *Heugl. Orn. N.O.-Afr.* p. 403; *Sharpe, Ibis,* 1870, p. 223; *id. Cat. Afr. B.* p. 53; *id. P. Z. S.* 1873, p. 714; *Bocage, Jorn. Lisb.* ii. p. 342; *Shelley, Ibis,* 1875, p. 75; *Barratt, Ibis,* 1876, p. 204.
Oriolus capensis, *Swains. Classif. B.* ii. p. 237; *id. B. W. Afr.* ii. p. 37.
Oriolus melanocephalus, *Des Murs in Lefebvr. Voy. Abyss. Zool.* p. 169 (*nec Linn.*).
Oriolus monachus, *Wagl. Syst. Av.* Oriolus, no. 7 (*nec Gm.*).
Oriolus arundinarius, *Burch. Trav. S. Afr.* i. p. 464.
Oriolus chloris, *Cuv. MS.*; *teste Hartl. Orn. Westafr.* p. 81.

Oriolus rolleti, *Salvad. Atti R. Accad. Torin.* vii. p. 151; *Heugl. Orn. N.O.-Afr.* i. p. 404.
Oriolus personatus, *Heugl. J. f. O.* 1867, p. 203.
Oriolus brachyrhynchus, *Schl. M. P.-B.* Coraces, p. 108 (*nec Swains.*).

Adult male. Whole of the head deep glossy black, extending down to the breast; nape and sides of the neck rich golden yellow; whole of the back and scapulars bright yellow, with a slight olive tinge on the latter, and brighter yellow on the rump; wing-coverts black, but so broadly edged with yellow that the black does not show; the outermost cubital coverts edged with grey; the primary-coverts black, tipped with white, forming a distinct speculum; quills black, the inner web paler, especially in the secondaries; the secondaries broadly edged with yellow on the outer web, those nearest the primaries having also a margin of white, which in the primaries themselves entirely takes the place of the yellow, so that these are edged and tipped with pure white; tail-feathers for the most part yellow, black at the base; as they approach the middle feathers the black colour occupies the most part of the feather, being, however, strongly tinged with olive on those nearest the middle feathers, which are entirely olive-green; under surface of the body brilliant golden yellow; bill brownish red; feet black. Total length 9 inches, bill from front 1·1, from gape 1·4, wing 5·4, tail 3·2, tarsus 0·9, middle toe 0·8, hind toe 0·4.

Young male. Similar to the old male, but the colours not nearly so pure, with indistinct brown stripes on the feathers of the upper part of the body; head dusky brown; quills brown, edged exteriorly with lighter brown; breast with black longitudinal markings; tail darker olive-green.

Hab. Southern and North-eastern Africa, the birds from the latter place being rather smaller, as is also the race from Angola.

a. Ad. st.	Cape of Good Hope.	Maj.-Gen. Hardwicke [P.].
b. Ad. sk.	South Africa.	J. Rocke, Esq. [P.].
c, d. ♂ ad. sk.	Eland's Post, S.E. Africa, May 20, 1870 (*T. C. Atmore*).	R. B. Sharpe, Esq.
e. ♂ juv. sk.	Eland's Post, June 8, 1870 (*T. C. Atmore*).	R. B. Sharpe, Esq.
f. ♂ juv. sk.	Knysna, January 2, 1865 (*Andersson*).	R. B. Sharpe, Esq.
g. Ad. sk.	Natal.	Thomas Ayres, Esq. [C.].
h. Juv. sk.	Natal.	
i. Ad. sk.	Rustenberg, S.E. Africa.	F. A. Barratt, Esq. [C.].
k. Ad. sk.	Angola.	J. J. Monteiro, Esq. [C.].
l. Juv. sk.	Angola.	
m. Ad. sk.	Rio Dande, Angola, June 24, 1869 (*Sala*).	R. B. Sharpe, Esq.
n. Juv. sk.	Mombas (*Wakefield*).	R. B. Sharpe, Esq. [P.].

28. **Oriolus brachyrhynchus.**

Oriolus brachyrhynchus, *Swains. B. W. Afr.* ii. p. 35; *Bp. Consp.* i. p. 347; *Hartl. Orn. W.-Afr.* p. 81; *id. J. f. O.* 1861, p. 162;

Sharpe, Ibis, 1870, pp. 57, 226, pl. viii. fig. 1; *id. Cat. Afr. B.* p. 54; *id. P. Z. S.* 1871, p. 611.
Oriolus baruffi, *Bp. Consp.* i. p. 347; *Schl. Mus. P.-B.* Coraces, p. 109; *Sharpe, Ibis*, 1869, p. 383; *id. Ibis*, 1870, p. 227, pl. viii. fig. 2.
Oriolus intermedius, *Hartl. Beitr. Orn.-Westafr.* p. 24, *et Orn. Westafr.* p. 24 (*ex Temm. MS. in Mus. Lugd.*); *id. J. f. O.* 1861, p. 163; *Cass. Pr. Philad. Acad.* 1859, p. 43.
Baruffius intermedius, *Bp. C. R.* xxxviii. p. 538.

Nestling. Bare on the head, rump, and underparts generally, excepting for some tracts of yellowish feathers along the sides of the body; the sprouting tail-feathers bright yellow, with dusky bases; plumes of the upper surface dusky olive-brown, with edgings of lemon-yellow; the wing-coverts and secondaries coloured like the dorsal plumes; primary-coverts and quills blackish, shaded with olive-yellow, and externally with grey, inclining to white on the outer margin.

Full-grown young bird. Above dull olive-yellow, brighter on the rump, duskier on the head; ear-coverts and feathers round the eye somewhat greyish; lores, cheeks, and throat light yellowish, the latter mottled with brighter yellow, all the plumes with dusky bases; rest of under surface of body bright yellow; wing-coverts olive-yellow, the least ones, as well as the inner greater coverts and the secondaries (which resemble the coverts), tipped with bright yellow; the rest of the wing-coverts blackish, externally shaded with grey and tipped with white, forming a speculum on the primary-coverts; rest of the quills blackish, the primaries externally grey, edged with white on the outer web; upper tail-coverts deep olive-yellow, with obsolete tips of brighter yellow; tail olive-green at base, the two centre feathers entirely of this colour, all the other feathers bright yellow at the tip, with a more or less distinct subterminal shade of black, the yellow increasing towards the outermost feathers, and extending chiefly along the outer webs.

Adult. Head glossy black, likewise the throat, extending a little way on the breast; upper surface of the body deep olive-yellow, brighter yellow on the nape, sides of the neck, and rump; wing-coverts of the same colour as the back; inner cubital coverts edged with olive-yellow, the two or three outer ones edged with grey; primary-coverts black, tipped with white, forming a distinct white alar speculum; quills brownish black, quite white on the inner web; the inner secondaries bright olive-yellow on the outer web, the outer ones grey, with a narrow outer edging of white; primaries edged with white on the outer web; the four middle tail-feathers olive-yellow, the outer ones olive-yellow at the base, bright golden yellow at the tip, preceded by a bar of black; entire under surface rich golden yellow; bill brownish red; feet black.

Female or young. The two middle rectrices olive-yellow, the two next with a black bar before the yellow tip.

Obs. The black head seems to be gradually assumed, several of the Museum specimens showing a tendency to it even in the young plumage.

Since I previously discussed the question of the identity of *O. baruffi* with *O. brachyrhynchus* ('Ibis,' 1870, pp. 226–228), the material at my disposition has been greatly increased, and I now believe them to constitute only one species. The chief difference pointed out by me (*l. c.*) was in the marking of the four tail-feathers, *O. baruffi* having the feather on each side next to the middle one marked with black. This, however, seems to vary with age, being present in the young birds and gradually disappearing with age, the yellow tips, however, increasing in extent as the bird gets older.

The present species in its adult stage is very closely allied to the small form of *O. larvatus*, called by Salvadori *O. rolleti*; but, independently of its smaller size and the totally different aspect of the young bird, the West-African species may be recognized by its much smaller alar speculum and by its olive-yellow secondaries, coloured like the back, with only a little of the inner web blackish; the secondaries in *O. larvatus* or *O. rolleti* are black externally, edged with olive-yellow or greenish.

Hab. West Africa, from Sierra Leone to Gaboon.

a. Ad. sk.	West Africa.	M. Verreaux.
b. Pull. sk.	Denkera (*Aubinn*).	R. B. Sharpe, Esq. [P.].
c. Juv. sk.	Denkera (*Aubinn*).	R. B. Sharpe, Esq. [P.].
d, e. Juv., *f, g.* Ad. sk.	Fantee (*Aubinn*).	R. B. Sharpe, Esq.
h, i, k. Ad. sk.	Fantee (*Ussher*).	R. B. Sharpe, Esq.
l. Ad. sk.	Denkera (*Ussher*).	R. B. Sharpe, Esq.
m. Juv. sk.	Accrá (*Haynes*).	R. B. Sharpe, Esq.
n. Ad., *o.* Imm. sk.	Accrá.	R. B. Sharpe, Esq. [P.].
p, q, r. Ad. sk.	Cameroons (*Crossley*).	R. B. Sharpe, Esq.
s. Ad. sk.	Gaboon (*Duchaillu*).	R. B. Sharpe, Esq.
t. Ad. sk.	River Danger, Gaboon.	Henry Ansell, Esq. [P.].

29. Oriolus nigripennis.

Oriolus nigripennis, *Verr. J. f. O.* 1855, p. 105; *Hartl. Orn. Westafr.* p. 85; *Cass. Pr. Philad. Acad.* 1859, p. 43; *Hartl. J. f. O.* 1861, p. 183; *Sharpe, Ibis,* 1869, p. 56; *id. Ibis,* 1870, p. 228, pl. vii. fig. 1; *id. Cat. Afr. B.* p. 54.

Entire head, throat, and upper part of the breast deep glossy black; back golden green, pure yellow on the nape and sides of the neck, and brighter yellow on the rump; wing-coverts blackish, broadly edged with olive-yellow; quills black, paler on the underside, the secondaries edged with olive-yellow, primaries edged with white towards the tip; middle tail-feathers deep glossy black, with a very narrow yellow edging at the tip, the other feathers black at the base, the outer ones for the most part yellow, and the inner ones for the most part black, those nearest the middle having most of the latter colour; under surface of the body very rich golden yellow; bill brownish red; feet black.

Hab. West Africa, from the Gold Coast to Gaboon.

a. Ad. st.	West Africa.	
b, c. Ad. sk.	West Africa.	Purchased of M. Verreaux. (Types of species.)

d. Ad. sk.	Enimill, Wassaw, March 1872 (*Blissett*).	R. B. Sharpe, Esq.
e. Ad. sk.	Fantee (*Ussher*).	R. B. Sharpe, Esq. (Spec. figured in Ibis, 1870, pl. vii. fig. 1.)

30. **Oriolus cruentus.**

Leptopteryx cruenta, *Wagler, Syst. Av. addit.* Leptopteryx, sp. 9.
Lanius cruentus, *Drap. Dict. Class.* xiii. p. 527.
Ocypterus sanguinolentus, *Temm. Pl. Col.* iii. pl. 499.
Analcipus sanguinolentus, *Swains. in Faun. Bor.-Amer.* p. 482; *id. Class. B.* ii. p. 222; *Bp. Consp.* i. p. 345; *Gray, Hand-l. B.* i. p. 290.
Artamia sanguinolenta, "*Isid. Geoffr.*," *Less. R. Z.* 1839, p. 197.
Philocarpus sanguinolentus, *Müll. Tijdschr. Nat. Gesch. Physiol. Amsterd.* ii. p. 331.
Erythrolanius rubropectus, *Less. R. Z.* 1840, p. 274.
Oriolus sanguinolentus, *Gray, Gen. B.* ii. p. 232.
Artamus sanguinolentus, *Low, Sarawak*, p. 410.
Analcipus cruentus, *Salvad. Ucc. Born.* p. 278.

Adult. General colour glossy blue-black, duller below, the centre of the breast deep crimson; wings and tail black, with the exception of the inner primary-coverts, which are crimson at the tip, forming a distinct alar speculum. Total length 9·8 inches, culmen 1·05, wing 5·35, tail 4·9, tarsus 0·9.

Hab. Java, Sumatra, Borneo.

a, b. Ad. st.	Indian archipelago.	Baron Laugier.
c. Ad. sk.	West Java.	A. R. Wallace, Esq. [C.].

31. **Oriolus ardens.**

Psaropholus ardens, *Swinh. Ibis*, 1862, p. 363, 1863, p. 293, 1866, pp. 297, 398; *id. P. Z. S.* 1871, p. 375.
Oriolus ardens, *Schl. M. P.-B.* Coraces, p. 115.
Analcipus ardens, *Gray, Hand-l. B.* i. p. 290.

Adult. General colour above and below deep crimson, the scapulars like the back, but more or less black externally; head all round black, this colour occupying also the entire fore neck; thighs blackish; under wing-coverts black, washed with crimson on the edge of the wing; wings above entirely black; tail crimson, not so bright as on the back; "bill light cobalt; skin round eye violet-grey; legs leaden grey; claws and soles dusky; iris crimson-brown" (*Swinhoe*). Total length 9·5 inches, culmen 1·1, wing 6·05, tail 4·35, tarsus 0·95.

Hab. Island of Formosa.

a. Ad. sk.	N. Formosa.	R. Swinhoe, Esq. [C.].

32. **Oriolus nigellicaudus.**

Psaropholus ardens, var. nigellicauda, *Swinh. Ibis*, 1870, p. 342; *id. P. Z. S.* 1871, p. 374.

Adult male. Plumage as in *O. ardens* of Formosa, except as regards the tail, which, instead of being entirely crimson, has black shafts to the feathers, the two middle ones being washed with black, and the rest with their outer webs for the most part black. Length of wing 5·7, tail 4·125. (*R. Swinhoe.*)

Young male. Head and tibiæ blackish brown; back brownish red; scapulars, wings, sides of breast and belly, and axillaries brown; throat, breast, and belly white, with long blackish brown spots; rump, vent, and a new feather or two on the breast crimson; tail pale crimson, the two middle feathers entirely washed with brown, the rest only on their outer webs; shafts of feathers black. Length of wing 5·6, tail 4·125. (*R. Swinhoe.*)

All the Formosan birds, in as mature plumage as the Hainan adult, have purely crimson tails. In the Formosan bird the black disappears; in the Hainan bird it intensifies with age. (*R. Swinhoe.*)

Hab. Hainan.

33. **Oriolus traillii.**

Pastor traillii, *Vigors, P. Z. S.* 1831, p. 175; *Gould, Cent. B. Himal. Mts.* pl. 35; *Hodgs. J. A. S. B.* v. p. 772.

Oriolus trailli, *M'Clell. P. Z. S.* 1839, p. 160; *Blyth, J. A. S. B.* xi. p. 797; *Gray, Gen. B.* i. p. 232; *Blyth, Cat. B. Mus. A. S. B.* p. 215; *Horsf. & Moore, Cat. B. Mus. E.I. Co.* i. p. 272; *Jerd. B. Ind.* i. p. 112; *Schl. Mus. P.-B.* Coraces, p. 114; *Hume, Str. F.* 1874, p. 477.

Psarophólus trailli, *Jard. & Selby, Ill. Orn.* iv. pl. 26; *Bp. Consp.* i. p. 345; *Gould, B. Asia,* part xxiii.; *Blyth & Wald. B. Burm.* p. 139.

Analcipus traillii, *Gray, Hand-l. B.* i. p. 290.

Young. Above dark brown, the head and hind neck blacker: rump and upper tail-coverts deep maroon, shaded with brown; wings blackish, with paler brown edgings; tail pale maroon, with more or less broad external edgings of dull brown; lores fulvous, with streaks of brown; ear-coverts blackish; cheeks and under surface of body buffy white, streaked with black, narrowly on the throat, more broadly on the breast, the sides of the upper breast being nearly uniform blackish, as are the thighs and under wing-coverts; under tail-coverts maroon.

Adult. General colour above and below shining maroon-red, the feathers with white bases; head all round glossy black, confined to the throat and jugulum below; wings black, glossed with greenish, some of the scapulars adjoining the wing externally black; tail maroon-red, but not so brilliant as the back; thighs and under wing-coverts black; bill bluish; legs dark plumbeous; iris pale yellow. Total length 10 inches, culmen 1·15, wing 5·9, tail 4·6, tarsus 0·95.

Obs. The very young birds have all the feathers of the upper surface broadly margined with rusty buff, the maroon upper tail-coverts mesially centred with black, while the central tail-feathers are dark maroon-brown, all the others being externally of this colour; the under surface is distinctly striped. In the next stage, which is fully described above, the rusty margins to the dorsal plumes are lost, and the under surface becomes gradually clouded until the throat and head are uniform black; at the same time a shade of maroon appears here and there, both on the upper and under surface, which becomes more and more developed till the full plumage is assumed.

Hab. Himalaya Mountains, extending into Assam, Arakan, and Tenasserim.

a. Juv. sk.	Darjiling (*Jerdon*).	C. Bygrave Wharton, Esq. [P.].
b. Ad. st.	Nepal.	B. H. Hodgson, Esq. [P.].
c,d. Ad., *e–k.* Juv. sk.	Nepal.	B. H. Hodgson, Esq. [P.].
l. Ad., *m–p.* Juv. sk.	Nynee Tal.	Capt. Stackhouse Pinwill [P.].
q. Juv. sk.	Bootan.	Secretary of State for India [P.].
r. Ad. sk.	Kumaon (*Strachey*).	Secretary of State for India [P.].

2. SPHECOTHERES.

	Type.
Sphecothera, *Vieill. N. Dict. d'Hist. Nat.* xxxii. p. 5 (1819)	S. viridis.
Picnorhamphus, *Von Rosenb. Tijdschr. Nederl. Ind.* xxix. p. 143 (1866)	S. flaviventris.

Range. Australia, New Guinea, Ké Islands, Timor.

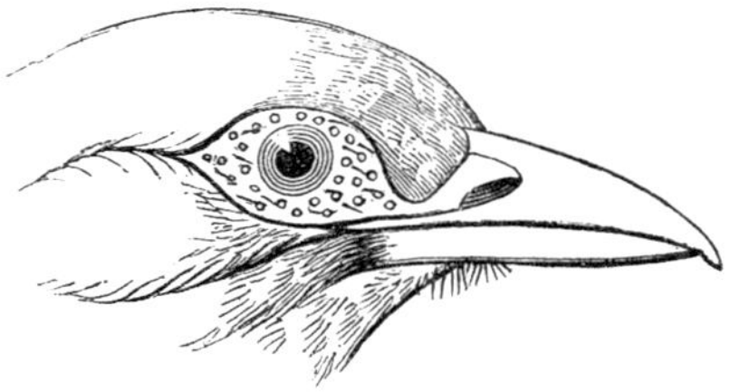

Head of *Sphecotheres flaviventris*, to show the bare face.

Key to the Species.

a. Outer tail-feather with external web white, as well as the terminal half or third of the inner web.
  a'. Throat and fore neck grey, like the sides of the neck.
    a''. Larger; wing 5·95; tarsus 1; middle of abdomen white .... *maxillaris*, p 224.
    b''. Smaller; wing 5·55, tarsus 0·9; middle of abdomen yellow .... *salvadorii*, p. 224.
  b'. Throat and fore neck yellow, like the breast .. *flaviventris*, p. 225.
b. Outer tail-feathers slightly tipped with white; chin black; throat and fore neck dull olivaceous .... *viridis*, p. 226.

1. Sphecotheres maxillaris.

Maxillary Thrush, *Lath. Gen. Syn. Suppl.* ii. p. 186.
Turdus maxillaris, *Lath. Ind. Orn. Suppl.* p. xliii.
Sphecotheres viridis, *Vig. & Horsf. Tr. Linn. Soc.* xv. p. 215 (*nec Vieill.*).
Sphecotheres virescens, *Jard. & Selby, Ill. Orn.* ii. pl. 79 (*nec Vieill.*); *Licht. Nomencl.* p. 25.
Sphecotheres australis, *Swains.* (*ubi?*); *Gould, B. Austr.* iv. pl. 15; *Diggles, Orn. Austr.* part 2.
Sphecothera grisea, *Less. Traité*, p. 351; *Gray, Gen. B.* i. p. 231.
Sphecotheres canicollis, *Swains. An. in Menag.* p. 320.
Sphecotheres maxillaris, *Gray, Gen. B.* i. p. 231; *Cab. Mus. Hein.* i. p. 212; *Bp. Consp.* i. p. 345; *Gould, Handb. B. Austr.* i. p. 467.
Oriolus maxillaris, *Schl. Mus. P.-B.* Coraces, p. 115.

Adult male. Above yellowish green, somewhat brighter on the lower back; wing-coverts uniform with the back, as also the innermost secondaries; quills black, with narrow greyish margins and tips, the secondaries externally yellowish green like the back, the bases of the outer webs black like the inner webs of these quills; tail black, the three outer feathers white at tip, the white increasing on the two external feathers, and extending along the outer web of the last one; crown of head, nape, and ear-coverts black, as also the fore part of the cheeks; hind neck, sides of neck, throat, and fore neck leaden grey; lores and region round the eye bare; rest of under surface dull yellowish green; thighs grey; vent and under tail-coverts white; under wing-coverts grey washed with yellowish green; under surface of quills light greyish; "orbits and a narrow space leading to the nostrils naked, and of a light buffy yellow; bill black; feet flesh-colour; iris very dark brown in some, red in others" (*Gould*). Total length 9·5 inches, culmen 1, wing 5·95, tail 4·4, tarsus 1.

Adult female. Differs from the male exactly as does the female of *S. salvadorii* or *S. flavigaster.*

Hab. Australia.

a, b. ♂ ad. st.	Australia.	Purchased.

2. Sphecotheres salvadorii. (Plate XII.)

Adult male. General colour above dull olive-yellow, the wing-coverts olive-yellow like the back, the outer greater coverts and the primary-coverts grey washed with olive-yellow; quills blackish, the primaries externally grey, the secondaries olive-yellow on the outer web like the back; four centre tail-feathers black, the next one tipped with white, the last three on each side very broadly tipped with white, increasing greatly in extent on the outermost; crown of head glossy black; the hinder neck grey, with which colour the mantle is also slightly washed; sides of the face grey, the ear-coverts blackish; throat and chest, as well as the sides of the neck, grey; rest of under surface olive-greenish, the feathers of the lower breast washed with grey; centre of the abdomen bright yellow; vent and under tail-coverts white; thighs grey; under wing-coverts and quill-lining grey, the former slightly washed with olive-yellow. Total length 9·5 inches, culmen 0·95, wing 5·55, tail 3·85, tarsus 0·9.

Adult female. General colour above brown, the head mottled with darker brown centres to the feathers; the lower back, rump, and upper tail-coverts strongly washed with greenish olive, rather yellower on the latter; wings brown, all the coverts and quills conspicuously margined with white on the outer web, the secondaries slightly washed with olive externally; tail brown, strongly shaded with olive-yellow on the centre feathers, the outer rectrices edged and tipped with white along the inner web; ear-coverts light ashy brown, streaked with darker brown; throat and fore neck light brown, with longitudinal streaks of darker brown; rest of under surface dull white, slightly washed with yellow on the breast and flanks, all the feathers with broad mesial streaks of brown, becoming narrower on the flanks and under tail-coverts; lower abdomen and under tail-coverts pure white, not streaked; under wing-coverts fulvous or ashy brown, with mesial streaks of darker brown. Total length 10 inches, culmen 1, wing 5·8, tail 4·1, tarsus 0·9.

Hab. South-eastern New Guinea.

a, b. ♂ ♀ ad. sk.	Near Port Moresby, S.E. New Guinea.	O. C. Stone, Esq. [C.].

3. Sphecotheres flaviventris.

Sphecotheres flaviventris, *Gould, P. Z. S.* 1849, p. 111; *id. B. Austr. Suppl.* pl. 37; *Bp. Consp.* i. p. 345; *Finsch, Neu-Guinea*, p. 173; *Diggles, Orn. Austr.* part 2; *Gould, Handb. B. Austr.* i. p. 468; *Ramsay, P. Z. S.* 1868, p. 385; *Gray, Hand-l. B.* i. p. 291; *Schl. N. T. D.* iii. p. 342.

Picnorhamphus cucullatus, *Rosenb. N. Tijdschr. Nederl. Ind.* xxix. p. 143.

Oriolus flaviventris, *Schl. Mus. P.-B.* Coraces, p. 116.

Adult male. Above yellowish green, the scapulars and wing-coverts uniform with the back, as also the innermost secondaries, the coverts and the last-named quills blackish on the inner webs; outer secondaries blackish, inclining to grey at the tips of the feathers, which are yellowish green externally; primaries and primary-coverts blackish, externally edged with greyish; upper tail-coverts yellowish green with black bases; tail-feathers black, externally margined with yellowish green, the three centre feathers on each side entirely black, the next with a white spot at the tip, increasing in size on the penultimate feather and on the outermost, both of which have the outer web almost entirely white; head and nape black, as also the cheeks; the lores and all the region of the eye bare; entire under surface of body bright yellow, the lower abdomen, thighs, and under tail-coverts white; sides of the upper breast green; under wing-coverts dull greyish with yellow edgings to the feathers; under surface of the quills blackish, inclining to greyish near the base of the inner web; "orbits and a narrow space leading to the nostrils naked and of a light buffy yellow or flesh-colour; bill black; feet flesh-colour" (*Gould*). Total length 10·2 inches, culmen 1·05, wing 5·65, tail 4·4, tarsus 0·95.

Adult female. General colour above brown, the wing-coverts and quills narrowly edged with greyish white; tail brown, all the feathers washed externally with olive-green; head light brown, the ear-coverts uniform, the crown, nape, and cheeks distinctly streaked with darker brown; under surface of body white, all the feathers with a distinct mesial streak of brown, decreasing in size on the abdomen and under tail-coverts; throat and fore neck shaded with brown, the feathers all centred with darker brown; under wing-coverts light brown, with dark brown shaft-streaks; lower surface of the quills brown, inclining to ashy brown towards the base of the inner web. Total length 10·5 inches, culmen 1, wing 6, tail 4·5, tarsus 1.

Hab. N.E. Australia, Ké Islands.

a, b. ♂ ♀ ad. st.	Cape York.	Capt. Stanley [P.].
c. ♂ ad. sk.	Cape York, Oct. 7, 1840.	Capt. Stanley [P.].
d. ♂ ad. sk.	Cape York.	Mr. Cockerell [C.].
e, f. ♂ ad. sk.	Port Albany.	Purchased.
g. ♂ ad. sk.	Albany Island.	J. R. Elsey, Esq. [P.].
h. Skull.		Capt. Stanley [P.].

4. Sphecotheres viridis.

Sphecothera viridis, *Vieill. N. Dict. d'Hist. Nat.* xxxii. p. 5.
Sphecothera virescens, *Vieill. et Oud. Gal. des Ois.* i. p. 238, pl. cxlvii.
Graucalus viridis, *Quoy et Gaim. Voy. de l'Uranie, Zool.* p. 103, pl. 21.
Sphecotheres viridis, *Gray, Gen. B.* i. p. 231; *Cab. Mus. Hein.* Th. i. p. 212; *Bp. Consp.* i. p. 345; *Wall. P. Z. S.* 1863, p. 485; *Finsch, Neu-Guinea*, p. 173; *Gray, Hand-l. B.* i. p. 291.
Oriolus timoriensis, *Schl. Mus. P.-B.* Coraces, p. 116.

Adult male. General colour yellowish olive above, the feathers of the upper surface rather more dusky in their centres, those of the mantle and upper back mesially streaked with black; wings black, all the feathers with greyish black bases and broadly margined with yellowish olive, especially on the secondaries, the primaries narrowly edged with yellow near the base, and with greyish white towards the tip; tail-feathers ashy brown, washed externally with yellow, almost imperceptibly on the two outermost ones; the three outer feathers with a whitish spot at the tips, the others pale brown at the extremities; head and nape black, as also the cheeks and ear-coverts; chin black; throat olive-greenish; rest of under surface olive-yellow, the abdomen yellowish white in the centre; under tail-coverts white, obscurely tipped with yellow; thighs greenish, with grey bases to the feathers; under surface of the wing light grey, with narrow olive-yellow edgings; the quills dark greyish black, with a very pale greyish shade on the inner web; bill and legs greyish black in skin. Total length 9 inches, culmen 0·9, wing 4·85, tail 4, tarsus 0·9.

Adult female. General colour above pale brown, with light greyish edgings to the feathers, the upper tail-coverts being faintly margined with olive-yellow: wings brown, the feathers margined with light greyish white or whity brown, the greater wing-coverts and inner secondaries faintly washed with olive-yellow externally; tail-feathers light brown, externally washed with an olive-yellow shade, absent on the outermost, all the feathers tipped with whitish; sides of face dusky pale brown; under surface dull white, the feathers mesially streaked with brown, not so distinctly on the throat; thighs pale brown, as also the under wing-coverts; under surface of quills brown, with the inner webs pale brown for the greater part of their extent; bill pale brownish (in skin), legs blackish. Total length 10·2 inches, culmen 0·9, wing 4·8, tail 4·45, tarsus 0·9.

Young female. Similar to the old female, but with more distinct olive-yellow margins to the feathers of the upper surface, the under surface being also somewhat washed with yellow; quills dark brown, edged with light olive-yellow, the coverts broadly margined with yellowish buff.

Young male. Similar at first to the young female; but afterwards the gradual disappearance of the stripes on the breast and the assumption of the black head is very distinctly indicated.

Hab. Timor and Semao.

a. Ad. st.	Timor.	Massena Collection.
b. Ad. st.	Timor.	Purchased.
c. ♂ ad., *d, e.* Imm. sk.	Timor.	A. R. Wallace, Esq. [C.].
f. [♂] ad., *g.* juv. sk.	E. Timor.	A. R. Wallace, Esq. [C.].
h. ♀ ad. sk.	E. Timor.	A. R. Wallace, Esq. [C.].
i, k. ♂ ♀ juv. sk.	Semao Island.	A. R. Wallace, Esq. [C.].
l, m. ♂ ♀ ad. sk.	Timor Coupong.	A. R. Wallace, Esq. [C.].

Family DICRURIDÆ.

Key to the Genera.

a. Tail more or less forked, generally very deeply so; plumage ordinary, generally black, with a more or less pronounced gloss.
  a'. Tail forked, but the outer feather not ending in a racket.
    a''. Fork of tail inconsiderable, not equal to the tarsus in length.
      a'''. Bristles strongly developed, but not reaching beyond the centre of the bill, which is long; the uncovered part of the maxilla from the nostril to the tip much broader than the bill at gape.
        *a*⁴. Tail nearly square, the outer feather not curved upwards .... 1. DICRURUS, p. 229.
        *b*⁴. Tail nearly square, but the outer feather recurved on itself towards the tip; head furnished in some species with long silky hairs .... 2. CHIBIA, p. 234.
      b'''. Bristles enormously developed, passing beyond the tip of the bill, which is short and stout, the uncovered part of the maxilla from the nostrils to the tip being about equal to the breadth of the bill at gape .... 3. CHÆTORHYNCHUS, p. 242.
    b''. Fork of tail strongly pronounced, the tarsus always less than the distance between the tips of the middle and outer feathers.
      c'''. Outer tail-feather not strikingly recurved at the tip, and not produced in a prominent manner, but approached by the successive graduation of the other tail-feathers.
        *c*⁴. Plumage scaly or spangled in appearance everywhere, the feathers with metallic tips; frontal plumes very dense and extending a good way on the upper mandible, the exposed part of the culmen not exceeding the hind toe in length . 4. CHAPTIA, p. 243.
        *d*⁴. Plumage ordinary black or grey, with metallic gloss; frontal plumes dense, but not extending as far as in the preceding genus, the exposed part of the bill exceeding the hind toe in length.
          *a*⁵. With full crest of frontal plumes,

1. DICRURUS.

	Type.
Dicrurus, *Vieill. N. Dict.* ix. p. 585 (1817)	D. balicassius.
Balicassius, *Bp. C. R.* xxxviii. p. 539 (1854)..........	D. balicassius.

Range. Indian region. From Nepal to Malacca. Philippine Islands. Africa: west coast, from Senegambia to Angola; Natal.

Key to the Species.

a. Size large: wing not less than 4·5 inches.
a'. Head and back uniform metallic colour.
a''. Silky feathers on the sides of the lower back white; colour brilliant green with purple shades.
a'''. Belly metallic purplish green, like the breast .... *balicassius*, p. 230.
b'''. Belly white .... *mirabilis*, p. 231.
b''. Silky feathers on the sides of the lower back blackish grey; colour dull metallic greenish.
c'''. Size large: wing 5·45 inches; tail very much forked .... *annectens*, p. 231.
d'''. Size small: wing 4·5 inches; tail less forked, nearly square .... *atripennis*, p. 232.
b'. Back deep purplish black of a velvety appearance, contrasting with the glossy crown and wings .. *modestus*, p. 232.
b. Size small: wing 3·9 inches; throat greyish black . *ludwigii*, p. 233.

The species contained in the present genus are by no means strictly of the same form, but exhibit slight differences. The type of the genus, *D. balicassius*, and its congener *D. mirabilis*, stand alone in the extreme brilliancy of their plumage. *D. annectens* has the coloration of a *Buchanga* without the strongly forked tail; but the outer feathers often show a tendency to turn up at the ends as in *Chibia*. The representative of *D. annectens* in Africa seems to be *D. atripennis*, but the smaller size and differently shaped tail distinguish it. *D. modestus* is in plumage a *Chibia*, resembling exactly the Malayan species of the genus without the chest-spangles: no specimen, however, as yet examined by me, exhibiting any tendency to have a twisted tail, or to show any hairs on the crown as in true *Chibia*, it has seemed best to keep it in *Dicrurus*. The last species, *D. ludwigii*, is easily recognizable on account of its small size, and appears to lead to *Chaptia*.

1. Dicrurus balicassius.

Le Choucas des Philippines, *Briss. Orn.* ii. p. 31, pl. 2. fig. 1; *Montb. Pl. Enl.* 603.
Corvus balicassius, *Linn. S. N.* i. p. 157 (*ex Briss.*).
Le Balicasse des Philippines, *Montb. Hist. Nat. Ois.* iii. p. 83.
Edolius furcatus, *Wagl. Syst. Av.* p. 322 (*ex Linn.*).
Edolius viridescens, *Gould, P. Z. S.* 1836, p. 6.
Dicrurus viridescens, *Blyth, Cat. B. Mus. A. S. B.* p. 202, App. xxii.
Balicassius philippensis, *Bp. C. R.* xxxviii. p. 539.
Balicassius furcatus, *Bp. t. c.* p. 539.
Dicrurus balicassius, *Gray, Hand-l. B.* i. p. 284; *Wald. Tr. Z. S.* ix. p. 180, pl. xxxi. fig. 1.

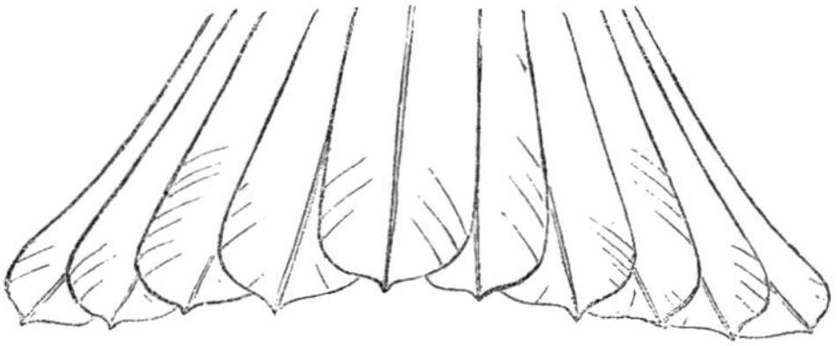

Tail of *Dicrurus balicassius.*

Adult. General colour above and below shining steel-green, with a strong metallic gloss of purple, the abdomen and under tail-coverts a little duller; the metallic endings to the feathers of the head, neck, and throat somewhat lanceolate in shape, giving a slightly spangled appearance to these parts; wings coloured like the back, with somewhat of a purplish shade under certain lights; quills black, the primaries glossed with green, the secondaries more broadly, the innermost being exactly like the back; middle tail-feathers resembling the back, the rest purplish black on the inner webs, glossy steel-green on the outer ones; flanks more or less mixed with grey, with a patch of silky white plumes on each side of the rump. Total length 10·5 inches, culmen 0·2, wing 5·5, tail 4·8, tarsus 1.

Hab. Philippine Islands: Luzon.

a. Ad. sk.	Manila.	Hugh Cuming, Esq. [C.].

2. Dicrurus mirabilis.

Dicrurus mirabilis, *Walden & Layard, Ibis*, 1872, p. 103, pl. 5; *Walden, Tr. Z. S.* ix. p. 181.

Adult. General colour above metallic greenish black, rather deeper and more steel-green on the head and neck; wing-coverts metallic green like the back, the quills and tail greenish black, externally of the same metallic green as the back, the innermost secondaries entirely of that colour; sides of face, throat, and neck black, glossed with metallic green, more conspicuously on the latter part; breast, abdomen, and under tail-coverts pure white; chest greenish black, spotted or streaked with white on the feathers adjoining the white breast; sides of body greyish; thighs and under wing-coverts greenish black; bill and legs black. Total length 10·5 inches, culmen 1·3, wing 5·35, tail 4·8, tarsus 1·05.

Hab. Island of Negros.

a. Ad. sk.	Negros.	Dr. J. B. Steere [C.].

3. Dicrurus annectens.

Buchanga annectans, *Hodgs. Ind. Rev.* i. p. 326.
Edolius annectans, *Blyth, J. A. S. B.* xi. p. 173.
Edolius affinis, *Blyth, t. c.* p. 174.
Edolius balicassius, *Blyth, t. c.* p. 800.
Dicrurus balicassius (*nec Linn.*), *Blyth, J. A. S. B.* xi. p. 175, xv. p. 298; *Gray, Cat. Mamm. &c. Hodgs. Coll. Nepal,* 1846, p. 98; *Blyth, Cat. B. Mus. A. S. B.* p. 202; *Bp. Consp.* i. p. 351; *Horsf. & Moore, Cat. B. Mus. E.I. Co.* i. p. 152; *Jerd. B. Ind.* i. p. 430; *Hume, S. F.* 1874, pp. 209, 474.
Dicrurus annectans, *Hodgs. in Gray's Zool. Misc.* p. 84; *Gray, Hand-l. B.* i. p. 285.
Dicrurus furcatus, *Gray, Hand-l. B.* i. p. 285.
Buchanga balicassius, *Ball, S. F.* 1873, p. 67.

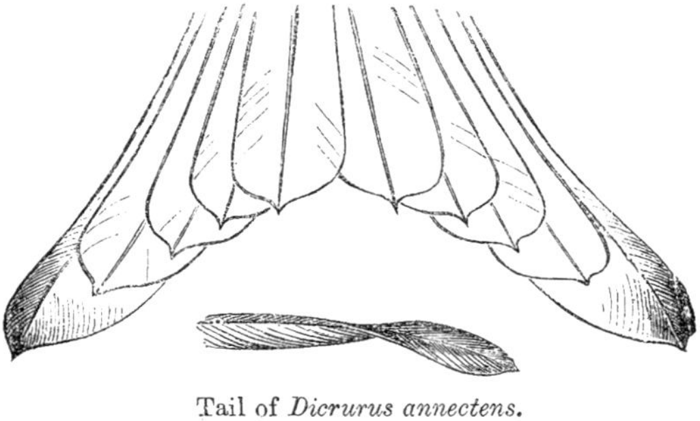

Tail of *Dicrurus annectens.*

Adult. General colour black, with a strong shade of metallic steel-green; wing-coverts and secondaries similarly coloured to the back, the primaries blackish, glossed with steel-green near the base of the outer web; tail black, glossed with steel-green on the outer webs of the feathers, this shade being almost absent on the outermost; lores and sides of face deep black; under surface of body black, glossed with deep metallic green, almost as strongly metallic as on the back, the under wing-coverts coloured like the breast; bill and legs black; iris brown. Total length 10 inches, culmen 1·1, wing 5·45, tail 4·1, to tip of outer feather 5·15, tarsus 0·75.

Young. Differs from the adult in having the feathers of the under surface edged with greyish white.

Obs. The persistence of these white edgings to the feathers, even when the bird is otherwise in fully adult and glossy plumage, is a special characteristic of this species.

Hab. Nepal, through the Burmese countries to Malacca and Sumatra.

a. Ad., *b, c.* Jun. sk.	Nepal.	B. H. Hodgson, Esq. [P.].
d. Ad. sk.	Oude Terai.	Capt. Stackhouse Pinwill [C.].
e. Imm. st.	Malacca.	F. Nightingale, Esq. [P.].
f, g. Vix ad. st.	Malacca.	Admiral Sir E. Belcher [P.].
h. ♀ ad. sk.	Malacca.	A. R. Wallace, Esq. [C.].
i, k. Imm. sk.	Malacca.	W. Harvey, Esq. [P.].
l. Ad. sk.	Malacca.	
m. Ad. sk.	Sumatra.	A. R. Wallace, Esq. [C.].

4. **Dicrurus atripennis.**

Dicrurus atripennis, *Swains. B. W. Afr.* i. p. 256; *Gray, Gen. B.* i. p. 287; *Bp. Consp.* i. p. 352; *Hartl. Orn. W.-Afr.* p. 101; *id. J. f. O.* 1861, p. 170; *Gray, Hand-l. B.* i. p. 286; *Sharpe, Ibis,* 1870, p. 481; *id. Cat. Afr. B.* p. 47.

Adult. General colour above glossy greenish black, the head uniform with the entire back and wing-coverts; quills and tail-feathers jet-black, externally glossed with greenish; lores and frontal plumes velvety black, the latter slightly crested; cheeks, throat, and breast black, glossed with green, more distinctly on the latter; belly and abdomen dull greyish black; under wing-coverts black, the feathers edged with a glossy green lustre; lower surface of quills entirely black, of the same shade as the wing-coverts. Total length 8·2 inches, culmen 0·85, wing 4·5, tail 4·1, tarsus 0·7.

Hab. West Africa.

a. Ad. sk.	W. Africa.	Capt. Sabine [P.].
b. Ad. sk.	W. Africa.	Purchased.
c, d, e. Ad. sk.	Fantee (*Aubinn*).	R. B. Sharpe, Esq.
f. Ad. sk.	River Gambia.	R. B. Sharpe, Esq.

5. **Dicrurus modestus.**

Dicrurus modestus, *Hartl. Rev. et Mag. de Zool.* 1849, p. 495; *id. Beitr. Orn. W.-Afr.* p. 50, Taf. 4; *id. Orn. W.-Afr.* p. 101; *Dohrn, P. Z. S.* 1866, p. 327; *Gray, Hand-l. B.* i. p. 286; *Sharpe, Cat. Afr. B.* p. 47.

Dicrurus coracinus, *J. & E. Verr. Rev. et Mag. de Zool.* 1851, p. 132; *Strickl. Contr. Orn.* 1851, p. 132; *Bp. C. R.* xxxviii. p. 540;

Hartl. Orn. W.-Afr. p. 101; *id. J. f. O.* 1861, p. 170; *Cass. Pr. Phil. Acad.* 1859, p. 53; *Gray, Hand-l. B.* i. p. 286; *Sharpe, Ibis,* 1869, p. 190; *id. Cat. Afr. B.* p. 47.
Dicrurus, sp., *Sharpe, P. Z. S.* 1870, p. 148.
Dicrurus atripennis, *Reichenow, J. f. O.* 1874, p. 26 (*nec Swains.*).
Musicus coracinus, *Bouvier, Cat. Ois. Marche &c.* p. 18.

Adult. Black: above blue-black, with a purplish lustre, but without any exact metallic shine on the back, sides of head and neck, and the underparts, all of which are of the same tint; frontal plumes, which are close-set and slightly curved forwards, and the crown of the head glossy steel-black, this coronal patch tapering off to a point on the occiput; wings glossy steel-black, the primaries black, externally glossed with metallic colour; upper and under tail-coverts glossy greenish black, all the feathers margined with metallic black; tail glossy steel-black, the central feathers entirely of this colour, the others black, glossed externally with metallic steel-black; under wing-coverts black, with margins of metallic purplish black. Total length 9 inches, culmen 0·95, wing 5·05, tail 4·6, tarsus 0·75.

Young. Differs from the adult in having the feathers of the under surface broadly margined with greyish white, the under wing- and tail-coverts included.

Obs. D. modestus, from Prince's Island, and *D. coracinus*, from Gaboon, are not to be separated except by the slightly larger size of the former, which is scarcely a sufficient character. Wing 5·1–5·45, tail 4·9–5·25, tarsus 0·8–0·85. The specimen described is a Gaboon bird. Examples from the Gold Coast are less typical, but are not to be specifically distinguished, though they are often more glossy above. Their forked tail prevents their being mistaken for the square-tailed *D. atripennis*.

Hab. West Africa, from the Gold Coast to Angola.

a, b. Ad. sk.	Ashantee.	John Gould, Esq.
c. Ad. sk.	Fantee.	Andrew Swanzy, Esq. [P.].
d, e. Ad. sk.	Fantee.	R. B. Sharpe, Esq.
f, g. Ad. sk.	Fantee (*Ussher*).	R. B. Sharpe, Esq
h. Ad. sk.	Yancoomassie, Fantee, Dec. 1871 (*Blissett*).	R. B. Sharpe, Esq.
i. Imm. sk.	Accra (*Haynes*).	R. B. Sharpe, Esq.
k. Ad. sk.	Gaboon.	M. Verreaux.
l, m. Ad. sk.	Gaboon (*Walker*).	R. B. Sharpe, Esq.
n. Ad., *o.* Imm. sk.	Prince's Island (*Dohrn*).	R. B. Sharpe, Esq.
p. Ad. sk.	Prince's Island.	Dr. A. Dohrn [C.].
q. Ad. sq.	Angola.	J. J. Monteiro, Esq.
r. ♀ juv. sk.	Rio Dande, Angola, April 25, 1869 (*Sala*).	R. B. Sharpe, Esq.

6. Dicrurus ludwigii.

Dicrurus ludwigii. *Smith, Ill. Zool. S. Afr.* pl. 34; *Gray, Gen. B.* i. p. 287; *Bp. Consp.* p. 352; *Gurney, Ibis,* 1864, p. 351; *Layard, B. S. Afr.* p. 154; *Sharpe, Cat. Afr. B.* p. 47; *Shelley, Ibis,* 1875, p. 75.
Melænornis ludwigii, *Gray, Hand-l. B.* i. p. 288.

Adult. General colour above greenish black, somewhat greyer on the rump; wing-coverts and quills black, the coverts and the secondaries narrowly edged with the same glossy green as the back; tail black, the feathers washed on their outer webs with glossy greenish black; lores and frontal feathers, as well as the sides of the face, black, the latter with a slight gloss; under surface of body greyish black, with a slight greenish gloss on the breast-feathers; "bill and legs black; iris deep orange." Total length 7 inches, culmen 0·8, wing 3·9, tail 3·7, tarsus 0·7.

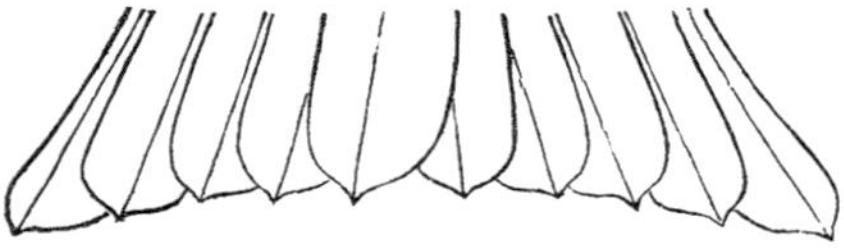

Tail of *Dicrurus ludwigii.*

Hab. South-eastern Africa.

a. Ad. sk.	Natal.	R. B. Sharpe, Esq.

2. CHIBIA.

	Type.
Chibia, *Hodgs. Ind. Review*, i. p. 324 (1837)	C. hottentotta.
Cometes, *Hodgs. J. A. S. B.* x. p. 29 (1841)	C. hottentotta.
Trichometopus, *Cab. Mus. Hein.* Th. i. p. 112 (1850) ..	C. hottentotta.

Range. India generally, ranging into Assam and Burmah, through the Malayan peninsula to the Moluccas, being found in all the principal groups of the Indo-Malayan and Austro-Malayan islands, and extending to N. Australia.

Key to the Species.

a. Head ornamented with several long silky hairs springing from the forehead, and falling backwards over the nape .... *hottentotta*, p. 235.

b. Silky hairs on the head either absent or only slightly developed.

  a'. Tail not 7 inches long.

    a''. Silky plumes on the sides of the lower back ashy whitish.

      a'''. Metallic shade on neck-hackles and on the spangles of the breast steel-green; loral plumes only moderately dense *bracteata*, p. 236.

      b'''. Metallic shade on the neck-hackles and spangles of breast purplish; loral plumes very full, and curved forwards in the form of a small drooping crest (see fig. p. 237) .... *bimaënsis*, p. 237.

    b''. Silky plumes on the sides of the lower back blackish.

      c'''. Metallic shade on neck-hackles purplish,

scarcely visible; spangles on breast minute .... *atrocærulea*, p. 238.
carbonaria, p. 238.
assimilis, p. 239.
amboinensis, p. 239.

d''. Neck-hackles metallic purple; breast-spangles very large.
a^4. Eye red.
a^5. Throat unspotted .... *pectoralis*, p. 240.
b^5. Throat spotted .... *læmosticta*, p. 241.
b^4. Eye white .... *leucops*, p. 241.
b'. Tail over 7 inches long.
c''. Larger; shade of body purple .... *megalornis*, p. 242.
d''. Smaller; shade of body green.......... *densa*, p. 241.

It is by no means certain that the numerous species included here should be really separated from *Dicrurus*. I follow Lord Tweeddale in affixing *D. balicassius* as the type of the latter genus; and *Chibia bimaensis* closely approaches it. The tendency to a turned-up outer tail-feather, so pronounced in *C. hottentotta*, is exhibited to a greater or less degree in all the species here united to that genus; but then *Dicrurus annectens*, most suitably named, also shows a tendency to have the outer tail-feathers recurved. At the same time, if *Chibia hottentotta* is to be kept separate from *Dicrurus* (and I do not see how they can be united), then all the Moluccan birds can hardly be kept out of the genus *Chibia*, for they all possess the same style of plumage, viz. a velvety black back, with strongly contrasting metallic wings and tail. The long silky hairs, which show so prominently in *C. hottentotta*, are also slightly developed in *C. pectoralis* (see fig. p. 240). In the above table of specific characters I endeavour to show that between *C. atrocærulea*, *C. carbonaria*, *C. assimilis*, and *C. amboinensis* no tangible differences can be drawn; they are nothing but races of one form, differing chiefly in size.

1. Chibia hottentotta.

Le Choucas du Cap de Bonne Espérance, *Briss. Orn.* i. p. 33, pl. ii. fig. 2.
Corvus hottentottus, *Linn. S. N.* i. p. 155 (*ex Briss.*).
Criniger splendens, *Tickell, J. A. S. B.* ii. p. 574.
Edolius crishna, *Gould, P. Z. S.* 1836, p. 5; *Blyth, J. A. S. B.* xi. pp. 169, 799.
Chibia casia, *Hodgs. Ind. Rev.* i. p. 324.
Chibia hottentotta, *Strickl. Ann. N. H.* xiii. p. 36; *Gray, Gen. B.* i. p. 287; *Blyth, J. A. S. B.* xv. p. 294; *id. Cat. B. Mus. A. S. B.* p. 200; *Bp. Consp.* i. p. 350; *Jerd. B. Ind.* i. p. 439; *Gray, Hand-l. B.* i. p. 287; *Hume, Str. F.* 1874, p. 474, 1875, p. 101; *Blyth & Wald. B. Burm.* p. 128.
Edolius barbatus, *Gray, Zool. Misc.* p.
Cometes crishna, *Hodgs. in Gray's Zool. Misc.* p. 84.
Trichometopus hottentottus, *Cab. Mus. Hein.* Th. i. p. 112.

Adult. Blue-black, velvet-like and with more or less of a metallic gloss on the back and lower parts; lores and frontal plumes velvety black, with a crest of long hairs springing from the base of the forehead and falling over the nape; sides of face purplish black; head and sides of neck ornamented with metallic steel-green feathers, those on the last-named part elongated and somewhat lanceolate; plumes of the throat and chest also with metallic steel-green tips, giving a spangled appearance; under tail-coverts glossed with green, as also the edge of the wing, the rest of the under wing-coverts being coloured like the breast; upper surface of the wing metallic bronzy green; the primaries purplish black, as also the greater portion of the secondaries; upper tail-coverts glossy steel-green; tail-feathers purplish black, steel-green on the outer webs, the centre feathers entirely glossed with the latter colour, the two outer feathers recurved inwards at the tips; bill and legs black; iris reddish brown. Total length 13 inches, culmen 1·55, wing 6·95, tail 6·5, tarsus 1.

Young. Dusky blackish, without any of the metallic feathers; wings glossy, but not so bright as in the adults; tail also less glossy than in the adult; under wing-coverts tipped with white.

Obs. *C. brevirostris*, from China, does not seem to me a distinct species, the differences of colour noticed by Dr. Cabanis not holding good; the bill is shorter than in Nepal specimens, measuring in a Chinese example before me 1·45 inch, but two specimens from the N.W. Himalayas have it exactly of this length also.

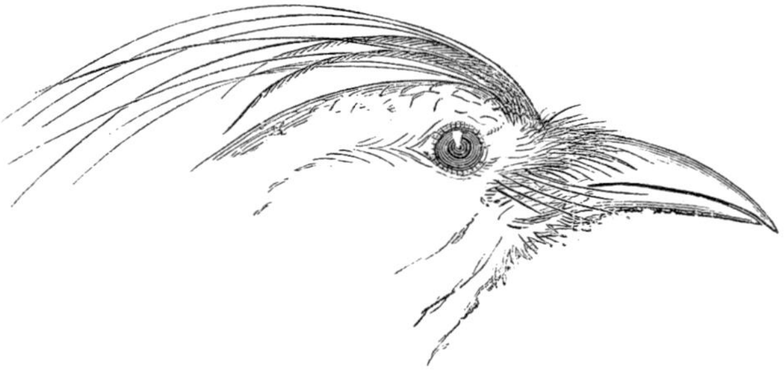

Head of *Chibia hottentotta.*

Hab. India, ranging through Assam and Burmah to China.

a. Ad. st.	India.	Capt. Honeywood [P.].
b, c. Ad. sk.	N.W. Himalayas.	Capt. Stackhouse Pinwill [P.].
d–g. Ad. sk.	Nepal.	B. H. Hodgson, Esq. [P.].
h. Juv. sk.	Nepal.	B. H. Hodgson, Esq. [P.].
i. Ad. sk.	N. China.	Mr. Fortune.
k. ♀ ad. sk.	Ningpo, Sept. 27, 1875.	Shanghai Museum [P.].

2. Chibia bracteata.

Dicrurus balicassius, *Vig. & Horsf. Tr. Linn. Soc.* xv. p. 211 (*nec Linn.*).
Dicrurus bracteatus, *Gould, P. Z. S.* 1842, p. 132; *id. B. Austr.* fol.

ii. pl. 82; *Gray, Gen. B.* i. p. 287; *Bp. Consp.* i. p. 352; *Finsch, Neu-Guinea,* p. 170; *Gray, Hand-l. B.* i. p. 285; *Diggles, Orn. Austr.* pl. 35.
Balicassius bracteatus, *Bp. C. R.* xxxviii. p. 539.
Chibia bracteata, *Gould, Handb. B. Austr.* i. p. 235.

Adult. General colour above black, of a velvety appearance, but with scarcely any metallic gloss, except on the outer scapulars; feathers of the crown tipped with metallic steel-green, the plumes of the rump also washed with greenish, the upper tail-coverts more conspicuously; wings metallic steel-green, with a slight purplish lustre on the greater series, the primaries black, only washed on the outer web with glossy green; tail black, glossed with green on the outer webs; lores, sides of face, and chin jet-black; the rest of the under surface black, with a slight greenish gloss, the throat and fore neck with a spot of metallic green at the tip of each feather; under wing-coverts distinctly spotted with white at the tip; bill and feet blackish brown; iris brownish red. Total length 10·8 inches, culmen 1·35, wing 5·8, tail 4·9, tarsus 0·9.

Young. Similar to adult, but browner, the tips to the feathers much less distinct.

Hab. Northern and North-eastern Australia.

a. Ad. st.	N. Australia.	J. Gould, Esq. [P.].
b. Ad. sk.	N. Australia.	J. W. E. Rishton, Esq. [P.].
c. Juv. sk.	N. Australia.	J. B. Jukes, Esq. [P.].
d. ♂ juv. sk.	Port Essington.	J. B. Jukes, Esq. [P.].
e. Ad. sk.	Port Essington.	Dr. Sibbald [P.].
f, g. Ad. st.	Port Essington.	Capt. Chambers [P.].
h. ♀ ad. sk.	Cape York.	Capt. Stanley [P.].
i. Ad. sk.	Cape York.	Capt. Stanley [P.].

3. Chibia bimaensis.

Dicrurus bimaënsis, *Bp. Consp.* i. p. 352 (*ex Temm. MS. in Mus. Lugd., descr. nulla*); *Wall. P. Z. S.* 1863, p. 494 (*descr. princeps*); *Finsch Neu-Guinea,* p. 171; *Gray, Hand-l. B.* i. p. 288.

Head of *Chibia bimaënsis.*

Adult. Above velvety black, with a very slight greenish gloss; wings glossy steel green, the secondaries exactly like the wing-coverts, the primaries black, externally green; crown of head glossy green; feathers of the neck long and hackle-like, glossy steel-green, and lanceolate in shape; upper tail-coverts and tail glossy green, the outer tail-feathers black on the inner web; lores, sides of face, and underparts purplish black, with a slight gloss, the plumes of the lower throat and fore neck tipped with steel-green; flanks silky grey. Total length 11 inches, culmen 1·35, wing 5·4, tail 5·2, tarsus 1·05.

Young. Much duller than the adult and with less gloss; underneath greyer, the feathers edged with dull whitish, the under wing-coverts also tipped with white.

Hab. Lombock and Flores.

a, b. ♂ ♀ ad. sk.	Lombock.	A. R. Wallace, Esq. [C.].
c. Ad. sk.	Lombock.	A. R. Wallace, Esq. [C.].
d. ♂ juv. sk.	Lombock.	A. R. Wallace, Esq. [C.].
e, f. Ad. sk.	Flores.	A. R. Wallace, Esq. [C.].

4. **Chibia atrocærulea.**

Dicrurus atrocæruleus, *Gray, P. Z. S.* 1860, p. 354; *Finsch, Neu-Guinea*, p. 171; *Gray, Hand-l. B.* i. p. 285; *Salvad. Ann. Mus. Civic. Genov.* vii. p. 772.

Adult male (type of species). General colour above velvety black, with only a very slight purplish gloss; crown of head metallic steel-black; neck-hackles not very pronounced, and only slightly glossed with metallic purple; wings metallic steel-black, with a purplish gloss on the lesser wing-coverts, the primaries black, with a slight metallic lustre; upper tail-coverts purplish black, with a very slight metallic gloss on the edge of the feathers; tail-feathers black, glossed on the outer web with metallic steel-black, scarcely perceptible on the outermost feather; lores, feathers round the eye, and sides of face deep velvety black, as also the under surface, on which a faint purplish lustre can be traced; the feathers of the lower throat very slightly, the fore neck and chest rather more conspicuously, tipped with dull metallic green; under wing-coverts black, with an indistinct gloss of metallic purple; bill and wings black; iris red. Total length 12·8 inches, culmen 1·3, wing 6·45, tail 6·3, tarsus 1.

Hab. Batchian and Gilolo.

a. ♂ ad. sk.	Batchian.	A. R. Wallace, Esq. [C.]. (Type of species.)
b. Ad. sk.	Gilolo.	A. R. Wallace, Esq. [C.].

Subsp. *a.* **Chibia carbonaria.**

Dicrurus carbonarius, *Bp. Consp.* i. p. 352 (*ex Müll. MS. in Mus. Lugd., descr. nulla*); *Gray, Cat. B. &c. New Guinea*, p. 33; *Finsch, Neu Guinea*, p. 171; *Gray, Hand-l. B.* i. p. 285; *Salvad. Ann. Mus. Civic. Genov.* vii. p. 772.

Adult. General colour above velvety black, with a purplish gloss, not very strongly developed, but visible on the neck-hackles and some of the dorsal plumes; wings metallic steel-black, with a very slight greenish cast; quills black, the primaries externally steel-green, the inner secondaries entirely so, like the wing-coverts;

upper tail-coverts and tail black, glossed on the outer webs with metallic steel-colour; crown of head metallic steel-black; lores, feathers round the eye, and sides of face deep velvety black; under

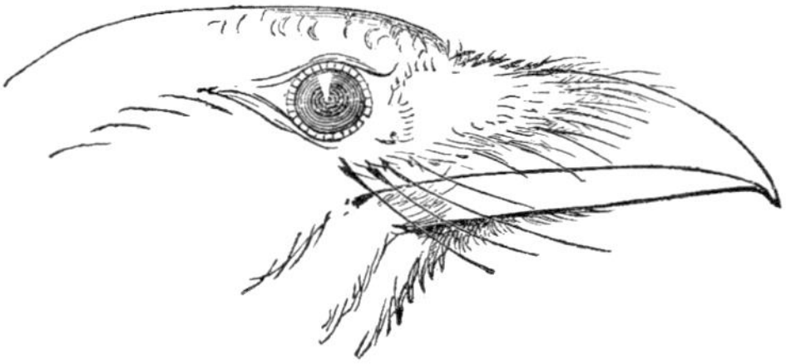

Head of *Chibia carbonaria.*

surface of body black, with a slight purplish gloss, the feathers ot the throat and breast slightly spangled with tips of metallic purple; bill and legs black; iris red. Total length 12 inches, culmen 1·25, wing 6, tail 5·55, tarsus 0·95.

Hab. Papuan group of islands.

a. Ad. sk.	New Guinea.	A. R. Wallace, Esq. [C.].
b, c. ♂ ad. sk.	Dorey, N. G.	A. R. Wallace, Esq. [C.].
d. Ad. sk.	Port Moresby, S.E. New Guinea.	O. C. Stone, Esq. [C.].
e. Ad. sk.	Mysol.	A. R. Wallace, Esq. [C.].
f. ♀ ad. sk.	Mysol.	A. R. Wallace, Esq. [C.].
g. ♀ ad. sk.	Waigiou.	A. R. Wallace, Esq. [C.].
h. ♂ ad. sk.	Waigiou.	A. R. Wallace, Esq. [C.].
i. Ad. sk.	Morty Island.	A. R. Wallace, Esq. [C.].

Subsp. β. **Chibia assimilis.**

Dicrurus assimilis, *Gray, P. Z. S.* 1858, pp. 179, 193; *id. Cat. B. &c. New Guinea,* p. 33; *Finsch, Neu-Guinea,* p. 171; *Gray, Hand-l. B.* i. p. 285.

Adult male. Similar to *C. carbonaria,* and only to be distinguished by its smaller size. Total length 10·4 inches, culmen 1·25, wing 5·3, tail 5, tarsus 0·9.

Adult female. Similar to the male. Total length 10·2 inches, culmen 1·2, wing 5·4, tail 4·9, tarsus 0·9.

Hab. Aru Islands.

a, b. ♂ ad. sk.	Aru Islands.	A. R. Wallace, Esq. [C.].
c. ♀ ad. sk.	Aru Islands.	A. R. Wallace, Esq. [C.].

Subsp. γ. **Chibia amboinensis.**

Dicrurus amboinensis, *Gray, P. Z. S.* 1860, p. 354; *Finsch, Neu-Guinea,* p. 171; *Gray, Hand-l. B.* i. p. 285.

Adult female (type of species). Similar to *C. carbonaria,* but differing in its long slender bill and in the more distinct spangles,

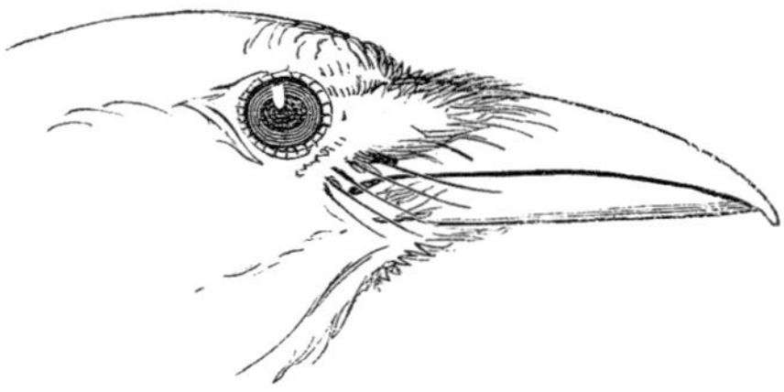

Head of *Chibia amboinensis.*

especially on the hinder neck. Total length 11·5 inches, culmen 1·3, wing 5·5, tail 5·5, tarsus 1.

Hab. Ceram and Amboyna.

a. ♀ ad. sk.	Amboyna.	A. R. Wallace, Esq. [C.]. (Type of species.)
b. Ad. sk.	Ceram.	A. R. Wallace, Esq. [C.].
c. Ad. sk.	E. Ceram.	A. R. Wallace, Esq. [C.].
d. Ad. sk.	N. Ceram.	A. R. Wallace, Esq. [C.].
e. Juv. sk.	N. Ceram.	A. R. Wallace, Esq. [C.].

5. Chibia pectoralis.

Dicrurus pectoralis, *Wallace, P. Z. S.* 1862, pp. 335, 342; *Finsch, Neu-Guinea,* p. 170; *Gray, Hand-l. B.* i. p. 285.

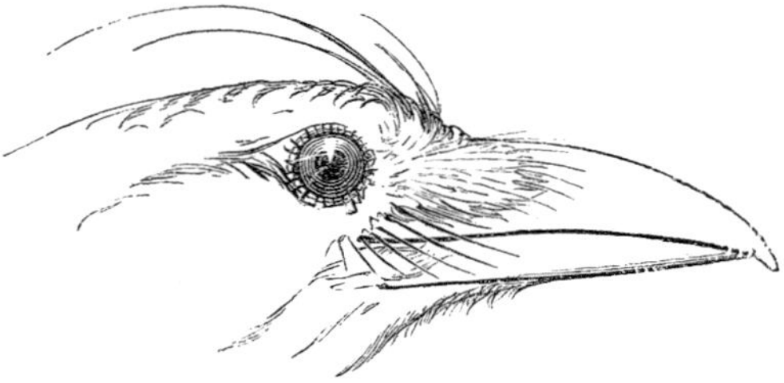

Head of *Chibia pectoralis.*

Adult male (type of species). General colour velvety black, with scarcely any gloss on the back; crown of head and long hackles on the side of the neck metallic steel-black, with a purplish shade; wings metallic steel-green, the primaries black externally, washed with a deeper shade of the same; upper tail-coverts and tail black, glossed with metallic steel-green on the outer webs of the feathers; lores, frontal plumes, and sides of face deep velvety black; on the forehead a few long silky hairs; rest of under surface of body deep velvety black without any perceptible gloss, the lower throat and breast very distinctly spangled with large tips of deep metallic green to all the feathers; bill and feet black; iris dark red. Total length 12 inches, culmen 1·3, wing 6·05, tail 5·75, tarsus 1·05.

Hab. Sula Islands.

a. ♂ ad sk.	Sula Islands.	A. R. Wallace, Esq. [C.]. (Type of species.)
b. Ad. sk.	Sula Islands.	A. R. Wallace, Esq. [C.].

6. Chibia læmosticta.

Dicrurus læmostictus, *Sclater, P. Z. S.* 1877.

Similar to *C. pectoralis*, but distinguished by having the throat, as well as the fore neck, spotted. Total length 11·3 inches, wing 6, tail 4·6, outer feather 5·4.

Hab. New Britain.

7. Chibia leucops.

Dicrurus leucops, *Wallace, P. Z. S.* 1865, p. 478; *Gray, Hand-l. B.* i. p. 285; *Wald. Tr. Z. S.* vii. p. 70; *Salvad. Ann. Mus. Civic. Genov.* vii. p. 658; *Brüggem. Abhandl. nat. Ver. Bremen*, v. p. 72.

Adult male (type of species). Very similar to *C. pectoralis*, but rather larger, with a stouter bill and less metallic brilliancy on the spangles of the breast-feathers; "iris always milky white" (*Wallace*). Total length 12·5 inches, culmen 1·35, wing 6·2, tail 5·9, tarsus 0·95.

Hab. Celebes.

a. ♂ ad. sk.	Tondano, Celebes.	A. R. Wallace, Esq. [C.].
b. ♂ ad. sh.	Macassar, Celebes.	A. R. Wallace, Esq. [C.].

8. Chibia densa.

Dicrurus densus, *Bp. Consp.* i. p. 352 (*ex Temm. MS. in Mus. Lugd.*, descr. nulla); *Finsch, Neu-Guinea*, p. 171; *Gray, Hand-l. B.* i. p. 285.

Adult male. Black, with a very decided greenish gloss on the head and back, the feathers of the head and neck metallic, the plumes on the sides of the latter long and lanceolate in shape, hackle-like, distinctly metallic green; back rather velvety in appearance, with a gloss of steel-green very distinct on the scapulars; tail very much forked, deep black, glossed with steel-green on the outer webs of all the feathers; lores velvety black; ear-coverts and sides of face black, with a very slight wash of green; under surface of body black, slightly washed with greenish, the throat and chest spangled with small spots of metallic steel-green at the tips of the feathers; sides of the body rather greyish, the bases of the feathers being of this colour; under wing-coverts black, washed with green and tipped with white spots; bill and feet black; "iris red" (*Wallace, MS.*). Total length 13·5 inches, culmen 1·35, wing 6, tail 7, tarsus 1·05.

Adult female. Similar to the male. Total length 13·5 inches, culmen 1·4, wing 5·85, tail 6·8, tarsus 1.

Hab. Timor group of islands.

a. ♂ ad. sk.	Semao Island.	A. R. Wallace, Esq. [C.].
b, c. ♀ ad. sk.	E. Timor.	A. R. Wallace, Esq. [C.].

9. Chibia megalornis.

Dicrurus megalornis, *Gray, P. Z. S.* 1858, pp. 179, 193; *id. Cat. B. &c. New Guinea*, p. 33; *Finsch, Neu-Guinea*, p. 171; *Gray, Hand-l. B.* i. p. 285.

Adult. General colour purplish black, the back being of a velvety appearance, and slightly glossed with metallic steel-black; head and feathers of the neck (the latter hackled) steel-black, with a slight tinge of greenish on the metallic portion of the neck-hackles; wings metallic steel-green, as also the outer webs of the scapulars, which resemble the secondary quills; primaries black, washed externally with greenish; upper tail-coverts black, glossed with steel-green; tail black, washed with metallic steel-green on the outer webs; lores, feathers round the eye, and sides of face blue-black; rest of under surface purplish black, the under tail-coverts slightly washed with steel-green; the feathers of the throat and breast very indistinctly spangled with tips of metallic steel-green; under wing-coverts black, glossed with steel-green, and all tipped with a white spot; bill and feet black; iris red. Total length 15 inches, culmen 1·5, wing 6·95, tail 7·6, tarsus 1·15.

Obs. Very like *C. densa* of Timor, but purplish black without the green shade. These differences are apparent, on comparing specimens, on the neck-hackles, wings, and upper tail-coverts, which have an *oily green* shade in *C. densa*.

Hab. Ké Islands and Goram.

a.	Ad. sk.	Ké Islands.	A. R. Wallace, Esq. [C.]. (Type of species.)
b, c.	Ad. sk.	Goram.	A. R. Wallace, Esq. [C.].

3. CHÆTORHYNCHUS.

Chætorhynchus, *Meyer, Sitz. Akad. Wien*, lxix. p. 493. . Type. C. papuensis.

Bill (from above) and profile of head of *Chætorhynchus.*

Range. Confined to New Guinea.

1. Chætorhynchus papuensis. (Plate XIII.)

Chætorhynchus papuensis, *Meyer, Sitz. Akad. Wien*, lxix. p. 493.

Adult female. General colour black, with a steely gloss, the head more brightly glossed with steel-green, the feathers rounded and somewhat scale-like in appearance; least wing-coverts steel-black.

the median, greater, and primary coverts black, narrowly edged with steel-green; quills black, the secondaries margined with steel-green,

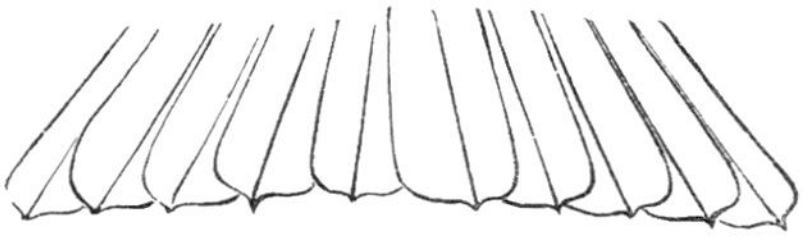

Tail of *Chætorhynchus papuensis.*

the innermost glossed with the latter colour; tail black, slightly glossed with steel-green on the edges of the feathers; sides of face and under surface of body black, the breast glossed with steel reflections; under wing-coverts greyish black; bill and feet black. Total length 7 inches, culmen 0·7 wing 4·2, tail 3·5, tarsus 0·7.

Hab. Arfak Mountains, New Guinea.

a.	♀ ad. sk.	Arfak Mountains, July 1873 (*Dr. A. B. Meyer*).	Dresden Museum [P.] (One of the typical specimens.)

4. CHAPTIA.

	Type.
Chaptia, *Hodgs. Ind. Rev.* i. p. 326 (1837)	C. ænea.
Prepopterus, *Hodgs. in Gray's Zool. Misc.* p. 84 (1844)......	C. ænea.
Entomoletes, *Sundev. Av. Meth. Tent.* p. 22 (1874)	C. ænea.

Range. The whole of India and Burmese countries, ranging throughout the Malayan peninsula; Sumatra, Borneo, Formosa.

Key to the Species.

a. General colour above rather uniform metallic greenish black, the rump inclining to greyish.
- *a'.* Smaller: wing not 5 inches .... *ænea*, p. 243.
- *b'.* Much larger: wing 5·15 inches, tail 5·2........ *brauniana*, p. 244.

b. Entire upper surface spangled with metallic plumes, the rump and upper tail-coverts being similarly metallic to the rest of the back .... *malayensis*, p. 244.

1. Chaptia ænea.

Le Drongo bronzé, *Levaill. Ois. d'Afr.* iv. p. 76, pl. 176; *Sundev. Crit. om Levaill.* p. 44.
Dicrurus æneus, *Vieill. N. Dict. d'Hist. Nat.* ix. p. 586 (*ex Levaill.*); *Jerd. Madr. Journ.* x. p. 240.
Dicrurus æratus, *Steph. Gen. Zool.* xiii. pt. 2, p. 138.
Chaptia muscipetoides, *Hodgs. Ind. Rev.* i. p. 327.
Prepopterus æneus, *Hodgs. in Gray's Zool. Misc.* p. 84.
Edolius æratus, *Blyth, J. A. S. B.* xi. p. 101.
Chaptia aenea, *Gray, Gen. B.* i. p. 258; *Blyth, J. A. S. B.* xv. p. 294; *id. Cat. B. Mus. A. S. B.* p. 200; *Bp. Consp.* i. p. 350; *Horsf. & Moore, Cat. B. Mus. E.I. Co.* i. p. 159; *Jerd. B. Ind.* i. p. 433; *Gray, Hand-l. B.* i. p. 287; *Hume, S. F.* 1874, p. 474, 1875, p. 100; *id. Nests and Eggs Ind. B.* p. 192; *Blyth & Wald. B. Burm.* p. 128.
Entomoletes aeneus, *Sundev. Av. Meth. Tent.* p. 22.

Adult. Upper surface of body greenish steel-black, highly metallic, the ground-colour of the plumage greyish, the rump and upper tail-coverts less distinctly glossed with green; wings metallic steel-colour, the primaries black, only washed externally with metallic colour; tail purplish black, glossed with metallic colour on the outer webs; lores and sides of face dull black; rest of under surface greyish black with a slight greenish gloss, the lower throat and chest spangled with metallic tips of steel-green; under wing-coverts black, slightly glossed with steel-colour; bill and feet black; iris deep brown*. Total length 9 inches, culmen 0·85, wing 4·75, tail 4·75, tarsus 0·6.

Hab. The whole of India, ranging into Assam and Burmah.

a. Ad. sk.	Travancore.	Capt. Biddulph [C.].
b, c. Ad. sk.	Malabar.	Purchased.
d, e, f, g. Ad. sk.	Nepal.	B. H. Hodgson, Esq. [P.].
h, i, k. Ad. sk.	Nazeerah, Upper Assam (*Dr. Reid*).	R. B. Sharpe, Esq. [P.].
l. Ad. sk.	Burmah.	Capt. Stackhouse Pinwill [P.].

2. Chaptia brauniana.

Chaptia brauniana, *Swinh. Ibis*, 1863, p. 269; *id. ib.* 1866, p. 399; *Gray, Hand-l. B.* i. p. 287; *Swinh. P. Z. S.* 1871, p. 378.

Adult. General colour above glossy greenish black, with a slight purplish tinge, this shade of colour spreading uniformly over the entire back, rump, and upper tail-coverts; wings uniform with the back, the primaries black, glossed with metallic; tail very much forked, purplish black, glossed on the outer webs with glossy steel-blue; lores and sides of face deep black; throat and breast black, spangled with metallic steel-blue, somewhat tinged with purple; the rest of the under surface dull grey, lighter on the lower flanks, and somewhat washed with metallic green near the breast, the under tail-coverts glossed with green; under wing-coverts black, glossed with purple; bill and legs black; iris blackish brown. Total length 9·7 inches, culmen 0·8, wing 5·15, tail 5·2, tarsus 0·55.

Hab. Island of Formosa.

a. Ad. sk.	Formosa.	Robert Swinhoe, Esq. [C.].

3. Chaptia malayensis †.

Chaptia malayensis, *Blyth, J. A. S. B.* xv. p. 294 (*ex A. Hay, MS.*); *id. Cat. B. Mus. A. S. B.* p. 200; *Gray, Gen. B.* iii. App. p. 13; *Bp. Consp.* i. p. 350; *Horsf. & Moore, Cat. B. Mus. E.I. Co.* p. 393; *Moore, P. Z. S.* 1854, p. 271; *Jerd. B. Ind.* i. p. 434;

* Mr. Oates gives the following note on the soft parts of specimens killed in Upper Pegu:—"Iris pinkish hazel; eyelids purplish grey; bill, legs, feet, and claws black."

† *Drymonax niger*, Less. Rev. Zool. 1839, p. 167, has been doubtfully referred by some authors to this species; but I cannot agree in the identification.

Sclater, P. Z. S. 1863, p. 218; *Gray, Hand-l. B.* i. p. 287; *Salvad. Ucc. Born.* p. 153.
Dicrurus picinus, *Bp. Consp.* i. p. 352 (*ex Müll. MS. in Mus. Lugd.*); *Gray, Hand-l. B.* i. p. 285; *Walden, B. Burm.* p. 128.
Chaptia malayana, *Sharpe, Ibis*, 1876, p. 45 (*lapsu*).

Adult male. Above brilliant metallic greenish black, continued over the entire back, rump, and upper tail-coverts; wings resembling the back, the primaries black, washed externally with metallic colour; tail purplish black, the outer webs dull metallic, the two centre feathers entirely so; nasal bristles, lores, and sides of face deep black; throat and entire breast spangled with metallic green, like the head, the abdomen and flanks greyish black, with only a very slight greenish gloss; under wing-coverts black, washed with purple; under surface of quills black; bill and feet black; iris dark blood-red. Total length 8·3 inches, culmen 0·85, wing 4·5, tail 4·1, tarsus 0·6.

Hab. Malayan peninsula, Sumatra, and Borneo.

a. ♂ ad. sk.	Mount Ophir, Malacca.	A. R. Wallace, Esq. [C.].
b. Ad. sk.	Borneo.	
c. ♀ ad. sk.	Sarawak.	A. R. Wallace, Esq. [C.].
d. ♂ ad. sk.	Jambusan, Sept. 15, 1874.	A. Everett, Esq. [C.].

5. BUCHANGA.

	Type.
Buchanga, *Hodgs. Ind. Rev.* i. p. 326 (1837)	B. macrocerca.
Musicus, *Reichenb. Syst. Av.* lxxxviii. fig. 9 (1850) ..	B. assimilis.

Range. The whole of Africa; India and Ceylon; Indo-Chinese region, China and Japan, Formosa, Hainan; Malayan peninsula, Java, Flores, Lombock; Mayotte.

Key to the Species.

a. Under tail-coverts deep black, with a gloss of steel-green, exactly like the under surface of the body .... { *atra*, p. 246. *assimilis*, p. 247.
  a'. Tail moderately long, not exceeding 7 inches; plumage blue-black.
  b'. Tail very long, 7·6 inches; plumage greenish black.............................. *waldeni*, p. 253.
b. Under tail-coverts slaty grey.
  c'. With a steel-green gloss on the tail-feathers *longicaudata*, p. 249.
  d'. With an ashy grey shade on the tail-feathers { *cineracea*, p. 250. *pyrrhops*, p. 251.
c. Under tail-coverts white.
  e'. General colour above light grey; lores, feathers round the eye, and cheeks white..... *leucogenys*, p. 251.
  f'. General colour above glossy greenish black; entire face blackish, like the throat.
    a''. Abdomen and under tail-coverts white.. { *cærulescens*, p. 252. *insularis*, p. 253.
    b''. Vent and under tail-coverts only white.. *leucopygialis*, p. 253.

1. Buchanga atra.

Le Drongolon, *Levaill. Ois. d'Afr.* iii. pl. 174.
Muscicapa atra, *Hermann, Obs. Zool.* p. 208.
Dicrurus macrocercus, *Vieill. N. Dict.* ix. p. 588; *Jerd. Madr. Journ.* xiii. pt. 2, p. 121; *Blyth, J. A. S. B.* xv. p. 298; *Gray, Gen. B.* i. p. 286; *id. Cat. Mamm. &c. Nepal Coll. Hodgs.* p. 98; *Bp. Consp.* i. p. 351; *Jerd. B. Ind.* i. p. 427; *Gray, Hand-l. B.* i. p. 286.
Edolius forficatus, *Horsf. Tr. Linn. Soc.* xiii. p. 144.
Muscicapa biloba, *Licht. Verz. Doubl.* p. 52.
Dicrurus indicus, *Steph. Gen. Zool.* xiii. p. 139 (*ex Levaill.*); *Hodgs. Asiat. Research.* xiii. p. 21, *cum fig.*
Dicrurus balicassius, *Sykes, P. Z. S.* 1832, p. 86 (*nec L.*); *Jerd. Madr. Journ.* x. p. 238; *M'Clell. P. Z. S.* 1839, p. 158; *Blyth, J. A. S. B.* xi. p. 175.
Buchanga albirictus, *Hodgs. Ind. Rev.* i. p. 326; *Hume, S. F.* 1873, p. 178; *Adam, t. c.* p. 377; *Ball, S. F.* 1874, p. 402; *Hume, t. c.* p. 474.
Dicrurus minor, *Blyth, Ann. & Mag. Nat. Hist.* xiii. p. 129; *id. Cat. B. Mus. A. S. B.* p. xxii; *id. Ibis*, 1867, p. 305.
Edolius fingah, *Blyth, J. A. S. B.* xv. p. 800.
Dicrurus fingah, *Hodgs. in Gray's Zool. Misc.* p. 84.
Dicrurus longus, *Bp. Consp.* i. p. 352; *Horsf. & Moore, Cat. B. Mus. E.I. Co.* i. p. 152.
Dicrurus himalayensis, *Tytler, Ibis*, 1868, p. 200 (descr. nulla); *Gray, Hand-l. B.* i. p. 286.
Dicrurus albirictus, *Gray, Hand-l. B.* i. p. 285; *Hume, S. F.* 1875, p. 97.
Dicrurus cathœcus, *Swinh. P. Z. S.* 1871, p. 377.
Buchanga minor, *Holdsw. P. Z. S.* 1872, p. 438.
Buchanga cathœca, *Walden, B. Burm.* p. 130.

Adult. General colour above glossy blue-black, the lesser and median wing-coverts uniform with the back; greater and primary coverts, as well as the inner secondaries, glossed with greenish, the primaries lighter brown, dusky at tip; lores, sides of face, and under surface of body black, with not quite so much metallic gloss as on the upper surface, and with somewhat of a greenish cast; under wing-coverts glossy black, resembling the breast; bill and feet black; iris red. Total length 11 inches, culmen 0·9, wing 5·4, tail 4·3, to tip of outer feather 6·2, tarsus 0·9.

Obs. In the Museum there are at present no Javan specimens; but Lord Tweeddale considers the Drongos from this island distinct, though he does not say how they differ. His latest conclusions (B. Burm. p. 129) are given in the following sentence:—"*Muscicapa ater* is Hermann's title for the South-Indian bird, which is invariably smaller than that of Northern and Eastern India. *D. macrocercus*, Vieillot, = *E. longus*, Temm., pertains to the Javan bird alone, a distinct form. Some Burman examples possess, while others want, the white rictal spot, an unstable character among the continental races, but never found, as far as at present recorded, in *B. macrocerca* or in *B. cathœca*. Adult Tonghoo birds agree best in the relative proportions of the rectrices with *B. cathœca*." I have seen no characters published to enable me to separate *Buchanga longa* (Temm.), of Java (N.B. I cannot understand why Vieillot's name of *macrocercus* applies particularly to the Javan bird) from *B. atra*, and I therefore unite the two. *B. cathœca* seems to me to be also indistinguishable as a species, Mr. Swinhoe's characters not holding good in the specimens which have fallen under my notice. A Chinese example in the Museum measures—total length 11·5 inches, culmen 0·9, wing 6·15, tail 4·7, to tip of outer feather 6·3, tarsus 0·85. Again, a Ceylonese specimen (*B. minor*) gives the following dimensions—total length 9·8 inches, culmen 0·9, wing 5·05, tail 3·4, to tip of outer feather 5·5, tarsus 0·85. It is much of the same size as *B. assimilis*, but has the large bill of *B. atra*.

Hab. The whole of India and Ceylon (smaller); Indo-Chinese region, China, and Formosa; Java.

a. B. macrocerca.

a. Ad. st.	India.	
b. ♂ ad. st.	India.	
c. ♀ juv. st.	India.	
d. Ad. sk.	India.	Dr. Burns [P.].
e. Jun. sk.	India.	Hon. E.I. Co. [P.].
f. Imm. sk.	Madras.	Rev. H. H. Baber [P.].
g, h. Ad. sk.	Kamptee, C. P.	Dr. B. Hinde [P.].
i. Juv. sk.	Kamptee, C. P.	Dr. B. Hinde [P.].
k, l. Ad. sk.	Behar.	B. H. Hodgson, Esq. [P.].
m–q. Ad. sk.	Nepal.	B. H. Hodgson, Esq. [P.].
r, s. Juv. sk.	Nepal.	B. H. Hodgson, Esq. [P.].
t. Ad. sk.	Gangaotri, Himalayas.	
u, v, w. Ad. sk.	N.W. Himalayas.	Capt. Stackhouse Pinwill [P.].
x, y. Juv. sk.	N.W. Himalayas.	Capt. Stackhouse Pinwill [P.].
z. ♂ ad. sk.	Bhamo, Upper Burmah, Sept. 9, 1868.	Dr. Anderson [C.].
a'. ♂ ad. sk.	Sanda, May 15, 1868.	Dr. Anderson [C.].
b'. ♀ ad. sk.	Mungla, Sanda valley, May 22, 1868.	Dr. Anderson [C.].
c', d'. Ad. sk.	Tenasserim.	J. C. D. Packman, Esq. [P.].
e'. Ad. sk.	Pegu.	W. T. Blanford, Esq. [P.].
f'. ♂ ad. sk.	Bangkok, Siam, May 30, 1872 (*Conrad*).	R. B. Sharpe, Esq. [P.].

β. B. cathœca.

g'. Ad., *h'.* Juv. sk.	China.	Capt. Blakiston [C.].
i', k'. Juv. sk.	Formosa.	R. Swinhoe, Esq. [C.].

γ. B. minor.

l'. ♂ imm. sk.	Jaffna, Ceylon, Jan. 1, 1873.	Capt. W. Vincent Legge, R.A. [P.].

Subsp. *a.* **Buchanga assimilis.**

Corvus afer, *Licht. Cat. Rev. Nat. Hamb.* p. 10 (*nec Linn.*).
Corvus adsimilis, *Bechst., Lath. Allg. Uebers. Vög.* ii. p. 562; *id. Kurze Uebers.* p. 117.

Le Drongear, *Levaill. Ois d'Afr.* p. 167; *Sundev. Crit. om Levaill.* p. 42.

Dicrurus musicus, *Vieill. N. Dict. d'Hist. Nat.* xi. p. 586; *Gray, Gen. B.* i. p. 287; *Bp. Consp.* i. p. 352; *Cab. Mus. Hein.* i. p. 111; *Hartl. Orn. W.-Afr.* p. 100; *id. J. f. O.* 1861, p. 170; *Grill, Zool. Anteck.* p. 33; *Layard, B. S. Afr.* p. 154; *Gray, Hand-l. B.* i. p. 283: *Sharpe, Cat. Afr. B.* p. 47; *Gurney, in Anderss. B. Dam. L.* p. 125; *Shelley, Ibis*, 1875, p. 75.

Muscicapa emarginata, *Licht. Verz. Doublett.* p. 52.

Muscicapa divaricata, *Licht. t. c.* p. 52.

Edolius lugubris, *Hempr. & Ehrenb. Symb. Phys.* Aves, fol. s. pl. viii. fig. 3; *Rüpp. N. Wirb.* p. 109.

Dicrurus canipennis, *Swains. B. W. Afr.* i. p. 254; *Bp. Consp.* i. p. 352.

Dicrurus lugubris, *Gray, Gen. B.* i. p. 287; *Rüpp. Syst. Uebers.* p. 61; *Bp. Consp.* i. p. 352; *Cab. Mus. Hein.* i. p. 110; *Heugl. Syst. Uebers.* p. 32; *Brehm, Reise Habesch.* pp. 215, 319; *Gray, Hand-l. B.* i. p. 286.

Dicrurus divaricatus, *Gray, Gen. B.* i. p. 287; *Cab. Mus. Hein.* i. p. 111; *Strickl. P. Z. S.* 1850, p. 217; *Horsf. & Moore, Cat. B. Mus. E.I. Co.* i. p. 155; *Hartl. Orn. W.-Afr.* p. 100; *Gray, Hand-l. B.* i. p. 286; *Finsch, Tr. Z. S.* vii. p. 249; *Blanf. Geol. & Zool. Abyss.* p. 344; *Finsch & Hartl. Vög. Ostafr.* p. 323; *Sharpe, Cat. Afr. B.* p. 47; *Heugl. Orn. N.O.-Afr.* pp. 422, cxii; *Antin. & Salvad. Viagg. Bogos*, p. 77.

Dicrurus emarginatus, *Bp. Consp.* i. p. 352.

Dicrurus aculeatus, *Cass. Pr. Philad. Acad.* 1851, p. 348.

Edolius erythrophthalmus, *Pr. Würt.*; *Heugl. J. f. O.* 1867, p. 294.

Dicrurus fugax, *Peters, J. f. O.* 1868, p. 132; *Cab. in Von der Decken's Reise*, iii. p. 29.

Dicrurus ludwigii, *Bianc. Spec. Zool. Mosamb.* fasc. xviii. (*nec Smith*); *Finsch & Hartl. Vög. Ostafr.* p. 156.

Dicrurus adsimilis, *Walden, Tr. Z. S.* ix. p. 181.

Musicus divaricatus, *Bouvier, Cat. Ois. Marche, &c.* p. 18.

Adult. General colour glossy black, with a steel-blue shade, more distinct on the secondaries and tail-feathers; quills brown, contrasting with the black of the body, especially underneath, the outer edges only glossed with greenish black; tail black, glossed with steel-green; under surface of body exactly like the upper, but not so glossy on the abdomen and under tail-coverts; no crest on the head, but the frontal feathers full; ear-coverts rather lighter than the head under certain lights, and somewhat of a deep oily greenish black; bill and feet black; iris red. Total length 9·5 inches, culmen 0·95, wing 5·55, tail 5·1, tarsus 0·9.

Obs. The difference between the large South-African Drongo and the smaller one of North-eastern Africa, *B. divaricata*, is simply one of size, and no other difference exists excepting a rather lighter under surface to the quills in the southern bird. This latter character is often approached in northern examples. For measurements of a large series, see Finsch and Hartlaub's 'Vögel Ostafrika's' (*l. c.*), in which work also Lord Tweeddale expresses his opinion that *B. adsimilis* is not to be specifically separated from *B. divaricata*. A comparison of a large series of specimens shows that the Indian *B. atra* is nothing but a larger race of the African Drongo.

Hab. The whole of Africa.

a. Ad. st.	South Africa.	
b. Ad. st.	South Africa.	J. Gould, Esq.
c. Ad. sk.	South Africa.	J. Gould, Esq.
d. Ad. st.	South Africa.	Sir A. Smith [C.].
e. Ad. sk.	South Africa.	F. Campbell, Esq. [P.].
f. Juv. sk.	South Africa (*Schomburgk*).	R. B. Sharpe, Esq.
g. Juv. sk.	Knysna.	R. B. Sharpe, Esq.
h, i, k. Ad. sk.	Near King-williamstown.	Herbert Trevelyan, Esq. [P.].
l. Ad. sk.	Port Elizabeth.	R. B. Sharpe, Esq.
m. Ad. sk.	Natal.	T. Ayres, Esq. [C.].
n. Ad. sk.	Near Rustenburg (*T. C. Atmore*).	F. A. Barratt, Esq. [C.].
o. ♀ ad. sk.	Graham's Town.	R. B. Sharpe, Esq.
p. Ad. sk.	Victoria Falls, Zambesi.	Purchased.
q, r. Ad. sk.	Tete, Zambesi.	Dr. Kirk [C.].
s. Juv. sk.	Otjimbinque, Damara Land, July 26, 1863 (*C. J. Andersson*).	R. B. Sharpe, Esq.
t, u. Ad. sk.	Damara Land (*Andersson*).	R. B. Sharpe, Esq.
v. Ad. sk.	Benguela.	J. J. Monteiro, Esq. [C.].
w, x. Imm. sk.	Benguela (*Anchieta*).	R. B. Sharpe, Esq.
y. Imm. sk.	River Niger (*Baikie*).	R. B. Sharpe, Esq.
z. Ad. st.	River Gambia.	Governor Rendall [C.].
a', b'. Ad. sk.	River Gambia.	Purchased.
c'. Ad. sk.	River Gambia.	R. B. Sharpe, Esq.
d', e'. Ad. sk.	Mombas (*Wakefield*).	R. B. Sharpe, Esq. [P.].
f'. Ad. sk.	N.E. Africa (*Kotschy*).	R. B. Sharpe, Esq.
g'. Var. st.	Abyssinia.	Dr. Rüppell [C.].
h'. Ad. sk.	White Nile.	Sir Samuel Baker [P.].
i'. ♂ ad. sk.	Near Magen, Senafé Pass, Feb. 18, 1868.	W. T. Blanford, Esq. [C.].
k'. ♀ imm. sk.	Near Magen, Feb. 18, 1868.	W. T. Blanford, Esq. [C.].
l'. ♀ ad. sk.	Anseba valley, Aug. 5, 1868.	W. T. Blanford, Esq. [C.].
m', n'. Ad. sk.	Bogos Land (*Esler*).	R. B. Sharpe, Esq.
o'. Imm. sk.	Bogos Land (*Esler*).	R. B. Sharpe, Esq.
p'. Ad. sk.	Shoa (*Harris*).	Secretary of State for India [P.].

3. Buchanga longicaudata.

Dicrurus macrocercus, *Jerd. Madr. Journ.* x. p. 240 (*nec V.*); *Blyth, Ann. & Mag. Nat. Hist.* xiv. p. 46.
Dicrurus longicaudatus, "*A. Hay,*" *Jerd. Mad. Journ.* xiii. pt. 2, p. 121; *Blyth, J. A. S. B.* xv. p. 298; *id. Cat. B. Mus. A. S. B.* p. 202; *Bp. Consp.* i. p. 351; *Horsf. & Moore, Cat. B. Mus. E.I. Co.* i. p. 152; *Jerd. B. Ind.* i. p. 430; *Gray, Hand-l. B.* i. p. 285; *Hume, Str. F.* 1875, p. 97.
Buchanga longicaudata, *Wald. Ibis,* 1868, p. 316; *Beav. t. c.* p. 497 *Holdsw. P. Z. S.* 1872, p. 438.
Buchanga waldeni, *Beav. Ibis,* 1868, p. 497 (*nec Schl.*).
Buchanga longicauda, *Hume, S. F.* 1874, p. 474.

Adult. General colour glossy greenish black, with somewhat of a greyish cast; quills and tail black, externally glossed with the same steel-green as the back; sides of face and under surface of body greyish, with a slight bluish gloss on the sides of the neck and chest, the lower abdomen and under tail-coverts dull greyish black; bill and feet black; iris brownish red. Total length 11·6 inches, culmen 1, wing 5·6, tail 4, outer tail-feathers 6·7, tarsus 0·8.

Hab. India and Ceylon.

a. ♂ ad. sk.	Colombo, Ceylon, Jan. 17, 1872.	Capt. Vincent Legge, R.A. [P.].
b, c. Ad. sk.	Madras.	Rev. H. H. Baber [P.].
d. Ad. sk.	Madras.	Dr. Jerdon [P.].
e. Ad. sk.	Kamptee, C. P.	Dr. B. Hinde [P.].
f–k. Ad. sk.	Nepal.	B. H. Hodgson, Esq. [P.].
l, m. Ad. sk.	N.W. Himalayas.	Capt. Stackhouse Pinwill [P.].
n. Ad. sk.	Ladakh.	Purchased.

4. **Buchanga cineracea.**

Edolius cineraceus, *Horsf. Tr. Linn. Soc.* xiii. p. 145.
Dicrurus leucophæus *Gray, Gen. B.* i. p. 287 (*nec V.*); *Bp. Consp.* i. p. 352; *Horsf. & Moore, Cat. B. Mus. E.I. Co.* i. p. 155.
Dicrurus intermedius, *Blyth, J. A. S. B.* xv. p. 299; *id. Cat. B. Mus. A. S. B.* p. 203.
Dicrurus cinerascens, *Gray, Gen. B.* i. p. 287.
Dicrurus cineraceus, *Bp. Consp.* i. p. 352; *Horsf. & Moore, Cat. B. Mus. E.I. Co.* i. p. 155.
Buchanga wallacii, *Walden, Ann. & Mag. Nat. Hist.* (4) v. p. 220.
Buchanga mouhoti, *Walden, Ann. & Mag. Nat. Hist.* (4) v. p. 220; *Swinh. P. Z. S.* 1870, p. 245; *id. P. Z. S.* 1871, p. 378.

Adult. General colour above ashy grey, slightly glossy; wing-coverts uniform with the back, the greater and primary-coverts with a faint subterminal shade of dusky, visible in certain lights; quills blackish, the primaries washed with grey near the base of the outer web; the secondaries externally grey, the innermost entirely of that colour; tail-feathers clear grey, with black shafts, the outer feathers rather more dusky blackish, the grey shade being confined to the inner web; nasal plumes blackish; lores and feathers in front of the eye dusky blackish; cheeks and ear-coverts rather more dingy ashy grey, and resembling more the shade of the throat; rest of under surface ashy grey, but rather duskier than the back, and only glossed on the breast; under wing-coverts resembling the breast; under tail-coverts rather lighter grey than the abdomen, but in no way whitish. Total length 10 inches, culmen 1, wing 5·05, tail 5·1, tarsus 0·7.

Obs. On comparing Lombock specimens (*B. wallacii*, Walden) with typical *B. cineracea* from West Java, I cannot find any specific differences; nor can I distinguish between *B. intermedia* and *B. cineracea*, judging from the few specimens I have compared. Mr. Hume (S. F. 1875, pp. 98–100) would unite *B. pyrrhops* and *B. intermedia* to *B. longicaudata*, as he has examples of all three forms from Dacca, where "they may be shot together off the same tree." Now although it is possible that at Dacca the ranges of these races may coalesce, I agree with Lord Tweeddale that they should not be specifically united; for I also have never seen a specimen of *D. longicaudata* that could be mistaken for *B. pyrrhops* or *B. cineracea* (*cf.* Walden, B. Burm. p. 131); the grey in the tail of the latter is a very good character, and easily recognizable. *B. pyrrhops* is nothing more than a large race of *B. cineracea.* The measurements of Mr. Blanford's Pegu skin are as follows:—Total length 9·6 inches, culmen 0·9, wing 5·2, tail 5·5, tarsus 0·75.

Hab. Java; Lombock; Pegu and Burmah.

a, b. Ad. sk.	W. Java.	A. R. Wallace, Esq. [C.].
c, d. ♂ ad. sk.	Lombock.	A. R. Wallace, Esq. [C.].
e. ♀ ad. sk.	Ampanam.	A. R. Wallace, Esq. [C.].
f. Ad. sk.	Pegu.	W. T. Blanford, Esq. [P.].

Subsp. *α.* **Buchanga pyrrhops.**

Dicrurus pyrrhops, *Hodgs. in Gray's Zool. Misc.* p. 14.
Dicrurus cinerascens, *Gray, Cat. Mamm. &c. Nepal coll. Hodgs.* p. 98.
Buchanga pyrrhops, *Jerd. Ibis,* 1872, p. 120; *Hume, S. F.* 1875, p. 98; *Wald. B. Burm.* p. 131.

Adult. General colour above ashy grey, with a light steel-blue lustre, rather lighter on the rump and upper tail-coverts; wing-coverts and scapularies more dusky, and glossed with darker steel-blue; quills dusky brown, glossed externally with a faint shade of the same steel-blue; tail greyish, lighter on the central feathers, and becoming more dusky on the outer ones, especially towards the tips, which are brownish; nasal plumes and lores dusky black, the latter slightly varied with greyish tips to the feathers; sides of face and entire under surface of body ashy grey, including the under wing- and tail-coverts, the breast slightly glossed with a dull metallic blue shade; bill and legs black; iris red. Total length 11 inches, culmen 1·1, wing 5·6, tail 4·8, to tip of outer feather 6·5, tarsus 0·8.

Hab. Himalayas, ranging to Dacca.

a, b. Ad. sk.	Nepal.	B. H. Hodgson, Esq. [P.].
c. Ad. sk.	N.W. Himalayas.	Capt. Stackhouse Pinwill [P.].

5. **Buchanga leucogenys.**

Dicrurus cineraceus, *Blyth, J. A. S. B.* xv. p. 299; *id. Cat. B. Mus. A. S. B.* p. 203.
Dicrurus leucophæus, *Swinh. P. Z. S.* 1863, p. 285 (*nec Vieill.*); *Hume, S. F.* 1874, p. 210, 1875, p. 99.
Buchanga leucogenys, *Walden, Ann. & Mag. Nat. Hist.* (4) v. p. 219; *Swinh. P. Z. S.* 1870, p. 245; *id. P. Z. S.* 1871, p. 378; *Hume, S. F.* 1875, p. 321.
Dicrurus leucogenys, *Gray, Hand-l. B.* i. p. 285.
Dicrurus leucophæus, *Gray, Hand-l. B.* i. p. 285.
Buchanga cinerascens, *Hume, S. F.* 1875, p. 321.

Adult. General colour above clear ashy grey, including the wings, the quills being blackish, washed externally with the same grey as the back, the primaries only near the base; upper tail-coverts and tail rather lighter grey than the back, gradually inclining to blackish at the tips and on the outermost feathers; nasal bristles black; lores, feathers round the eye, and ear-coverts pure white; cheeks and under surface of body light pearly grey, shading off into pure white on the abdomen and under tail-coverts; under wing-coverts light grey, with whitish edgings; bill and legs black; iris red. Total length 10·2 inches, culmen 0·95, wing 5·4, tail 4·5, to tip of outer feather 5·6, tarsus 0·7.

Hab. Malacca, ranging through the Indo-Chinese region to China and Japan.

a. Ad. sk.	[Malacca.]	Purchased.
b. Imm. sk.	Malacca.	A. R. Wallace, Esq. [C.].
c. Ad. sk.	Penang.	Purchased.
d, e. Ad. sk.	Tenasserim.	J. C. D. Packman, Esq. [P.].
f. Ad. sk.	Foochow, June 1, 1875.	R. Swinhoe, Esq. [C.].

6. Buchanga cærulescens.

The Fork-tailed Indian Butcher-bird, *Edwards, B.* ii. p. 56, pl. 56.
La Pie-Grièche à queue fourchue de Bengale, *Briss. Orn.* ii. p. 189.
Lanius cærulescens, *Linn. S. N.* i. p. 134; *Gm. S. N.* i. p. 297.
Lanius cæruleus, *Müll. S. N. Suppl.* p. 72.
Le Fingah, *Buff. H. N. Ois.* i. p. 108.
Lanius fingah, *Shaw, Gen. Zool.* vii. p. 291.
Dicrurus cærulescens, *Vieill. N. Dict. d'Hist. Nat.* ix. p. 587; *Jerd. Madr. Journ.* x. p. 239; *Blyth, Ann. & Mag. Nat. Hist.* xiv. p. 47; *id. J. A. S. B.* xv. p. 298; *Gray, Gen. B.* i. p. 286; *Blyth, Cat. B. Mus. A. S. B.* p. 203; *Bp. Consp.* i. p. 351; *Horsf. & Moore, Cat. B. Mus. E.I. Co.* i. p. 154; *Jerd. B. Ind.* i. p. 432; *Ball, S. F.* 1874, p. 403.
Drongo cærulescens, *Tickell, J. A. S. B.* ii. p. 573.
Dicrurus fingah, *Hodgs. in Gray's Zool. Misc.* p. 84.
Dicrurus cæruleus, *Cass. Pr. Philad. Acad.* 1864, p. 238; *Gray, Hand-l. B.* i. p. 285.

Adult. General colour above slaty grey, with a slight metallic gloss of steel blue, more apparent on the head, which is rather darker than the back; rump and upper tail-coverts clearer grey than the back; wings deep ashy brown, the coverts all glossed with the same slightly metallic steel-blue as the back; quills glossed only on the outer web, and dusky at tips; the inner secondaries entirely glossy; tail ashy brown, only very faintly glossed with steel-blue, the feathers rather more dusky towards their tips; lores and frontal plumes blackish; sides of face, throat, and chest ashy brown, glossed with steel-blue on the sides of the neck and sides of the upper breast; rest of the under surface, from the breast downwards, and including the under tail-coverts, white, the flanks somewhat washed with greyish; thighs glossy black; under wing-coverts deep ashy brown like the breast, with a slight metallic gloss; bill and legs black; iris lake-red. Total length 9·6 inches, culmen 0·95, wing 5·05, tail 3·65, to tip of outer feather 5, tarsus 0·8.

Hab. India.

a. Ad. sk.	Nepal.	B. H. Hodgson, Esq. [P.].
b. Ad. sk.	Behar.	B. H. Hodgson, Esq. [P.].
c. Ad. sk.	N.W. Himalayas.	Capt. Stackhouse Pinwill [P.].
d. Ad. sk.	Gondul, Katiawar, Nov. 26, 1871.	Major J. Hayes Lloyd [P.].

Subsp. *α*. **Buchanga insularis.**

Buchanga cærulescens, *Holdsw. P. Z. S.* 1872, p. 439 (*nec L.*).

Adult male. Similar to *B. cærulescens*, but much smaller and darker, especially on the throat and chest, which are slaty brown with a distinct gloss of steel-blue; "bill and legs black, the edges of the scales on the latter greyish; iris reddish brown" (*Legge, MS.*). Total length 9 inches, culmen 0·95, wing 4·6, tail 3·7, to tip of outer feather 4·5, tarsus 0·75.

Hab. The whole of the east of Ceylon and central province, eastern district (*Legge, MS.*).

a. ♂ ad.	Ceylon, Sept. 20, 1875.	Capt. W. Vincent Legge, R.A. [P.].
b, c. Ad.	Uva district.	Mr. E. Boate [C.].
d. Juv. sk.	Uva district.	Mr. E. Boate [C.].

7. **Buchanga leucopygialis.**

Dicrurus leucopygialis, *Blyth, J. A. S. B.* xv. p. 298; *id. Cat. B. Mus. A. S. B.* p. 203.
Buchanga leucopygialis, *Holdsw. P. Z. S.* 1872, p. 439.

Very similar to *B. insularis*, if, indeed, really separable, the only differences being that the entire abdomen is brownish, only slightly mixed with white, the under tail-coverts being white: this gives a darker appearance to the bird underneath. "Bill and feet black, iris brown" (*Legge, MS.*). Total length 9 inches, culmen 0·85, wing 4·7, tail 3·5, to tip of outer feather 4·5, tarsus 0·7.

Hab. Ceylon.

a. ♂ ad. sk.	Ceylon, Jan. 26, 1872.	Capt. W. Vincent Legge, R.A. [P.].

8. **Buchanga waldeni.**

Dicrurus waldeni, *Schl. N. T. D.* 1865, p. 86; *id. P. Z. S.* 1866, p. 423; *id. & Poll. Faun. Madag.* p. 80, pl. 23; *Gray, Hand-l. B.* i. p. 286.

Adult. Above and below glossy greenish black, deeper black on the lores and sides of the face; quills and tail black, externally glossed with green; bill and feet black; iris brown. Total length 12·6 inches, culmen 1·2, wing 5·6, tail 4·9, outer tail-feathers 7·6, tarsus 1·05. (*Mus. Marquis of Tweeddale.*)

6. EDOLIUS.

Type.
Edolius, *Cuvier, Règn. Anim.* i. p. 365 (1829).......... E. forficatus.
Range. Confined to Madagascar.

1. Edolius forficatus.

Le Grand Gobe-mouche noir hupé de Madagascar, *Briss. Orn.* ii. p. 388, pl. xxxvii. fig. 4.
Gobe-mouche hupé de Madagascar, *Buff. Pl. Enl.* 189.
Lanius forficatus, *Linn. S. N.* i. p. 134.
Fork-tailed Crested Shrike, *Lath. Gen. Syn.* i. p. 158.
Muscicapa galatea, *Bodd. Tabl. Pl. Enl.* p. 11.
Le Drongo, *Levaill. Ois. d'Afr.* iv. pl. 166; *Sundev. Crit. om Levaill.* p. 43.
Dicrurus lophorhinus, *Vieill. N. Dict.* ix. p. 587; *Gray, Gen. B.* . p. 286.
Dicrurus cristatus, *Vieill. N. Dict.* ix. p. 587, pl. D. 3. fig. 2; *id. et Oud. Gal. Ois.* i. p. 228, pl. cxli.
Edolius forficatus, *Cuvier, Règne Anim.* i. p. 365.
Dicrurus forficatus, *Steph. Gen. Zool.* xiii. part 2, p. 138; *Gray, Gen. B.* i. p. 286; *Bp. Consp.* i. p. 351; *Hartl. Faun. Madag.* p. 49; *Roch & E. Newt. Ibis,* 1862, p. 274; *E. Newt. Ibis,* 1863, p. 348; *Scl. P. Z. S.* 1863, p. 163; *A. Newt. P. Z. S.* 1865, p. 834; *Schl. P. Z. S.* 1866, p. 423; *Grand. Rev. et Mag. de Zool.* 1867, p. 386; *Schl. & Pollen, Faun. Madag. Ois.* p. 79; *Gray, Hand-l. B.* i. p. 286; *Sharpe, P. Z. S.* 1870, p. 390; *id. Cat. Afr. B.* p. 47. no. 842.

Adult. Black, everywhere glossed with steel- or blue-black; quills and tail black, glossed on the outer web with greenish steel-black; lores glossy black; on the forehead a very large erect crest, composed of glossy black feathers 1·45 inch long; under surface of body glossy steel-black, rather duller on the abdomen and under tail-coverts; bill and legs black. Total length 10·8 inches, culmen 1, wing 5·35, tail 5·55, tarsus 0·9.

Hab. Madagascar, and Johanna Island in the Comoro group.

a, b. Ad. sk.	Madagascar.	Purchased.
c. Ad. sk.	Madagascar.	J. Caldwell, Esq. [P.].
d. Ad. sk.	Mohambo.	E. Gerrard, Esq. [C.].
e. Ad. sk.	Nossi Vola (*Crossley*).	R. B. Sharpe, Esq.

7. DISSEMUROIDES.

Type.
Dissemuroides, *Hume, S. F.* 1873, p. 408.......... D. dicruriformis.

Range. Ceylon; Andaman Islands and Nicobars.

Key to the Species.

a. Larger; forehead with a few hairy bristles .. { *andamanensis,* p. 255. / *dicruriformis,* p. 255.
b. Smaller; frontal crest composed of plumes, not hairs .... *edoliiformis,* p. 256.

1. Dissemuroides andamanensis.

Dicrurus andamanensis, *Tytler & Beavan, Ibis,* 1867, p. 322; *Gray, Hand-l. B.* i. p. 285; *Ball, J. A. S. B.* xli. p. 282.
Buchanga (Dicrurus) andamanensis, *Ball, S. F.* 1873, p. 66.
Dissemuroides andamanensis, *Hume, S. F.* 1874, p. 211.

Adult male. General colour black, with a slight purplish gloss on the back; wings glossed with metallic steel-green, as also the secondaries; quills brown, with reddish brown shafts, and slightly glossed with greenish at the base of the outer web; tail black, with a steel-green gloss on the outer webs; lores, sides of face, and under surface deep black, with a slight purplish gloss on the breast; under wing-coverts like the breast, but tipped with white. "Bill and legs black; iris hair-brown, sometimes very deep and almost blackish. Total length 11·25–12·0, wing 5·0–5·25, tail 7·0–8·25, bill from gape 1·2–1·35, tarsus 0·8–0·9" (*Hume*).

Adult female. Rather smaller than the male (*Hume*).

Obs. The single skin which we possess in the Museum is in bad condition and moulting; but the metallic gloss both on the upper and under surface seems to me more purple and less of a greenish cast than *D. dicruriformis.* Mr. Hume, who had a large series, does not notice this difference, which perhaps may not be observable when a quantity of specimens are examined.

Hab. Andaman Islands.

a. ♂ ad.	Port Blair, South Andamans, June 23, 1873 (*Wimberley*).	R. B. Sharpe, Esq. [P.].

Subsp. *α.* Dissemuroides dicruriformis.

Dissemuroides dicruriformis, *Hume, S. F.* 1873, p. 408; *id. S. F.* 1874, p. 211.

Adult female. General colour above black, with a greenish gloss deepening to steel-green on the head and neck; wing-coverts uniform with the back, as also the inner secondaries; primaries brown, with reddish brown shafts, slightly glossed with greenish near the base of the outer web, and inclining to dusky brown at tips; tail deep black, with a steel-green gloss on the outer webs; lores and sides of face deep black; under surface of body black, with a slight gloss of steel-green on the feathers of the lower throat and breast; under wing-coverts black, with a slight metallic gloss, and all tipped with white; bill and feet black; irides dark brown. Total length "13·25–14·6, wing 5·6–5·9, bill from gape 1·4–1·5, tarsus 0·9–1·0" (*Hume*).

Obs. Mr. Hume has kindly presented to the Museum an example of this bird; but as the tail is not fully grown, I have preferred to quote his own comparative measurements. As he claims a specific rank for this bird only on account of its larger size, I am induced to look upon it merely as a race of the ordinary *D. andamanensis.*

Hab. Andamans: Great Coco Island and Table Island.

a. ♀ ad. sk.	Great Coco Island.	A. O. Hume, Esq. [P.].

2. Dissemuroides edoliiformis.

Le Drongup, *Levaill. Ois. d'Afr.* pl. 173; *Sundev. Crit. om Levaill.* p. 43.
Dicrurus edoliiformis, *Blyth, J. A. S. B.* xv. p. 297; *id. Cat. B. Mus. A. S. B.* p. 202; *id. Ibis,* 1867, p. 305.
Dicrurus lophorhinus, *Gray, Hand-l. B.* i. p. 285 (*nec V.**).
Dissemurus lophorhinus, *Holdsw. P. Z. S.* 1872, p. 439.
Dissemuroides lophorhinus, *Hume, S. F.* 1874, p. 408.

Adult female. General colour above black, with a metallic green shade, rather more purplish on the head and on the neck-hackles; wings metallic steel-green, the inner secondaries also; the rest of the quills black, glossed externally with steel-green, the primaries only at the base; tail-feathers black, glossed with steel-green on the outer webs; forehead crested, the plumes deep black, as also those of the lores and sides of the face; under surface of body black, glossed with steel-green, uniformly distributed on the abdomen, but rather more metallic on the breast, where it is chiefly confined to the tips of the breast-feathers, producing a slightly spangled appearance; under wing-coverts black, glossed with metallic green like the breast; "bill and legs black; iris dull brownish red" (*Legge*). Total length 12·3, culmen 1·25, wing 5·6, tail 5·2, to tip of outer feather 7·2, tarsus 0·95.

Adult male. Similar to the female, but a little larger. Captain Legge says that the wing of a male bird, shot with the female above described, measured 6 inches in length.

Hab. Ceylon.

a. ♀ ad. sk.	Opate Forests, South Province, Ceylon.	Capt. W. Vincent Legge, R.A. [P.].

8. DICRANOSTREPTUS.

	Type.
Dicranostreptus, *Reichenb. Syst. Av.* pl. lxxxviii. fig. 12 (1850)	D. megarhynchus.

Range. New Ireland; New Guinea (?)†, Ké Islands‡.

1. Dicranostreptus megarhynchus.

Edolius megarhynchus, *Quoy et Gaim. Voy. d'Astrol. Zool.* p. 184, pl. 6.§
Dicrurus megarhynchus, *Gray, Gen. B.* i. p. 286; *Bp. Consp.* i.

---

* Vieillot's name of *D. lophorhinus* (N. Dict. d'Hist. Nat. ix. p. 587) is generally applied to this species; but the description seems to me to be that of *D. forficatus* of Madagascar, if, indeed, recognizable at all.

† Very doubtful. The bird is probably confined to New Ireland.

‡ *Teste* Dr. O Finsch, 'Neu-Guinea,' p. 171.

§ The habitat is given as Dorey, New Guinea; but probably this locality is erroneous. The Museum specimens are from Carteret Harbour, New Ireland, a locality visited by the 'Astrolabe' (*cf. t. c.* p. 181).

p. 352; *Gray, Cat. B. New Guinea*, p. 33; *Finsch, Neu Guinea*, p. 171; *Gray, Hand-l. B.* i. p. 287.
Dicronostreptus mezorhynchus (lapsu), *Jerd. B. Ind.* i. p. 430, note.

Adult. Above blue-black, with a very slight steel gloss on the back, rump, and upper tail-coverts; head and sides of neck metallic steel-purple, the feathers on the latter elongated and somewhat lanceolate; frontal plumes, lores, and sides of face dull purplish black; under surface of body dusky purplish black, with faintly indicated tips of glossy green on some of the feathers of the throat and chest; under wing-coverts purplish black like the breast; wings above dull glossy steel-green; quills purplish black, the secondaries externally dull steel-green, not so metallic as the coverts; tail-feathers purplish black, glossed with metallic steel-green on the outer edge of both webs, more distinctly and broadly on the centre ones, the outermost feather elongated and twisted in an inward curl at the tip; bill and feet black. Total length 20·5 inches, culmen 1·5, wing 7, tail 6, outer feather 14, tarsus 1·05.

Hab. The same as that of the genus.

a. Ad. st.	Carteret Harbour.	Capt. Lambrick, R.N. [P.].
b. Ad. sk.	Carteret Harbour.	Capt. Lambrick, R.N. [P.].

9. BHRINGA.

	Type.
Bhringa, *Hodgson, Ind. Rev.* i. p. 325 (1837)............	B. remifer.
Melisseus, *Hodgson, in Gray's Zool. Misc.* p. 84 (1844)....	B. remifer.

Range. Himalayas, Burmese provinces; Tenasserim; Java.

1. Bhringa remifer.

Edolius remifer, *Temm. Pl. Col.* iii. pl. 178; *Blyth, J. A. S. B.* xi. pp. 169, 800.
Bhringa tectirostris, *Hodgs. Ind. Rev.* i. p. 325; *Hume, Str. F.* 1874, p. 474, 1875, p. 101; *id. Nests and Eggs Ind. B.* p. 193.
Bhringa remifer, *Gray, Gen. B.* i. p. 288; *Blyth, J. A. S. B.* xv. p. 294; *id. Cat. B. Mus. A. S. B.* p. 200; *Bp. Consp.* i. p. 350; *Horsf. & Moore, Cat. B. Mus. E.I. Co.* i. p. 159; *Jerd. B. Ind.* i. p. 434; *Gray, Hand-l. B.* i. p. 287; *Godw.-Aust. J. A. S. B.* xxxix. p. 268; *Blyth & Wald. B. Burm.* p. 128.
Melisseus remifer, *Hodgs. in Gray's Zool. Misc.* p. 84; *Cab. Mus. Hein.* Th. i. p. 111.

Adult. Black; upper surface of body with a rich steel-green metallic gloss; forehead, lores, and sides of face purplish black; below, the breast is metallic greenish like the back, the throat also spangled with steel-green tips to the feathers; abdomen blackish grey, with a slight steel gloss, the under tail-coverts distinctly washed with steel-green; under wing-coverts purplish black, with a few steel reflections near the edge of the wing, some few of the greater series tipped with white; wings above glossy greenish like the back, but a little duller; quills blue-black on the inner web, the secondaries externally glossed like the back, the primaries only narrowly washed with the same on the outer web; tail-feathers blue-black, externally washed with steel-green, the two outermost feathers produced into a long shaft, with an elongated black spatule at the end. Total length 9·6 inches, culmen 0·95, wing 5·25, tail 5·1, outer tail-feathers 17·2, tarsus 0·75.

Hab. The same as the genus.

a. Ad. sk.	Java.	Purchased.
b. Ad. sk.	W. Java.	A. R. Wallace, Esq. [C.].
c, d, e, f. Ad. sk.	Nepal.	B. H. Hodgson, Esq. [P.].
g. Juv. sk.	Nepal.	B. H. Hodgson, Esq. [P.].
h. Ad. sk.	Nepal.	Purchased.
i. Juv. sk.	N.W. Himalayas.	Capt. Stackhouse Pinwill [P.].

10. DISSEMURUS.

Type.

Dissemurus, *Gloger, Handb. Naturg.* p. 347 (1842) .... D. paradiseus.

Range. Indian peninsula, Ceylon, and Andaman Islands; Indo-Chinese countries; Malayan peninsula; Java; Borneo.

1. Dissemurus paradiseus.

Le Coucou vert hupé de Siam, *Briss. Orn.* iv. p. 151, pl. xiv. A. fig. 1.
Cuculus paradiseus, *Linn. S. N.* i. p. 172.
Paradise Cuckoo, *Lath. Gen. Syn.* i. p. 529.
Malabar Shrike, *Lath. Gen. Syn. Suppl.* p. 56, pl. 108; *id. Gen. Hist. B.* ii. p. 57, pl. xviii.
Lanius malabaricus, *Lath. Ind. Orn.* i. p. 66; *Raffl. Tr. Linn. Soc.* xiii. p. 306.
Drongo à raquettes, *Levaill. Ois. d'Afr.* iv. p. 73, pl. 175.
Dicrurus platurus, *Vieill. N. Dict. d'Hist. Nat.* ix. p. 588; *Gray, Hand-l. B.* i. p. 287.
Edolius malabaricus, *Horsf. Tr. Linn. Soc.* xiii. p. 145.
Edolius rangoonensis, *Gould, P. Z. S.* 1836, p. 5; *Blyth, J. A. S. B.* xi. p. 172.
Edolius grandis, *Gould, P. Z. S.* 1836, p. 5; *Blyth, J. A. S. B.* xi. p. 172; *Bp. Consp.* i. p. 351.
Chibia malabaroides, *Hodgs. Ind. Rev.* i. p. 325.
Dicrurus retifer, *Jerd. Madr. Journ.* x. p. 241.
Dicrurus rangoonensis, *M'Clell. P. Z. S.* 1839, p. 158; *Jard. & Selby, Ill. Ind. Orn.* iv. pl. 38.
Dicrurus grandis, *M'Clell. P. Z. S.* 1839, p. 158; *Gray, Gen. B.* i. p. 286.
Edolius cristatellus, *Blyth, J. A. S. B.* xi. p. 171.
Edolius dentirostris, *A. Hay, apud Jerd. Madr. Journ.* x. p. 121.
Edolius orissæ, *A. Hay, apud Jerd. l. c.*
Dicrurus paradiseus, *Gray, Gen. B.* i. p. 286; *Pelz. Voy. Novara, Vög.* p. 82; *Horsf. & Moore, Cat. B. Mus. E.I. Co.* i. p. 155; *Gray, Hand-l. B.* i. p. 287.
Dicrurus malabaricus, *Gray, Gen. B.* i. p. 286; *Blyth, J. A. S. B.* xv. p. 296; *Horsf. & Moore, Cat. B. Mus. E.I. Co.* i. p. 157; *Gray, Hand-l. B.* i. p. 286.
Edolius malabaroides, *Bp. Consp.* i. p. 350.

Edolius malabaricus, *Bp. Consp.* i. p. 351; *Jerd. B. Ind.* i. p. 437.
Edolius paradiseus, *Bp. Consp.* i. p. 351; *Jerd. B. Ind.* i. p. 435.
Edolius brachyphorus, *Bp. Consp.* i. p. 351 (*ex Temm. MS. in Mus. Lugd.*); *Schl. Handl. Dierk.* p. 479, pl. 2. fig. 4; *Scl. P. Z. S.* 1863, p. 218.
Dissemurus paradiseus, *Cab. Mus. Hein.* Th. i. p. 112.
Dissemurus malabaricus, *Cab. t. c.* p. 112, note; *Holdsw. P. Z. S.* 1872, p. 439.
Dissemurus grandis, *Cab. t. c.* p. 112, note.
Dissemurus brachyphorus, *Cab. t. c.* p. 112; *Salvad. Ucc. Born.* p. 154.
Dissemurus formosus, *Cab. t. c.* p. 111; *Bp. C. R.* xxxviii. p. 539.
Dissemurus setifer, *Cab. t. c.* p. 111.
Edolius brachyptorus, *Schl. Handl. Dierk.* i. p. 287 (*lapsu*).
Dicrurus malayensis, "*Blyth*," *Jerd. B. Ind.* i. p. 438.
Edolius formosus, *Jerd. t. c.* p. 438.
Dicrurus brachyphorus, *Pelz. Reis. Novara, Vög.* p. 83.
Dicrurus setifer, *Pelz. t. c.* p. 83.
Edolius affinis, *Tytler & Beav. Ibis*, 1867, p. 323.
Dicrurus malabaroides, *Gray, Hand-l. B.* i. p. 286.
Dicrurus formosus, *Gray, t. c.* p. 286.
Dicrurus singularis, *Gray, t. c.* p. 287.
Dissemurus affinis, *Wald. Ibis*, 1873, p. 310.

Adult. General colour above blue-black, with steel-coloured reflections at the ends of the feathers, the plumes of the sides of the neck elongated and lanceolate; the metallic gloss on the wings rather brighter than on the back; primaries black, slightly glossed externally on the outer web, the secondaries metallic like the coverts; tail black, the feathers glossed with steel-colour on the outer webs, the two outermost elongated and furnished with a spatula, which takes half a turn at the end; head blue-black, with an enormous crest, the frontal plumes erect and attenuated, the next ones broader and curved backward, the hindermost very broad and recurved over the nape; lores, sides of face, and under surface of body blue-black, with a steel gloss, somewhat of a greenish tinge, confined to a spot at the extremity of the breast-feathers; bill and feet black; iris red. Total length 14 inches, culmen 1·3, wing 6·6, tail 7, outer tail-feathers 13, tarsus 1·15. (*Nepal.*)

Younger birds seem to have the crest less strongly developed, and have the under wing-coverts tipped with white.

The following is a description of a typical *D. brachyphorus*.

Adult male. Black; the upper surface with metallic steel lustre, less pronounced on the rump, where there is somewhat of a greyer shade, the feathers being grey washed with steel-green; wings metallic steel-black like the back, the primaries black slightly washed with metallic steel reflections, the secondaries externally metallic steel-black like the back; tail black, with metallic steel reflections on both webs, the outer feather with a small elongate racket, slightly twisted and recurved; lores, sides of face, and underparts blue-black, with a gloss of steely purple on the fore neck and breast, rather shaded with greenish on the abdomen, the ground-colour of which is grey; under wing-coverts purplish black; bill and feet black; iris red. Total length 10·6 inches, culmen 1·25, wing 5·55, tail 5·4, tarsus 0·9, long outer tail-feathers 12.

Obs. I am unable to separate *D. brachyphorus* from Sumatra and Singapore from those of Borneo: the latter, however, have much smaller rackets at the end of their tails, which are also shorter. The measurements of a pair in the Museum are as follows:—

	Total length. in.	Culmen. in.	Wing. in.	Tail. in.	Outer tail-feathers. in.	Tarsus. in.
a. ♂. Singapore (*Wallace*)	10·5	1·3	5·8	5·4	13·7	0·9
b. ♀. Sumatra (*Wallace*) .	11·0	1·25	5·4	5·3	10·7	0·9

Young birds appear to be more dusky in colour than the adults, and to have the under wing-coverts tipped with white.

After a careful study of what has been written on the subject of these racket-tailed Drongos, and after comparison of the series contained in the British Museum, I have arrived at the conclusion that between *D. malabaroides* of the Eastern Himalayas and *D. brachyphorus* of Borneo, an unbroken chain of connecting races exists; and this I believe I can show by the accompanying observations. This conclusion is very much the same as that adopted by Mr. Blyth in 1849; and I do not see that the labours of subsequent ornithologists have much tended to upset this, though many more species have recently been recognized. Of these the late Mr. G. R. Gray admitted in the 'Hand-list' no less than eight, as follows:—

4240. *malabaricus*, Lath. India, Malabar, Tenasserim.
4241. *brachyphorus*, Temm. Borneo.
4242. *malabaroides*, Hodgs. India, Himalayas, &c.
4243. *formosus*, Cab. Java, Banda.
4244. *singularis*, Gray, n. sp. Malabar.
4245. *platurus*, Vieill. Java, Malacca.
4246. *paradiseus*, L. Siam, Burmah.
4247. *affinis*, Tickell. Andamans.

These species may at once be reduced to seven by suppressing the supposed new species, *D. singularis*, which is nothing but the young of the Malabar bird with a smaller crest and racket. It is quite certain that the size of both of these portions of the plumage varies, even in specimens from the same locality; and although the development of the crest is greater in old birds than in young ones, it must be remembered also that the preparation of the skin has a great deal to do with the appearance of the crest; for if the latter has been flattened when the bird was skinned, it will be very difficult to restore it to its original proportions; this is more especially the case with the short-crested forms, the full-crested species having the crest always remarkably developed.

Taking the remainder of Mr. Gray's species in somewhat different order, the most distinct is, of course, the Nepalese *D. malabaroides*, with an enormous crest (fig. 1).

Fig. 1.

Head of *Dissemurus malabaroides.*

D. paradiseus, from Siam (fig. 2), is a full-crested form, but, of course not to be compared in this respect with Nepalese examples; and the Burman birds are intermediate (fig. 3). The Tenasserim and Cambodian specimens cannot be separated from the Siamese.

Fig. 2.

Head of *Dissemurus paradiseus.* Siam.

Fig. 2.

Head of *Dissemurus paradiseus.* Siam.

D. platurus, from Java, is a strictly intermediate race between

D. paradiseus and *D. brachyphorus*, having a small crest but a tolerably large and very twisted racket (fig. 4). In the previously mentioned birds the racket does not take more than a half turn; but in Javan specimens it is twisted over so as nearly to meet the opposite web.

Fig. 4.

Racket of *Dissemurus platurus.*

Malaccan specimens are again intermediate, having a very small crest and racket, which makes them referable to *D. brachyphorus*, with which species I keep them, though the racket is much twisted, as in *D. platurus*.

Figs. 5 & 6.

Head and racket of *Dissemurus formosus*.

D. formosus is included in the 'Hand-list' on the authority of a Bornean specimen received from Verreaux. It must be kept with the Javan bird, on account of the crest (fig. 5), but the racket (fig. 6) is very flat and large. It is a question, however, whether this last-named character does not change its form a good deal with the growth of the feather.

D. brachyphorus is a small, almost crestless race (fig. 7), with a very small racket (fig. 8), and would have to be regarded as a distinct species on this account, if it were not approached by the intermediate forms in Malacca &c.

Fig. 7.

Head of *Dissemurus brachyphorus.*

Fig. 8.

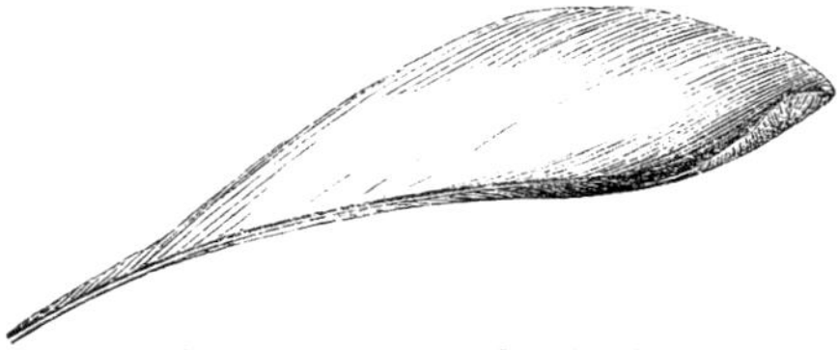

Racket of *Dissemurus brachyphorus.*

D. affinis (fig. 9) is a small-crested form of the Malabar bird, the crest, such as it is, being full and the racket large.

Fig. 9.

Head of *Dissemurus affinis.*

D. malabaricus of the 'Hand-list' must be called *D. cristatellus*, Blyth; for the former name, founded on Sonnerat's utterly irrecognizable plate (Voy. Ind. ii. pl. cxi.), ought to be discarded, as there is no chance of the figure ever being referred with certainty to any species. The crest in this form is very close-set; but considerable variation appears to exist even in birds from different parts of Southern India.

Lastly, the form from Ceylon (fig. 10) is quite as distinct as any of the foregoing races. It has quite as small a racket (fig. 11) as in *D. brachyphorus*, with a crest like *D. paradiseus*; and if the others are separated, this must be called *D. ceylonensis*.

Fig. 10.

Head of *Dissemurus ceylonensis*.

Fig. 11.

Tail feather of *Dissemurus ceylonensis*.

Finding it, however, quite impossible to define exact specific characters for these supposed species, I have united them under one name, and merely keep the specimens arranged under different headings; for I fully believe that a larger series will only show more connecting links, and that it will be found more and more impossible to give specific characters for those differentiated under the various titles mentioned.

Hab. India and Ceylon, Indo-Chinese region; Malacca; Sumatra; Java; Banda; Borneo.

α. *D. malabaroides*.

a. Ad. st.	India.	W. Sotheby, Esq. [P.].
b, *c*. Ad. sk.	Nepal.	B. H. Hodgson, Esq. [P.].

β. *D. paradiseus*.

d, *e*, *f*, *g*. Ad. sk.	Burmah.	Capt. Stackhouse Pinwill* [P.].

* Not typical, but at the same time not separable as a species from the foregoing. *D. rangoonensis*, Gould, is a short, nearly crestless specimen, and very doubtfully recognizable.

h. Ad. st.	Siam.	Hon. E.I.Co. [P.].
i. ♂ ad. st.	Bangkok, April 3, 1872 (*Conrad*).	R. B. Sharpe, Esq. [P.].
k. Ad. sk.	Cambodia.	M. Mouhot [C.].
l. Ad. sk.	Tenasserim.	J. C. D. Packman, Esq. [P.].

γ. *D. platurus.*

m. ♂ ad. sk.	W. Java.	A. R. Wallace, Esq. [C.].
n. Ad. sk.	E. Java.	A. R. Wallace, Esq. [C.].
o. Ad. sk.	Borneo.	Purchased.

δ. *D. cristatellus.*

p, *q*. Ad. sk.	Malabar.	Purchased.
r. Juv. st.	Malabar.	Mrs. Spinks.
s. Ad. sk.	Moalmaza.	
t. Ad. sk.	Travancore.	Capt. Biddulph [C.].
u. Ad. st.	India.	J. Gould, Esq. [P.].

ε. *D. ceylonensis.*

v. ♀ jun. sk.	N.E. Province, Ceylon, Jan. 7. 1873.	Capt. Vincent Legge [P.].

ζ. *D. affinis.*

w, *x*. Ad. sk.	Andaman Islands.	Purchased.

η. *D. brachyphorus.*

y, *z*. ♂ ad. sk.	Sarawak, Borneo, 1855.	A. R. Wallace, Esq. [C.].
a'. ♂ ad. sk.	Sibu, Sarawak.	A. Everett, Esq. [C.].
b'. Ad. sk.	Borneo.	
c'. Ad. sk.	Borneo.	Hugh Low, Esq. [C.].
d'. Ad. sk.	Banjermassing.	J. Motley, Esq.
e'. ♀ Ad. sk.	Sumatra, 1861.	A. R. Wallace, Esq. [C.].
f'. ♂ Ad. sk.	Singapore, 1854.	A. R. Wallace, Esq. [C.].
g'. Jun. sk.	Malacca.	Capt. Stackhouse Pinwill [C.].
h', *i'*. Ad. sk.	Malacca.	W. Harvey, Esq. [P.].

11. IRENA.

	Type.
Irena, *Horsf. Tr. Linn. Soc.* xiii. p. 153	I. turcosa.

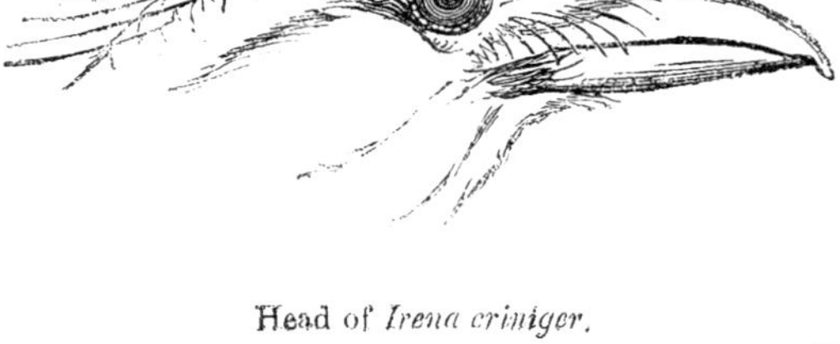

Head of *Irena criniger*.

Range. Indian peninsula, Indo-Chinese countries, Malayan peninsula; Java; Sumatra; Borneo; Philippine Islands.

Key to the Species.

a. Underneath deep purplish blue.
  a'. Mantle deep purplish blue .... *cyanogastra*, p. 266.
  b'. Mantle velvety black .... *melanochlamys*, p. 266.
b. Underneath black.
  c'. Under tail-coverts reaching to the tip, or even beyond the tip of the tail .... { *turcosa*, p. 267. / *criniger*, p. 267.
  d'. Under tail-coverts not reaching to the tip of the tail.
    a''. Under tail-coverts falling short of the tip of the tail by more than length of tarsus.
      a'''. Enamelled blue feathers of a deep purplish shade .... *puella*, p. 268.
      b'''. Enamelled blue feathers of a dull verditer .... *tweeddalii*, p. 268.
    b''. Under tail-coverts reaching nearly to the tip of the tail, not falling short of it by as much as the length of the tarsus .... *cyanea*, p. 269.

1. Irena cyanogastra.

Irena cyanogastra, *Vigors, P. Z. S.* 1831, p. 97; *Gray & Mitch. Gen. B.* i. p. 288, pl. 70; *Bp. Consp.* i. p. 349; *Cass. U.S. Expl. Exp.*, Aves, p. 143; *Gray, Hand-l. B.* i. p. 288; *Walden, Tr. Z. S.* ix. p. 190.

Adult male. General colour above deep purplish blue, slightly varied on the rump with ashy brown, where the bases of the feathers show through; the scapulars resembling the back; least and median wing-coverts deep cobalt-blue, the greater series black, shading off into deep cobalt at the tip, very slightly on the outermost, conspicuously on the inner ones; primary-coverts black, the quills black, the inner secondaries externally deep cobalt like the wing-coverts, the greater upper tail-coverts deep cobalt, brighter than the lower back; tail black, washed with deep purple, more plainly on the centre feathers; crown of head and nape deep cobalt-blue, a little brighter on the fore part of the crown; lores, sides of face and of neck, entire throat and chest, black; rest of under surface deep purplish blue like the back; under tail-coverts deep cobalt-blue; thighs and under wing-coverts brown, washed with purplish blue; iris red. Total length 10 inches, culmen 1·15, wing 5·35 tail 4·4, tarsus 0·85.

Sexes alike (*Meyer*; *cf. Walden, l. c.*).

Hab. Philippine Islands.

a, b. Ad. st. Manila. Hugh Cuming, Esq. [C.].

2. Irena melanochlamys.

Adult male (type of species). Crown of head and nape glistening purplish cobalt, extending to the middle of the hind neck; sides of neck, mantle, and scapulars deep velvety black; lower back and rump dull purplish cobalt, much brighter on the upper tail-coverts; wings black, the lesser and median series purplish cobalt, the greater series black, tipped with the same cobalt, the innermost, as well as the inner secondaries, externally of the same colour; tail black, shaded with purple on the centre feathers and the outer webs of the others; lores, feathers above and round the eye, throat, and fore neck deep velvety black, the rest of the under surface dull purple, the under tail-coverts bright purplish cobalt; under wing-coverts dull purple; bill black; legs black; iris carmine. Total length 10 inches, culmen 1·05, wing 4·8, tail 4·1, tarsus 0·7. (*Coll. J. B. Steere.*)

Hab. Island of Basilan, Philippines.

3. Irena turcosa.

Irena puella, *Horsf. in Linn. Trans.* xiii. p. 153; *id. Zool. Research. in Java, Birds*, pls. 13, 14; *Steph. Gen. Zool.* xiii. pt. 2, p. 142, pl. 48; *Gray, Gen. B.* i. p. 288; *Cab. Mus. Hein.* Th. i. p. 110.
Edolius puellus, *Temm. Pl. Col.* pls. 70, 476, 225.
Irena turcosa, *Walden, Ann. N. H.* (4) v. p. 417.

Adult male. General arrangement of colour as in *I. puella*; but the blue is of a turquoise shade, inclining to silvery cobalt on the head; under tail-coverts reaching a little beyond the tip of the tail. Total length 8·5 inches, culmen 1·1, wing 4·95, tail 3·6, tarsus 0·8.

Hab. Java.

a. Ad. st.	[Java.]	Hon. E.I. Company [P.].
b. Juv. sk.	Java.	Secretary of State for India [P.].

4. Irena criniger.

Coracias puella, *Raffl. Tr. Linn. Soc.* xiii. p. 302 (*nec Lath.*).
Irena puella (*nec Lath.*), *Motley & Dillw. N. Hist. Labuan*, p. 23; *Scl. P. Z. S.* 1863, p. 217.
Irena cyanea, *Salvad. Ucc. Born.* p. 151 (*nec Begbie*).
Irena turcosa, *Sharpe, Ibis*, 1876, p. 44 (*nec Wald.*).

Adult male. Blue and black, as in *I. puella*, the distribution of colours being the same; the hue of the blue colour intense cobalt, very rich on the head; the under tail-coverts produced to the very end of the tail, the upper coverts also nearly as far; bill and legs black; iris red. Total length 9·3 inches, culmen 0·95, wing 4·6, tail 3·25, tarsus 0·65.

Adult female. Differs from the male exactly as does the female of *I. puella* from the male.

Obs. Sumatran birds seem to be exactly the same as the Bornean, a pair from Mr. Wallace's collection measuring as follows:—

		Total length. in.	Culmen. in.	Wing. in.	Tail. in.	Tarsus. in.
a. ♂ ad.	Sumatra....	9·3	1	4·65	3·55	0·65
b. ♀ ad.	Sumatra....	9·0	1	4·5	3·25	0·7

Hab. Borneo and Sumatra.

a. [♂] ad. sk.	Borneo.	Rajah Brooke [P.].
b. [♂] ad. sk.	Banjermassing.	J. Motley, Esq. [C.].
c. ♂ ad. sk.	Jambusan, Sept. 11, 1874.	Alfred Everett, Esq. [C.].
d. ♂ ad. sk.	Sibu.	Alfred Everett, Esq. [C.].
e. [♂] ad. sk.	Labuan.	Hugh Low, Esq. [C.].
f, g. ♂ ♀ ad. sk.	Sumatra.	A. R. Wallace, Esq. [C.].
h. [♀] ad. st.	Sumatra.	Purchased.

5. Irena puella.

Fairy Roller, *Lath. Gen. Syn. Suppl.* p. 87.
Coracias puella, *Lath. Ind. Orn.* i. p. 171; *Daud. Traité*, ii. p. 264.
Irena puella, *Jerd. Madr. Journ.* xiii. p. 262; *M'Clell. P. Z. S.* 1839, p. 160; *Gray, Gen. B.* i. p. 288, pl. 70. fig. 3; *Horsf. & Moore, Cat. B. Mus. E.I. Co.* p. 273; *Jerd. B. Ind.* ii. p. 105; *Gray, Hand-l. B.* i. p. 288; *Stoliczka, J. A. S. B.* xxxix. pt. 2, p. 318; *Wald. Ibis*, 1871, p. 170; *Holdsw. P. Z. S.* 1872, p. 452; *Wald. Ibis*, 1873, p. 306.
Irena indica, *A. Hay in Blyth's Report, J. A. S. B.* xv. p. 309; *Bp. Consp.* i. p. 349.

Adult male. General colour above deep purplish cobalt, of a purer blue shade somewhat on the forehead; scapulars as well as the least and median coverts coloured exactly like the back, the greater series blackish, with a cobalt-blue spot at the tips of some of them; quills and tail black, with a slight bluish shade on the secondaries and central tail-feathers; lores, feathers over the eye, sides of face, sides of neck, and entire under surface, including the thighs and under wing-coverts, black; under tail-coverts deep-purplish cobalt; bill and legs black; iris ruby-red. Total length 10 inches, culmen 0·95, wing 5, tail 4·45, tarsus 0·75, distance between tip of tail-coverts and tip of tail 1·6.

Adult female. Different in colour from the male, being of a dull Prussian blue, a little brighter and inclining to cobalt on the rump and upper and under tail-coverts; wings and tail dark brown, the quills narrowly edged, and the secondaries and tail-feathers washed with dull blue, the latter more conspicuously.

Hab. Malabar and Travancore; Assam and Burmese provinces to Tenasserim.

a, b. [♂] ad. sk.	Travancore.	Capt. Biddulph [C.]
c. [♂] ad. sk.	Moalmaza.	
d, e. [♂] ad. sk.	Tenasserim.	J. C. D. Packman, Esq. [P.].
f. [♂] ad. sk.	Siam.	M. Mouhot [C.].

6. Irena tweeddalii.

Adult male (type of species). Crown of head, nape, and hind neck, as well as the entire back, scapulars, rump, and upper tail-coverts brilliant turquoise-blue, the head more silvery; lesser and median wing-coverts turquoise-blue like the back; rest of the wing-coverts black, some of the innermost of the greater series with a blue spot at the tips; quills and tail black; lores, feathers in front of and over the eye, sides of face, sides of neck, and entire under surface of body, from chin to vent, rich velvety black, including the under wing-coverts; under tail-coverts bright turquoise-blue, a little more purplish than the upper surface; bill and legs black. Total length 8 inches, culmen 1·05, wing 4·85, tail 3·85, tarsus 0·7. (*Coll. J. B. Steere.*)

Adult female. General colour above and below dull verditer-blue, brighter and more washed with cobalt on the rump and upper tail-coverts; least and median wing-coverts exactly resembling the back, the greater series, quills, and tail blackish brown, externally washed with blue; bill and legs black; "iris carmine" (*Steere*). Total length 9·5 inches, culmen 1·05, wing 4·85, tail 3·8, tarsus 0·75. (*Coll. J. B. Steere.*)

Hab. Island of Balabac, Philippines,

7. Irena cyanea.

Irena puella, *Blyth, J. A. S. B.* xv. p. 308 (*nec Lath.*); *id. Cat. B. Mus. A. S. B.* p. 214 (pt.); *Bp. Consp.* i. p. 349.

Irena malayensis, *Moore, in Horsf. & Moore, Cat. B.* i. p. 274; *Jerd. B. Ind.* ii. p. 106; *Gray, Hand-l. B.* i. p. 288; *Wald. Ibis,* 1871, p. 170.

Adult male. Similar to *I. criniger* in distribution of colour and shade of blue, but having the under tail-coverts falling short of the tip of the tail by half an inch. Total length 9·3 inches, culmen 1, wing 4·6, tail 3·5, tarsus 0·7.

Adult female. Similar to the female of *I. criniger.* Total length 8·5 inches, wing 4·6, tail 3·7, tarsus 0·7.

Hab. Malacca.

a. Imm. st.	Malacca.	Hugh Cuming, Esq. [C.].
b, c. [♂ ♀] ad. sk.	Malacca.	
d. [♀] ad. sk.	Malacca.	W. Harvey, Esq. [C.].
e. ♂ ad. sk.	Malacca.	A. R. Wallace, Esq. [C.].
f, g. ♂ ♀ ad. sk.	Penang.	Secretary of State for India [P.].
h. ♂ imm. sk.	Penang.	Secretary of State for India [P.].

Family PRIONOPIDÆ.

a. Bill broader than it is high .. *Subfam. PRIONOPINÆ,* p. 270.
b. Bill higher than it is broad .. *Subfam. EURYCEROTINÆ,* p. 326.

Subfam. PRIONOPINÆ.

Key to the Genera.

a. Distance between the primaries and the secondaries greater than the length of the tarsus.
  a'. Wings short or moderate, falling short of the tip of the tail by more than the length of the culmen; no eye-wattle.
    a''. Tail square or slightly rounded, shorter than the wing.
      a'''. First primary moderate; second primary *longer than* the secondaries .... 1. GRALLINA, p. 272.
      b'''. First primary moderate; *second* primary *equal to* the secondaries .... 2. TEPHRODORNIS, p. 274.
      c'''. *First* primary equal to the secondaries .... 3. EUROCEPHALUS, p. 279.
    b''. Tail graduated, longer than the wing .... 4. PŒOPTERA, p. 281.
  b'. Wings nearly reaching to the end of the tail, the distance between the tip of the wing and the tip of the tail not so long as the culmen; round the eye a fleshy wattle .... 5. LEPTOPTERUS p. 282.
b. Distance between the tips of the primaries and the tips of the secondaries less than the length of the tarsus.
  c'. No fleshy wattle round the eye.
    c''. Without an occipital crest of greater length than the head.
      d'''. Wing a little longer than the tail, the former not exceeding the latter by as much as the length of the culmen.
        a⁴. Inner toe (without claw) longer than the hind toe (without claw).
          a⁵. Second primary equal to the secondaries in length.
            a⁶. Head with a crest of

elongated plumes; rictal bristles not extending beyond the nostrils: bill simply notched at the end, but serrated very finely along the edge of the upper mandible .... 6. RECTES, p. 283.

b^6. Head with a slight crest, the plumes not elongated; bill simply notched at the end and not serrated; rectal bristles as in the preceding...... 7. PSEUDORECTES, p. 287.

c^6. Head with a slight crest, the plumes not elongated; bill with a strong circular notch; rictal bristles and nasal bristles strongly developed, and extending far beyond the middle of the culmen.......... 8. MELANORECTES, p. 289.

b^5. Second primary longer than the secondaries; head not crested........ 9. COLLYRIOCINCLA, p. 289.

b^4. Inner toe (without claw) shorter than, or equal to, the hind toe (without claw).

c^5. Second primary equal to the secondaries in length.

d^6. Nostrils somewhat exposed.

a^7. Culmen much longer than the hind toe and claw.............. 10. PINAROLESTES, p. 293.

b^7. Culmen not so long as the hind toe and claw.............. 11. CUPHOPTERUS, p. 302.

e^6. Nostrils entirely hidden by plumes.

c^7. Bill moderately strong; feet very strong; culmen not so long as the middle toe and claw.............. 12. FRASERIA, p. 303.

d^7. Bill very stout; feet extremely weak; culmen considerably longer than the middle toe and claw.... 13. HEMIPUS, p. 305.

d^5. Second primary longer than the secondaries.... 14. BRADYORNIS, p. 308.

e'''. Tail longer than the wing.

c^4. First primary rather long, extending beyond the primary-coverts .... 15. MELÆNORNIS, p. 315.

d⁴. First primary very small, not extending as far as the primary-coverts .... 16. HYPOCOLIUS, p. 316.
d''. With a long occipital crest of dilated plumes, equalling the head in length.................... 17. PLATYLOPHUS, p. 317.
d'. With a fleshy wattle round the eye.
e''. Wing rather short, the quills falling short of the tail by more than the length of the first primary; head crested, the frontal plumes extending forwards, so as to completely hide the nostrils; the exposed part of the culmen not exceeding the diameter of the eye 18. PRIONOPS, p. 318.
f''. Wing rather long, the distance between the wings and tail not so much as the length of the first primary; head moderately crested, the nostrils not entirely hidden by the nasal plumes; the exposed part of the culmen exceeding the diameter of the eye. 19. SIGMODUS, p. 319.

1. GRALLINA.

Type.
Tanypus*, *Oppel, Münch. Denkschr.* 1811-12, p. 159 .... G. picata.
Grallina, *Vieill. N. Dict. d'Hist. Nat.* xiii. p. 401 (1817).. G. picata.

Range. Australia and New Guinea.

Key to the Species.

a. Larger; sexes different; the black colour on the throat occupying the breast also; underparts pure white .... *picata*, p. 272.
b. Smaller; sexes alike; the black colour confined to the throat; underparts tinged with buff .... *bruijni*, p. 273.

1. Grallina picata.

Gracula picata, *Lath. Ind. Orn., Suppl.* p. 29.
Tanypus australis, *Oppel, Münch. Denkschr.* 1811-12, p. 159.
Cracticus cyanoleucus, *Vieill. N. Dict.* v. p. 356; *Less. Traité*, p. 39 pl. 39.
Grallina melanoleuca, *Vieill. N. Dict. d'Hist. Nat.* xiii. p. 401; *id. et Oud. Gal. Ois.* i. pl. 150; *Cab. Mus. Hein.* i. p. 60.
Grallina bicolor, *Vig. & Horsf. Tr. Linn. Soc.* xv. p. 233; *Less. Traité*, p. 398.
Corvus cyanoleucus, *Lath. Gen. Hist. B.* iii. p. 49.
Grallina australis, *Gray, List Gen. B.* 1841, p. 33; *Gould, B. Austr.* fol. ii. pl. 54.
Grallina cyanoleuca, *Gray & Mitch. Gen. B.* i. p. 204; *Bp. Consp.* i. p. 251.
Grallina picata, *Strickl. Ann. & Mag. Nat. Hist.* xi. p. 335; *Gould, Handb. B. Austr.* i. p. 188; *Gray, Hand-l. B.* i. p. 248.

* Preoccupied in entomology.

Adult. General colour above blue-black, the scapulars uniform with the back; all the wing-coverts pure white, excepting the outermost of the greater series, which are black, with a white tip to a few of them; primary-coverts and quills black, the secondaries tipped with white, the innermost with white bases extending for more than half of the quills, and forming a longitudinal wing-bar; feathers of the lower back mottled with white, each feather having a broad subterminal bar of this colour; rump, upper tail-coverts, and tail pure white, the latter with a broad terminal bar of blue-black, some of the outer feathers tipped with white; lores, feathers in front of the eye, and a broad band extending from the eye above the ear-coverts to the sides of the neck blue-black; above the eye to the sides of the nape a broad streak of white, forming a very distinct eyebrow; ear-coverts and sides of neck white, forming a large oblong patch; below the eye a small half-circle of white feathers; cheeks, throat, and breast blue-black; rest of under surface of body pure white, including the under wing- and tail-coverts; quills black below, with concealed white bases; "bill yellowish white; feet black; iris straw-yellow" (*Gould*). Total length 11 inches, culmen 0·95, wing 7·2, tail 5·2, tarsus 1·7.

Adult female. Similar to the male, but having a frontal band and the throat white, and no white line under the eye; the general shade of the black colour greener than in the male, where it is blue-black. Total length 10 inches, culmen 0·95, wing 6·95, tail 4·95, tarsus 1·6.

Young. Similar to the adult female, but more dingy and browner; forehead blackish brown like the crown; all the quills and tail-feathers distinctly tipped with white.

Hab. Australia.

a. ♀ ad. st.	Australia.	
b. [♀] ad. st.	Australia.	Linnean Society [P.]. (Type of *G. bicolor.*)
c. [♀] ad. sk.	South Australia.	Sir George Grey [P.].
d, e. ♂ ♀ ad. sk.	Murray River.	Sir George Grey [P.].
f. Pull. st.	Swan River.	Purchased.
g, h. ♂ ♀ ad., *i.* Pull. sk.	N.W. Australia.	J. R. Elsey, Esq. [P.].

2. Grallina bruijni.

Grallina bruijnii, *Salvad. Ann. Mus. Civic. Genov.* vii. p. 929.

Head above, neck, back, throat, cheeks, quills, and the apical half of the tail-feathers black; lores, a superciliary band drawn backwards as far as the sides of the neck, upper part of the breast, upper and under wing-coverts white; rump, upper tail-coverts, and basal half of the rectrices white, slightly tinged with buff; the base of the quills white; bill lead-coloured in the skin, the edges of the mandibles and the tip pale; feet black. Total length 7·3 inches, culmen 0·7, wing 4·3, tail 3·15, tarsus 1·05. (*Salvadori.*)

Hab. Arfak Mountains in North-western New Guinea.

2. TEPHRODORNIS.

	Type.
Tephrodornis, *Swains. Faun. Bor.-Amer. Birds, App.* p. 482 (1831)	T. virgatus.
Keroula, *J. E. Gray, in Hardw. Ill. Zool.* ii. pl. 33 (1833–34)................................	T. pondicerianus.
Tenthaca, *Hodgs. Ind. Review*, i. p. 447 (1837)	T. pelvicus.
Creurgus, *Hodgs. J. A. S. B.* x. p. 29 (1847)	T. pelvicus.
Tentheca, *G. R. Gray, List Gen. B.* 1841, p. 48....	T. pelvicus.
Tenthera, *G. R. Gray, Gen. B.* i. p. 289 (1847)....	T. pelvicus.
Tephrolanius, *Bp. C. R.* xxxix. p. 536 (1854)	T. gularis.

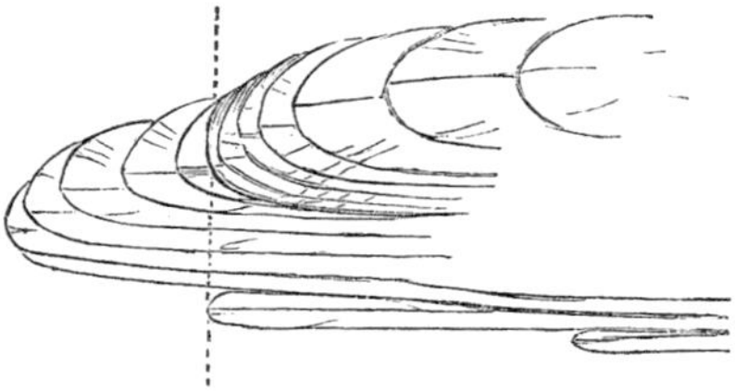

Wing of *Tephrodornis gularis.*

Range. India and Ceylon; Indo-Chinese countries to Hainan; Malayan peninsula; Java; Sumatra; Borneo.

Key to the Species.

a. Two outer tail-feathers white, blackish at base, and sometimes spotted near the tip with the same.
  a′. General colour above ashy brown, with a very distinct buffy white eyebrow ...... *pondicerianus*, p. 275.
  b′. General colour above dull cinereous, with only a faintly indicated eyebrow of lighter grey .... *affinis*, p. 276.
b. Outer tail-feathers uniform with the rest of the tail, or sometimes a little whitish on the inner web.
  c′. Head grey; back ashy brown; size large; wing 4·75 inches .... *pelvicus*, p. 276.
  d′. Head dark grey or slaty brown like the back.
    a″. Larger; wing 4·55 inches; upper tail-coverts dark brown.................. *sylvicola*, p. 277.
    b″. Smaller; wing 3·6–3·9 inches; tail-coverts black.
      a‴. Cheeks and throat white like the rest of the under surface; the breast slightly tinged with vinaceous grey .. *virgatus*, p. 278.
      b‴. Cheeks white, contrasting strongly with the throat and breast, which are ashy grey .... *gularis*, p. 278.

1. Tephrodornis pondicerianus.

Gobe-mouche de Pondichéry, *Sonn. Voy. Ind.* ii. p. 198.
Muscicapa pondiceriana, *Gm. S. N.* i. p. 939.
The Keroula Shrike, *Lath. Gen. Hist. B.* ii. p. 29.
Lanius muscicapoides, *Frankl. P. Z. S.* 1831, p. 117; *id. J. A. S. B.* i. p. 265.
Keroula indica, *J. E. Gray in Hardw. Ill. Zool.* pl. 33 (1833–34).
Lanius griseus, *Tickell, J. A. S. B.* ii. p. 573.
Lanius sordidus, *Less. Compl. Buff.* ii. p. 422.
Tenthaca leucurus, *Hodgs. Ind. Review*, i. p. 447.
Taphrodornis superciliosus, *Jerd. Madr. Journ.* x. p. 237; *Blyth, J. A. S. B.* xi. p. 799.
Tephrodornis superciliosa, *Swains. An. in Menag.* p. 283.
Tephrodornis pondiceriana, *Blyth, J. A. S. B.* xv. p. 305; *id. Cat. B. Mus. A. S. B.* p. 153; *Bp. Consp.* i. p. 358; *Horsf. & Moore, Cat. B. Mus. E.I. Co.* i. p. 169; *Jerd. B. Ind.* i. p. 410; *Gray, Hand-l. B.* i. p. 394; *Hume, Str. F.* 1873, p. 177; *Adam, t. c.* p. 376; *Ball, S. F.* 1874, p. 399; *Hume, S. F.* 1875, p. 92; *id. Nests and Eggs Ind. B.* p. 176; *Blyth. & Wald. B. Burm.* p. 122.
Tephrodornis indica, *G. R. Gray, Gen. B.* i. p. 290; *id. Cat. Mamm. &c. Nepal, Hodgs.* p. 99.

Adult. Above ashy brown, the rump rather greyer; across the lower rump a white band; upper tail-coverts blackish; tail dark brown, the two centre feathers much lighter brown, the two outermost white, dark brown at the base, and spotted at the tip with a mark of the same colour, confined to the outer web of the last feather; wing-coverts ashy brown like the back, the greater series browner, as also the quills, these having paler brown margins and being washed externally with grey; lores and nasal plumes fulvescent, produced into a well-defined superciliary band of fulvous white; feathers in front of and round the eye ashy brown, darker than the back; cheeks, throat, abdomen, and under tail-coverts white, the breast greyish brown, extending onto the throat; thighs ashy brown; under wing-coverts dull cinereous, the edge of the wing whitish; bill dusky horny; legs plumbeous brown; iris greenish yellow. Total length 6·4 inches, culmen 0·75, wing 3·15, tail 2·55, tarsus 0·75.

Obs. The black markings at the ends of the two outer feathers seem to disappear with age, leaving the whole feather white. Capt. Pinwill's specimens from the North-west Himalayas are very much paler and greyer than those from Southern India; they are also rather larger; wing 3·5. The bird described is from Travancore.

Hab. India, ranging into the Burmese and Indo-Chinese countries; Hainan.

a, b. Ad. st.	India.	J. Gould, Esq.
c, d. Ad. sk.	Travancore.	Capt. T. Biddulph [C.]
e. Ad. sk.	Madras.	Rev. H. H. Baber [C.].
f, g. Ad. sk.	Kamptee, C. P.	Dr. Hinde [P.].
h, i. Ad. sk.	Behar.	B. H. Hodgson, Esq. [P.].
k, l, m. Ad. sk.	Nepal.	B. H. Hodgson, Esq. [P.].
n–q. ♂ ♀ ad. sk.	N.W. Himalayas.	Capt. Stackhouse Pinwill [P.].
r. Ad. sk.	Pegu.	W. T. Blanford, Esq. [P.].

2. Tephrodornis affinis.

Tephrodornis affinis, *Blyth, J. A. S. B.* xvi. p. 473; *id. Cat. B. Mus. A. S. B.* p. 153; *Bp. Consp.* i. p. 358; *Gray, Hand-l. B.* i. p. 394; *Hume, Str. F.* 1873, p. 434.
Tephrodornis pondiceriana, *Holdsw. P. Z. S.* 1872, p. 437.

Adult [*male*]. Above slaty grey, with only a slight trace of a paler eyebrow; wing-coverts grey like the back, the greater series ashy brown like the quills, which are externally washed with grey; lower rump white; upper tail-coverts black; tail-feathers blackish brown, the two outermost white, with blackish bases and a black spot at the tip of the feather; nasal bristles and lores greyish brown, the feathers above and below the eye and the ear-coverts deeper black; cheeks and chin white; rest of under surface ashy, with a pinkish shade, gradually becoming white on the abdomen and under tail-coverts; under wing-coverts cinereous, the edge of the wing white; "bill dusky; feet dusky lead-colour; iris dull yellow" (*Holdsworth*). Total length 6·4 inches, culmen 0·8, wing 3·35, tail 2·45, tarsus 0·65.

Adult female (Colombo: *Legge*). Similar to the supposed male above described, but duller in colour, browner above, and below washed with ashy brown, extending well onto the throat; the eyebrow rather whiter and more distinct. Total length 6 inches, culmen 0·75, wing 3·3, tail 2·5, tarsus 0·65.

Obs. Mr. Holdsworth says that, after comparing Ceylonese specimens with a series of Indian ones, he cannot separate them; but my experience is directly the opposite, and I consider that *T. affinis* is quite distinct, and approaches *T. gularis* of Malacca.

Hab. Ceylon.

a, b. Ad. sk.	Ceylon.	E. L. Layard, Esq. [C.].
c. Ad. sk.	Ceylon.	T. Thwaites, Esq. [C.].
d, e. Ad., *f.* juv. sk.	Near Newara Elia.	Mr. E. Boate [C.].
g. ♀ ad. sk.	Colombo, Jan. 1, 1869.	Capt. Vincent Legge [P.].

3. Tephrodornis pelvicus.

Tentheca pelvica, *Hodgs. Ind. Review*, i. p. 447.
Tentheca petrica, *Gray, List Gen. B.* 1841, p. 48.
Tephrodornis pelvica, *Gray, Cat. Mamm. &c. Nepal, Hodgs.* p. 99; *Blyth, J. A. S. B.* xv. p. 304; *Gray, Gen. B.* i. p. 290; *Blyth, Cat. B. Mus. A. S. B.* p. 153; *Bp. Consp.* i. p. 357; *Horsf. & Moore, Cat. B. Mus. E.I. Co.* i. p. 171; *Jerd. B. Ind.* i. p. 409; *Gray, Hand-l. B.* i. p. 394; *Swinh. P. Z. S.* 1871, p. 377; *Godwin-Aust. J. A. S. B.* xxxix. p. 99; *Hume, Str. F.* 1874, p. 474, 1875, pp. 13, 92; *Blyth & Wald. B. Burm.* p. 122.

Adult. General colour above pale ashy brown, the head cinereous grey, occupying also the nape and hind neck, the forehead slightly whitish; feathers of rump greyish white at base, barred with brown near the tip, upper tail-coverts white, the long ones brown like the tail-feathers, which have light-brownish shafts and narrow white tips; wings pale brown, the coverts narrowly margined with whity brown, the inner secondaries washed and margined with the latter colour, the primaries sandy brown externally; nasal plumes, lores, feathers above and below the eye, and ear-coverts blue-black, forming a broad band on each side of the head; sides of face, sides of neck, and entire underparts white, with a slight pinkish shade on the breast; thighs and under wing-coverts light brown; bill dusky blackish; legs plumbeous brown; iris brownish yellow. Total length 8·5 inches, culmen 0·95, wing 4·75, tail 3·4, tarsus 0·8.

Young. Differs from the adult in being light brown, with the head of the same colour as the back, the feathers of the upper surface having all of them light margins to the feathers, rather rufous on the wing-coverts and quills, the upper tail-coverts and tail strongly washed with rufous; underneath creamy white, with an ashy shade on the throat and breast; band along the side of the head ashy brown instead of black.

Hab. Eastern Himalayas, ranging into Assam and Burmah.

a. Ad. st.	Nepal.	B. H. Hodgson, Esq. [P.].
b. Ad. sk.	Nepal.	B. H. Hodgson, Esq. [P.].
c, *d*. Juv. sk.	Nepal.	B. H. Hodgson, Esq. [P.].
e. Juv. sk.	Sikhim.	W. T. Blanford, Esq. [P.].
f. Juv. sk.	Darjiling (*Dr. Jerdon*).	C. Bygrave Wharton, Esq. [P.].
g, *h*. Ad. sk.	N.W. Himalayas.	Capt. Stackhouse Pinwill [P.].
i, *k*. Ad., *l*, *m*. Juv. sk.	Nazeerah, Upper Assam (*Dr. Reid*).	R. B. Sharpe, Esq. [P.].

4. Tephrodornis sylvicola.

Tephrodornis sylvicola, *Jerd. Madr. Journ.* x. p. 236; *Blyth, J. A. S. B.* xv. p. 304; *id. Cat. B. Mus. A. S. B.* p. 153; *Bp. Consp.* i. p. 357; *Jerd. B. Ind.* i. p. 408; *Gray, Hand-l. B.* i. p. 394.

Adult. Above dark cinereous brown, the lower back greyer at the base of the feathers and dark grey towards their tips; upper tail-coverts white, the longer ones blackish brown; wing-coverts slaty like the back, the greater series and the quills and tail brown; nasal bristles, lores, feathers above and below the eye and ear-coverts blue-black; cheeks and entire under surface of body white, with an ashy pink shade on the breast; thighs ashy brown; under wing-coverts dull cinereous; bill blackish; legs plumbeous; iris wax-yellow. Total length 8·5 inches, culmen 0·8, wing 4·55, tail 4·4, tarsus 0·8.

Obs. The single specimen in the Museum being in poor condition, and the description being, therefore, perhaps a little inaccurate, I subjoin the following remark of Dr. Jerdon's respecting it:—"This species differs from *T. pelvica* in being smaller, in the purer and darker cinereous hue of the upper plumage, and in the white beneath; also in the more conspicuous and darker eye-streak."

Hab. Southern parts of the Indian peninsula.

a. Ad. sk.	[India.]

5. Tephrodornis virgatus.

Lanius virgatus, *Temm. Pl. Col.* 256. fig. 1.
Tephrodornis virgata, *Swains. Classif. B.* ii. p. 221; *Gray, Gen. B.* i. p. 290.
Tephrodornis gularis, *Cab. Mus. Hein.* i. p. 67.

Adult male. Above blue-grey, the forehead white, as also the lower rump; wing-coverts blue-grey like the back, with concealed blackish bases; primary-coverts and quills blackish, externally edged with blue-grey like the back, the secondaries so broadly as to appear entirely of the same colour as the latter; upper tail-coverts and tail black; lores, nasal bristles, and feathers round the eye and ear-coverts blue-black; cheeks, sides of neck, and entire under surface of body white, with a slight blush of greyish pink on the breast; under wing-coverts blue-grey; bill black; feet blackish. Total length 7·1 inches, culmen 0·8, wing 3·65, tail 2·75, tarsus 0·7.

Adult female. Blue-grey, but much duller than the male; the forehead scarcely perceptibly whitish; rump white; upper tail-coverts and tail black, some of the outer tail-feathers marked with white on the outer web, and tipped on the inner with the same; wings as in the male, but paler; under surface of body white, with a pinkish blush on the breast; the blackish band on the sides of the head much paler than in the male. Total length 7·2 inches, culmen 0·75, wing 3·6, tail 2·85, tarsus 0·7.

Hab. Java.

a, b. Ad. st.	Java.	Purchased.
c, d. ♂ ♀ ad. sk.	E. Java.	A. R. Wallace, Esq. [C.].

6. Tephrodornis gularis.

Lanius gularis, *Raffl. Tr. Linn. Soc.* xiii. p. 304.
Tephrodornis gularis, *Blyth, J. A. S. B.* xv. p. 304; *id. Cat. B. Mus. A. S. B.* p. 153; *Bp. Consp.* i. p. 357; *Horsf. & Moore, Cat. B. Mus. E.I. Co.* p. 171; *Moore, P. Z. S.* 1854, p. 272; *Gray, Hand-l. B.* i. p. 394; *Salvad. Uc. Born.* p. 157.
Tephrolanius gularis, *Bp. C. R.* xxxix. p. 536.
Tephrodornis obscura, *Stolicz. J. A. S. B.* xxxix. p. 320.

Adult. Dull blue-grey above, the forehead paler and slightly whitish; rump-feathers tipped with white; upper tail-coverts pure white, the longer ones black; quills and tail brown, the wing-coverts and secondaries grey like the back, the primaries also washed externally with grey; nasal bristles, lores, and feathers above and below the eye and ear-coverts blue-black; cheeks white; under surface of body pale vinaceous grey, shading off into white on the abdomen and under tail-coverts; under wing-coverts greyish like the inner lining of the wing. Total length 6·6 inches, culmen 0·8, wing 3·9, tail 2·65, tarsus 0·7.

Young. Browner than the adult, especially the quills and tail-feathers, which have reddish shafts; head rather greyer than the back; feathers of the lower back, wing-coverts, and quills obscurely edged with lighter brown; upper tail-coverts white, mottled with remains of brown spots; band along the sides of the head browner than in adults; under surface strong vinaceous grey; the cheeks, chin, lower abdomen, and under tail-coverts white.

Obs. Although very closely allied to *T. gularis*, the present species appears nevertheless to be distinct, its chief character being the vinaceous grey colour of the under surface, which, by extending up onto the throat, causes the white cheeks to stand out with the appearance of a bold moustache; in *T. gularis* the cheeks and throat are uniform pure white. I have taken for granted that the Sumatran bird is the same as the Malaccan, and for that reason have referred the specimens from these localities to *T. gularis*, restricting the name of *T. virgatus* to the Javan bird. The specimens from Borneo examined by me are similar to the Malaccan; and I do not find them agreeing with Temminck's plate, as Count Salvadori states (*l. c.*).

Hab. Malacca; Borneo; ? Sumatra.

a, b.	Ad. sk.	Malacca.	A. R. Wallace, Esq. [C.].
c, d.	Jun. sk.	Malacca.	A. R. Wallace, Esq. [C.].

3. EUROCEPHALUS.

	Type.
Eurocephalus, *Smith, Report Expl. Exp. C. Afr., App.* p. 52 (1836)	E. anguitimens.
Chætoblemma, *Swains. An. in Menag.* p. 282 (1837)	E. anguitimens.

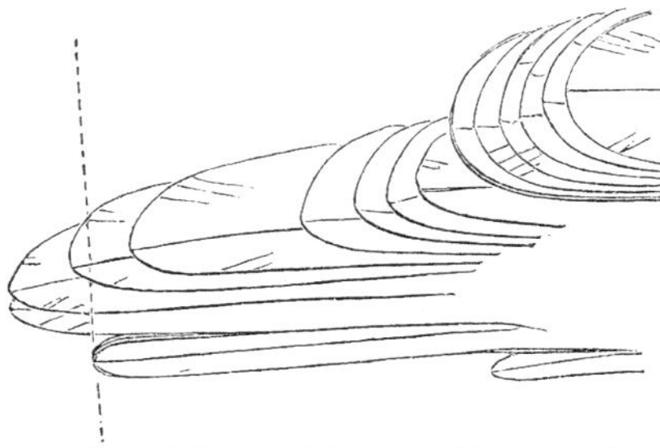

Wing of *Eurocephalus anguitimens*, reduced.

Range. Southern and North-eastern Africa.

Key to the Species.

a. Rump pale ashy brown like the back; abdomen and under tail-coverts pale brown .... *anguitimens*, p. 279.

b. Back brown; rump pure white; abdomen and under tail-coverts white .... *rueppelli*, p. 280.

1. Eurocephalus anguitimens.

Eurocephalus anguitimens, *Smith, Rep. Expl. Exp. C. Afr., App.* p. 53; *Gray, Gen. B.* i. p. 293; *Bp. Consp.* i. p. 365; *Strickl. & Scl. Contr. Orn.* 1852, p. 145; *Bp. Rev. et Mag.* 1853, p. 440;

Layard, B. S. Afr. p. 162; *Gray, Hand-l. B.* i. p. 395; *Sharpe, P. Z. S.* 1870, p. 143; *id. Cat. Afr. B.* p. 51; *Gurney in Anderss. B. Dam. Ld.* p. 140; *Bocage, Jorn. Lisb.* 1874, pp. 7, 24; *Buckley, Ibis*, 1874, p. 377.

Chætoblemma leucocephala, *Swains. An. in Menag.* p. 282.

Adult male. Head creamy white above, extending onto the nape; lores and feathers below the eye black; ear-coverts, hinder part of cheeks, and sides of neck blackish brown, shading off into deep chocolate-brown on the latter; fore part of cheeks white, like the throat; entire back clear light ashy brown, many of the feathers shading off into whitish on their edges; wings dark brown, the coverts and quills very narrowly edged with lighter brown, the quills darker brown than the coverts; tail dark brown, the feathers with very narrow edgings of lighter brown; throat and entire breast white; abdomen and under tail-coverts light ashy brown, the thighs and under wing-coverts rather darker ashy brown; "bill black horn-colour; legs and toes a livid brown; iris brown" (*Andersson*). Total length 9 inches, culmen 0·8, wing 5·35, tail 4·4, tarsus 1·05.

Adult female. Similar to the male. Total length 8·5 inches, wing 5·35, tail 4·4, tarsus 1.

Young. Much paler on the back than the adult; lores and anterior part of cheeks very pale brown; sides of neck white; ear-coverts whitish and mottled with brown; under surface of body whitish, with a wash of pale brown on the abdomen and thighs.

Hab. South Africa.

a. Ad. st.	South Africa (*Smith*).	South-African Museum.
b. ♀ ad. st.	South Africa.	Sir A. Smith [C.].
c. Juv. sk.	South Africa (*Smith*).	Zoological Society. (Type of species.)
d, e. ♂ ad. sk.	Transvaal (*T. Ayres*).	R. B. Sharpe, Esq.
f. ♀ ad. sk.	Otjimbinque, Damara Land, May 30, 1866 (*C. J. Andersson*).	R. B. Sharpe, Esq.
g, h. ♂ ♀ ad. sk.	Otjimbinque, July 30, 1866 (*Andersson*).	R. B. Sharpe, Esq.
i. ♀ ad. sk.	Katenbella, Benguela, November 1 (*Sala*).	R. B. Sharpe, Esq.

2. **Eurocephalus rueppelli.**

Eurocephalus anguitimens, *Rüpp. Syst. Uebers.* p. 53, Taf. 27 (*nec Smith*); *Horsf. & Moore, Cat. B. Mus. E.I. Co.* i. p. 169; *Antinori, Cat. Descr. Ucc.* p. 57; *Salvad. Atti R. Accad. Torino*, 1870, p. 738.

Eurocephalus rueppelli, *Bp. Rev. et Mag. de Zool.* 1853, p. 440; *Heugl. Syst. Uebers.* p. 34; *id. Orn. N.O.-Afr.* i. pp. 487, cxxii; *Gray, Hand-l. B.* i. p. 395.

Adult. Head and nape pure white; back brown, rather darker on the lower part; rump and upper tail-coverts pure white; wings and tail brown, the quills rather darker than the coverts; tail dark brown, white at base, like the upper tail-coverts; lores, ear-coverts, hinder part of cheeks, and sides of neck very deep chocolate-brown; fore part of cheeks, throat, and entire under surface of body white, the breast browner; thighs and under wing-coverts brown; under tail-coverts white, like the abdomen. Total length 9·5 inches, culmen 0·8, wing 5, tail 4·1, tarsus 1.

Hab. North-east Africa.

a. Ad. sk.	Abyssinia.	Sir W. C. Harris [C.].

4. PŒOPTERA.

	Type.
Pœoptera, *Bp. C. R.* xxxviii. p. 381	P. lugubris.
Myiopsar, *Cabanis, J. f. O.* 1876, p. 93	P. lugubris.

Range. Confined to Western Africa.

1. Pœoptera lugubris *.

Pœoptera lugubris, *Bp. C. R.* xxxviii. p. 381; *Hartl. Orn. W.-Afr.* p. 69.
Thamnobia lugubris, *Gray, Hand-l. B.* i. p. 212.
Myiopsar cryptopyrrhus, *Cabanis, J. f. O.* 1876, p. 93.

Adult male. General colour above glossy purple, the least and median coverts coloured like the back; the greater series of wing-coverts light brown, edged with whity brown; the primary-coverts and bastard wing darker brown, with slight purple gloss; quills rather darker brown, slightly shaded with purple towards their tips; the secondaries externally whity brown; tail strongly graduated, blackish, glossed with greenish purple on the edges of the feathers; entire under surface of body glossy purple, including the under wing- and tail-coverts; lower surface of quills dusky brown, inclining to ashy brown on the inner webs, and slightly washed with rufous near the base of the primaries; bill and legs black. Total length 8·5 inches, culmen 0·6, wing 3·55, tail 4·45, tarsus 0·7.

Adult female. Different from the male. Greyish above, with a very distinct steel-blue gloss; wing-coverts like the back, the greater series dark brown, glossed with steel-blue on the outer webs; quills blackish brown, the inner webs rufous for the greater part of their extent; tail blackish, glossed with steel-green on both webs; under surface of body ashy grey, with a gloss of steel-green; the wing-coverts like the breast; quills blackish below, the rufous inner webs showing very plainly. Total length 7·6 inches, culmen 0·55, wing 3·3, tail 3·8, tarsus 0·75.

Hab. West Africa: from the Gold Coast to the Congo.

a, b. ♂ ♀ ad. sk.	Forests of Denkera, interior of Fantee.	Mr. St. Thomas D. Aubinn [C.].

---

* Prince Bonaparte seems to have named this species without due consideration, and it is quite open to question whether this name should be accepted. He has confused it with *Muscicapa lugubris* of Müller, to which it bears no resemblance at all. Dr. Hartlaub has adopted Bonaparte's name, and I follow him, but not without doubt as to the reception of the title by future students.

5. LEPTOPTERUS.

Type.

Leptopterus, Bp. C. R. xxxix. p. 538 (1854) .... L. chabert.

Head of *Leptopterus chabert.*

Range. Confined to Madagascar.

1. Leptopterus chabert.

La petite Pie-grièche verte de Madagascar, *Briss. Orn.* ii. p. 195, pl. 15. fig. 3.
Green Shrike, *Lath. Gen. Syn.* i. p. 179.
La Tcha-bert, *Buff. Pl. Enl.* 32. fig. 1.
Lanius chabert, *P. L. S. Müll. Syst. Nat.* Anhang, p. 72.
Lanius violaceus, *Bodd. Tabl. Pl. Enl.* p. 3.
Lanius viridis, *Gm. S. N.* i. p. 306 (*ex Lath.*).
Leptopteryx viridis, *Wagler, Syst. Av.* Leptopteryx, sp. 7.
Artamia viridis, *Lafr. Dict. Univ. d'Hist. Nat.* ii. p. 166; *Grand. Rev. et Mag. de Zool.* 1867, p. 386; *Schl. & Poll. Faun. Madag. Ois.* p. 84, pl. 27; *Sharpe, P. Z. S.* 1870, p. 389; *id. Cat. Afr. B.* p. 52.
Analcipus hirundinaceus, *Sw. An. in Menag.* p. 284.
Leptopterus viridis, *Bp. C. R.* xxxix. p. 538; *Hartl. Faun. Madag.* p. 48; *Roch & E. Newt. Ibis*, 1862, p. 273; *E. Newt. Ibis*, 1863, p. 348; *A. Newt. P. Z. S.* 1865, p. 835.
Artamus viridis, *Vieill. N. Dict. d'Hist. Nat.* xvii. p. 299; *Gray, Gen. B.* i. p. 285; *id. Hand-l. B.* i. p. 290.

Adult male. Above shining greenish black, the hind neck and mantle mottled with subterminal white spots to the feathers; head and sides of face shining blue-black; wing-coverts, quills, and tail-feathers greenish black, with distinct edges of shining green to the feathers; cheeks and entire under surface of body pure silky white, the flank-feathers long and greyish; thighs black; under wing-coverts greenish black, the feathers edged with shining green; "beak greyish pearl; feet black; iris brown; eye-wattle clear blue" (*Grandidier*). Total length 6·2 inches, culmen 0·75, wing 3·75, tail 2·1, tarsus 0·75.

Obs. A young bird, which has not quite completed the full plumage, has the rump greyish, with green edges to the feathers; the secondaries edged with grey; on the head the feathers are brown, minutely spotted and streaked with dull white.

Hab. North-western Madagascar.

a, b. Ad., *c.* Juv. sk.	Madagascar.	E. Wilson, Esq. [P.].
d. Ad. sk.	Madagascar.	R. B. Sharpe, Esq.
e. Ad. sk.	Mohambo, Madagascar.	Mr. Gerrard [C.].

6. RECTES.

Type.

Pitohui, *Lesson, Traité*, i. p. 375 (1831) .... R. cirrhocephalus.
Rectes, *Bp. C. R.* xxxi. p. 563 (1850).

Head, foot, and wing of *Rectes uropygialis*.

Range. New Guinea and Papuan Islands; Aru Islands; Mysol; Waigiou; Jobi.

Key to the Species.

a. With a very distinct cap, contrasting with the colour of the back.
  a'. Upper tail-coverts rufous like the rest of the back .... *cirrhocephalus*, p. 284.
  b'. Upper tail-coverts black; back chestnut.
    a''. Black colour ending abruptly on the breast.
      a'''. Black colour occupying the whole of the lower back, as well as the rump and upper tail-coverts; thighs chestnut .. *uropygialis*, p. 285.
      b'''. Black colour confined to the lower rump and upper tail-coverts.......... *tibialis*, p. 285.
    b''. Black colour extending down the centre of the breast and abdomen; upper tail-coverts only black, not the rump .... *aruensis*, p. 285.
b. With no distinct cap, the head being of the same colour as the mantle, or nearly so.
    c''. Throat ashy, contrasting with the breast.. *cerviniventris*, p. 286.
    d''. Throat rufous, like the rest of the under surface.
      c'''. Brownish red above, inclining to olive; bill black .... *cristatus*, p. 286.
      d'''. Foxy red above, like the under surface; bill light horn-colour................ *jobiensis*, p. 287.

1. Rectes cirrhocephalus.

Vanga kirrhocephalus, *Less. Voy. Coquille, Zool.* p. 633, pl. xi.
Pitohui, *Less. Traité*, p. 375.
Rectes kirrocephalus, *Bp. C. R.* xxxi. p. 563; *Scl. Journ. Linn. Soc.* 1858, p. 161; *Gray, P. Z. S.* 1859, p. 157; *id. Cat. Mamm. &c. N. Guin.* p. 33; *Finsch, Neu-Guinea*, p. 170; *Scl. P. Z. S.* 1873, p. 696.
Timalia poliocephala, *Bp. C. R.* xxxi. p. 563 (*ex Müll.*).
Rectes dichrous, *Bp. C. R.* xxxi. p. 563; *Scl. Journ. Linn. Soc.* 1858, p. 161.
Rectes bicolor, *Finsch, Neu-Guinea*, p. 170 (*ex Müll. MS. in Mus. Lugd.*).
Colluricincla kirrocephala, *Gray, Hand-l. B.* i. p. 385.
Rectes cirrhocephala, *Meyer, Sitz. Akad. Wien*, lxix. pp. 205, 391; *Salvad. Ann. Mus. Civic. Genov.* vii. p. 772.

Adult (*R. dichrous*). Above maroon-chestnut, including the upper tail-coverts; head crested; sides of face, throat, and fore neck black; rest of under surface of body rich orange fawn-colour, including the under wing- and tail-coverts; least wing-coverts chestnut like the back; greater and median series black; quills and tail black; bill black. Total length 10 inches, culmen 1·2, wing 4·5, tail 4·5, tarsus 1·55.

Adult female. Similar to the male, but having the black of the head and throat somewhat obscured by olive-brown; quills also washed externally with brownish, and the outer tail-feather brownish on the outer web. Total length 10·8 inches, culmen 1·25, wing 4·8, tail 4·7, tarsus 1·5.

Young (*R. cirrhocephalus*). General colour above bright maroon-chestnut, the least wing-coverts uniform with the back; greater and median wing-coverts ashy brown, slightly washed with olive, and margined on the edge of the feathers with olive-brown; quills dark brown, externally ashy brown, with olive margins, the secondaries distinctly washed with the latter colour; tail ashy brown, somewhat washed with olive on the edges of the feathers; head crested, grey, forming a cap, the sides of the face and entire throat also grey like the head; sides of neck maroon-chestnut like the back; rest of under surface of body fawn-colour, including the thighs and under wing- and tail-coverts. Total length 10·9 inches, culmen 1·1, wing 4·6, tail 4·6, tarsus 1·3.

Younger. Similar to the foregoing, but everywhere paler in colour.

Obs. I could not at first, from a study of the two birds in the British Museum, receive Dr. Meyer's conclusions as to *Rectes cirrhocephalus* being the young of *R. dichrous*, and I am therefore greatly indebted to him for sending over for my inspection his fine series of these birds, after examining which there is no doubt of the fact of their being one and the same species. The adult male is black-headed, and the adult female is similar, but has the black cap and throat browner. In the young birds a perfect gradation is seen from the pale ashy-headed specimens to the dark-brown ones of the young males, when they resemble the *adult* females before finally completing their perfect dress. In the first plumage both males and females seem to be perfectly alike. Dr. Meyer's specimens were all collected in May and June, and exhibit every stage from the very young bird, through successive changes, to the perfectly adult.

Hab. New Guinea.

a, b. ♂ imm. sk.	Dorey, New Guinea.	A. R. Wallace, Esq. [C.].
c. Imm. sk.	Dorey, New Guinea.	A. R. Wallace, Esq. [C.].
d. Juv. st.	New Guinea.	Leiden Museum.
e. [♀] ad. sk.	New Guinea.	Purchased *.

2. Rectes uropygialis.

Rectes uropygialis, *Gray, P. Z. S.* 1861, p. 430; *Finsch, Neu-Guinea,* p. 170; *Meyer, Sitz. Akad. Wien,* lxix. p. 208.
Colluricincla uropygialis, *Gray, Hand-l. B.* i. p. 385.

Adult male (type of species). Head, which is strongly crested, and nape black, as also the entire sides of the face, sides of neck, throat, fore neck, and chest; hind neck, mantle, and upper back rich maroon-chestnut, rather more orange on the hind neck; lower back, rump, and upper tail-coverts black; wings and tail black; remainder of under surface of body, from the chest downwards, including the thighs, under wing- and tail-coverts, rich maroon-chestnut; bill and legs black. Total length 11 inches, culmen 1·1, wing 4·7, tail 4·55, tarsus 1·36.

Another male, from Mysol, is slightly larger. Total length 11 inches, culmen 1·25, wing 4·85, tail 4·7, tarsus 1·45.

Adult female. Similar to the adult male, and quite as richly coloured. Total length 11 inches, culmen 1·25, wing 4·95, tail 4·75, tarsus 1·35.

Hab. Mysol.

a. ♂ ad. sk.	Mysol.	A. R. Wallace, Esq. [C.]. (Type of species.)
b, c. ♂ ♀ ad. st.	Mysol.	A. R. Wallace, Esq. [C.].

3. Rectes tibialis.

Similar in all respects to *Rectes uropygialis,* but rather smaller; the black confined to the rump and upper tail-coverts, and not extending on to the lower back; hinder part of thighs black, and not maroon as in *R. uropygialis.* Total length 10 inches, culmen 1·2, wing 5·1, tail 4·85, tarsus 1·25.

Hab. New Guinea.

a. Ad. sk.	New Guinea.	A. R. Wallace, Esq. [C.].

4. Rectes aruensis.

Rectes dichrous (*nec Bp.*), *Gray, P. Z. S.* 1858, p. 179; *id. List Mamm. &c. N. Guin.* p. 33.

---

* The first four are in the grey-headed plumage, the only stage obtained by Mr. Wallace: the last one mentioned is in the black-headed dress.

Colluriocincla dichroa, *Finsch, Neu-Guinea*, p. 170.
? Rectes dichrous ceramensis, *Meyer, Sitz. Akad. Wien*, lxix. p. 208.

Adult male. Above very bright chestnut, the upper tail-coverts black; head crested, black all round, with the sides of face, fore neck, and chest also entirely black; rest of the under surface of the body deep tawny buff, the whole of the breast shaded with black; under wing-coverts deep ochraceous, those near the edge of the wing black; wings and tail deep black; bill and legs black. Total length 10 inches, culmen 1·2, wing 4·3, tail 4·1, tarsus 1·35.

Young. Paler ochre-rufous below, not mixed with black on the breast; the cap and black throat almost as strongly defined as in the adult, but both are much washed with brown; rump dingy ochraceous brown, washed with chestnut, the tail-coverts inclining to blackish; quills and tail blackish, the outer webs of the quills dingy ochraceous brown; tail-feathers brown, slightly washed with rufous, and inclining to black near the base and upwards along the centre of the feathers. Total length 10 inches, culmen 1·2, wing 4·45, tail 4·3, tarsus 1·3.

Hab. Aru Islands.

a, b. ♂ ad. sk.	Aru Islands, 1857.	A. R. Wallace, Esq.
c, d. ♂ ♀ juv. st.	Aru Islands, 1857.	A. R. Wallace, Esq.
e. Juv. sk.	Aru Islands (*Cockerell*).	R. B. Sharpe, Esq.

5. Rectes cerviniventris.

Rectes cerviniventris, *Gray, P. Z. S.* 1861, p. 430; *Finsch, Neu-Guinea*, p. 176; *Meyer, Sitz. Akad. Wien*, lxix. p. 208.
Colluricincla cerviniventris, *Gray, Hand-l. B.* i. p. 465.

Adult male. General colour above ashy olive, inclining to olive on the rump and upper tail-coverts; the head crested, ashy grey, this colour pervading the mantle and back; sides of face and sides of neck rather darker ashy; wings olive-brown, the inner webs of the feathers darker brown, the outer webs washed with reddish olive; tail-feathers brown, washed with olive on the edges of the feathers, the outer ones reddish brown, edged with deep fawn-colour on the outer web; throat and fore neck ashy, washed with ochre; rest of under surface of body chestnut fawn-colour, including the thighs, under wing- and tail-coverts; under surface of body dark ashy brown, the inner webs rufous fawn-colour. Total length 8·6 inches, culmen 1·05, wing 3·75, tail 3·5, tarsus 1·25.

Hab. Waigiou.

a, b. ♂ ad. sk.	Waigiou.	A. R. Wallace, Esq. [C.]. (Type of species.)
c, d. ♂ ad. st.	Waigiou.	A. R. Wallace, Esq. [C.].

6. Rectes cristatus.

Rectes cristata, *Salvad. Ann. Mus. Civic. Genov.* vii. p. 930.

Above brownish red, rather inclining to olive; underneath bright rufous, the breast and centre of abdomen paler; head conspicuously crested; upper tail-coverts and tail brownish ruddy; eyebrow dusky; bill black; feet very strong, lead-coloured in skin. Total length 9·1 inches, culmen 0·9, wing 5·15, tail 3·95, tarsus 1·65. (*Salvadori.*)

Hab. Mount Morait, N.W. New Guinea.

7. Rectes jobiensis.

Rectes jobiensis, *Meyer, Sitz. Akad. Wien*, lxix. p. 205; *Salvad. Ann. Mus. Civic. Genov.* vii. p. 773.

Adult male (type of species). General colour above bright foxy red, paler on the head; wing-coverts like the back; quills dark brown, externally rufous, the inner secondaries entirely rufous; tail-feathers chestnut, the shafts rufous; lores, sides of face, and under surface of body deep rich tawny buff, darker on the throat and fore neck, and inclining to chestnut on the lower flanks and under tail-coverts; under wing-coverts bright tawny, like the breast; quills dusky brown below, rufous on the outer web and along the inner web; bill light yellowish horn-colour; feet blackish, claws pale. Total length 9·2 inches, culmen 11·5, wing 4·65, tail 4·35, tarsus 1·35.

Adult female. Similar to the male. Total length 10·5 inches, wing 4·6, tail 4·3, tarsus 1·25.

Young male. Similar in colouring to the adult, but rather duller; the bill shorter and duller brown.

Hab. Island of Jobi, in the Bay of Geelvink, New Guinea.

a, b. ♂ ♀ ad. sk. Jobi Island. Dr. A. B. Meyer [C.].

7. PSEUDORECTES.

Rectes (pt.), *auct. recent.*

Range. New Guinea and adjacent Papuan Islands. Aru Islands; Waigiou; Mysol.

Key to the Species.

a. Smaller: bill black; underneath fawn-buff.... *ferrugineus*, p. 287.
b. Much larger: bill yellowish white; below chestnut fawn-colour........................ *leucorhynchus*, p. 288.

1. Pseudorectes ferrugineus.

Pie-Grièche bruyante, *Hombr. et Jacq. Voy. Pôle Sud, Atlas*, pl. 6. fig. 1.
Garrulax ferrugineus, *S. Müll. in Mus. Lugd. undè*
Rectes ferrugineus, *Bp. C. R.* xxxi. p. 563; *Salvad. Ann. Mus. Civic. Genov.* vii. p. 823.
Rectes strepitans, *Hombr. et Jacq. Voy. Pôle Sud, Texte*, iii. p. 60; *Gray, Cat. Mamm. &c. New Guinea*, p. 34; *Finsch, Neu-Guinea*, p. 170.
Colluricincla strepitans, *Gray, Hand-l. B.* i. p. 385.

Adult female (Dorey: *Wallace*). Above dark reddish brown; rump and upper tail-coverts rather more rufous, especially the latter; tail rufous; ear-coverts rufous brown, streaked along the shaft with whitish buff; rest of under surface of body fawn-colour, the under wing- and tail-coverts uniform with the breast; upper wing-coverts a little darker than the back; quills dark brown, externally fawn-brown, very slightly washed with olive; "bill black; feet ashy; iris white or pale yellow" (*D'Albertis*). Total length 11·5 inches, culmen 1·2, wing 5·25, tail 4·9, tarsus 1·45.

Adult male (Dorey). Similar to the female. Total length 11 inches, culmen 1·15, wing 5·4, tail 4·7, tarsus 1·45.

Obs. The specimens from different localities vary very little in size, the following being a synopsis of the measurements of the Museum series:—

	Total length. in.	Culmen. in.	Wing. in.	Tail. in.	Tarsus. in.
a. Mysol (*Wallace*)	10·5–11·5	1·25	5·5 –5·65	4·8–5·2	1·4 –1·5
b. New Guinea (*Wallace*)	11·0–11·5	1·15–1·2	5·35–5·4	4·7–4·9	1·45
c. Aru Islands (*Wallace*)	10·0–10·5	1·05–1·2	4·8 –5·3	4·4–4·7	1·35

Thus it will be observed that the Mysol birds are rather the largest, the Aru ones the smallest, while the New-Guinea ones are intermediate.

Hab. New Guinea; Aru Islands; Mysol.

a. ♂ ad. st.	Dorey, New Guinea.	A. R. Wallace, Esq. [C.].
b, c. ♀ ad. sk.	Dorey, New Guinea.	A. R. Wallace, Esq. [C.].
d. ♂ ad. sk.	Nicura, S.E. New Guinea, Dec. 1875.	Dr. James [C.].
e, f. ♂ ♀ ad. sk.	Aru Islands.	Dr. James [C.].
g. ♂ ad. st.	Aru Islands.	Dr. James [C.].
h. Ad. sk.	Aru Islands (*Cockerell*).	R. B. Sharpe, Esq. [P.].
i, k. ♀ ad. sk.	Mysol.	A. R. Wallace, Esq. [C.].

2. Pseudorectes leucorhynchus.

Rectes leucorhynchus, *Gray, P. Z. S.* 1861, p. 430; *Finsch, Neu-Guinea,* p. 170.
Colluricincla leucorhyncha, *Gray, Hand-l. B.* i. p. 386.

Adult male. Above rufous brown, the rump and the tail a little brighter, the crown and sides of the head slightly darker; wing-coverts like the back, but a little darker brown; quills dark brown, externally dark rufous brown; under surface of body deep rufous fawn-colour, the throat paler; under wing- and tail-coverts uniform with the breast; bill yellowish white. Total length 12 inches, culmen 1·15, wing 5·7, tail 5·2, tarsus 1·5.

A second male, collected by Mr. Wallace, measures—total length 11 inches, culmen 1·15, wing 6, tail 5·1, tarsus 1·15.

Hab. Waigiou.

a. ♂ ad. sk.	Waigiou.	A. R. Wallace, Esq. [C.].
b. Ad. st.	Waigiou.	A. R. Wallace, Esq. [C.].

8. MELANORECTES.

Rectes (pt.), *auct recent.*

Range. North-western New Guinea.

1. Melanorectes nigrescens.

Rectes nigrescens, *Schl. N. T. D.* iv. p. 46; *Meyer, P. Z. S.* 1874, p. 419; *id. Sitz. Akad. Wien*, lxix. p. 390.
Rectes bennetti, *Sclater, P. Z. S.* 1873, p. 692; *Meyer, Sitz. Akad. Wien*, lxx. p. 126.

Adult male. General colour above slaty black, deeper and more jet-black on the crown and sides of the head; wing-coverts like the back; quills and tail blackish brown; under surface of body black, inclining to slaty black on the flanks and under tail-coverts; under wing-coverts resembling the breast; quills blackish brown below, margined along the inner web with lighter brown; "bill black; legs dull lead-colour; iris chestnut-brown" (*D'Albertis*). Total length 9 inches, culmen 0·95, wing 5, tail 4·1, tarsus 1·2. (*Dr. A. B. Meyer: Mus. Dresden.*)

Adult female. Above olive-brown, a little redder on the rump and upper tail-coverts; wing-coverts rather more reddish brown, the greater series somewhat olivaceous; quills brown, reddish on the outer webs; tail chestnut; head above ashy brown, slightly greyer on the forehead and lores; sides of face, sides of neck, and under surface of body fulvous brown, the throat somewhat ashy; under wing-coverts rich fawn-colour; quills light brown below, chestnut along the inner web. (*Dr. A. B. Meyer: Mus. Dresden.*)

Hab. North-western New Guinea.

9. COLLYRIOCINCLA.

	Type.
Colluricincla, *Vig. & Horsf. Tr. Linn. Soc.* xv. p. 214 (1825)	C. harmonica.
Collurisoma, *Swains. Classif. B.* ii. p. 10 (1837)	?
Pnigocichla, *Cab. Mus. Hein.* Th. i. p. 66 (1850)......	C. harmonica.

Head and wing of *Collyriocincla harmonica.*

Range. Australia and Southern New Guinea.

Key to the Species.

Bill dark blackish or deep brown.
  a'. Back umber-brown, contrasting with the grey head and rump.
    a''. Bill stout, 0·9 inch long; white loral spot very distinct; no eyebrow in adult...... *harmonica*, p. 290.
    b''. Bill very long, 1·2 inch; white loral spot very faintly indicated; a distinct eyebrow. *rectirostris*, p. 291.
  b'. Back ashy brown, like the rest of the upper surface.
    a'''. Under tail-coverts white .... *brunnea*, p. 291.
    b'''. Under tail-coverts bright fawn-buff.... *rufiventris*, p. 292.
b. Bill light yellowish horn-colour; eyebrow indistinct .... *pallidirostris*, p. 293.

1. Collyriocincla harmonica.

Harmonic Thrush, *Lath. Gen. Syn. Suppl.* p. 182.
Turdus harmonicus, *Lath. Ind. Orn. Suppl.* p. xli.
Colluricincla cinerea, *Vig. & Horsf. Tr. Linn. Soc.* xv. p. 214; *Jard. & Selby, Ill. Orn.* ii. pl. 71; *Sw. Classif. B.* ii. p. 221.
Lanius saturninus, *Nordm. in Erman's Reis.* p. 14.
Colluriocincla harmonica, *Gray, Gen. B.* i. p. 295; *Gould, B. Austr.* ii. pl. 7; *Bp. Consp.* i. p. 366; *Gould, Handb. B. Austr.* i. p. 220; *Gray, Hand-l. B.* i. p. 385; *Diggles, Orn. Austr.* pl. 34. fig. 1.
Pnigocichla harmonica, *Cab. Mus. Hein.* Th. i. p. 66.

Adult. Head and nape grey; hind neck, back, and scapulars clear umber-brown; least and median coverts uniform with the back, the greater series ashy grey; quills dark brown, externally ashy grey, the secondaries slightly washed with olivaceous; lower back, rump, and upper tail-coverts ashy grey; tail ashy grey, the feathers slightly tipped with whitish, the shafts brown above, white below; lores and feathers in front of the eye white; round the eye a ring of dusky grey plumes; ear-coverts ashy grey, slightly washed with brown; cheeks whitish, grey on their hinder margin; under surface of body light ashy grey, the throat, centre of abdomen, and under tail-coverts white; under wing-coverts ashy whitish; quills light brown below, whitish along the inner webs; "bill blackish brown; feet dark greenish grey; iris dark brown" (*Gould*). Total length 9·8 inches, culmen 0·9, wing 5·15, tail 4·0, tarsus 1·25.

Young. Distinguished by its rufous eyebrow and by the stripes on the breast; the whole colour of the upper surface is darker, the ashy grey colour of the head and back being shaded with brown.

Hab. Australia.

a, b.	Ad. juv. sk.	Australia.	Linnean Society. (Types of *C. cinerea.*)
c.	Juv. st.		Ronald Gunn, Esq. [P.].
d.	♂ ad. sk.	S. Australia.	Sir George Grey [P.].
e, f.	Jun. sk.	S. Australia.	Sir George Grey [P.].
g.	Juv. sk.	Australia.	Sir T. Mitchell [P.].

h, *i*. Ad. juv. sk.	Australia.	Sir T. Mitchell [P.].
k. Imm. sk.	River Darling, lat. 32°.	Capt. Sturt [P.].
l. Ad. sk.	[New Zealand.]	New-Zealand Co. [P.].

2. Collyriocincla rectirostris.

Colluricincla rectirostris, *Jard. & Selby, Ill. Orn.* iv. pl. xxxi.; *Gray, Gen. B.* i. p. 295.

Colluricincla strigata, *Swains. An. in Menag.* p. 283.

Colluricincla selbii, *Gould, B. Austr.* ii. pl. 77*; *Bp. Consp.* i. p. 366; *Gould, Handb. B. Austr.* i. p. 224; *Gray, Hand-l. B.* i. p. 385.

Adult. Head grey; hind neck, back, and scapulars umber-brown, contrasting with the head; rump and upper tail-coverts grey; wings brown, the least and median coverts umber-brown like the back, the greater series rather greyer; primary coverts and primaries dark brown, externally washed with dull grey, the inner secondaries rather more olive-brown; tail dull grey, the outer feathers browner on the inner web, the shafts of the quills reddish brown; lores dull white; over the eye an obscurely indicated eyebrow of whitish; round the eye a ring of dusky greyish feathers; ear-coverts grey like the head; cheeks dull grey, whiter on the fore part; under surface of body greyish white, the flanks, thighs, and under tail-coverts grey; under wing-coverts ashy grey, darker near the edge of the wing; quills ashy brown below, greyish along the inner webs; "bill black; feet light lead-colour; iris brown" (*Gould*). Total length 9·5 inches, culmen 1·2, wing 5, tail 4·3, tarsus 1·3.

Adult female. Similar to the male; "bill horn-colour at the base, black at the tip" (*Gould*).

Young. Duller than the adult, the head washed with ashy brown; over the eye a distinct rufous eyebrow; primary-coverts externally rufous, the quills olivaceous on the outer web, the inner webs to the quills also fawn-coloured; under surface of body dull white, everywhere distinctly streaked with shaft-stripes of dark brown.

Hab. Van Diemen's Land.

a. Ad. st.	Van Diemen's Land.	J. Gould, Esq.
b. Ad., *c*, *d*. Juv. st.	Van Diemen's Land.	Ronald Gunn, Esq. [P.].
e. ♂ ad., *f*. ad. sk.	Van Diemen's Land.	Antarctic Expedition.
g. ♂ juv. sk.	Van Diemen's Land.	J. Gould, Esq.
h. Juv. sk.	Van Diemen's Land.	

3. Collyriocincla brunnea.

Colluricincla brunnea, *Gould, P. Z. S.* 1840, p. 164; *Gray, Gen. B.* i. p. 295; *Gould, B. Austr.* ii. pl. 76; *Bp. Consp.* i. p. 366; *Gould, Handb. B. Austr.* i. p. 223; *Gray, Hand-l. B.* i. p. 385.

Colluricincla superciliosa, *Masters, Tr. Linn. Soc. N. S. W.* i. p. 50.

---

* Mr. Gould refers us to a note to pl. 71 of the 'Illustrations of Ornithology' for this name, by which the species is generally known. In the copies of the work examined by me I find no mention of such a name.

Adult. Above greyish brown, clearer on the head, which has faintly indicated shaft-lines of brown; wing-coverts brown, externally washed with grey, clearer on the greater series; quills dark brown, externally grey, much clearer on the secondaries; upper tail-coverts much lighter greyish brown than the back; tail-feathers ashy brown, with light greyish margins, all the feathers narrowly whitish at their tips; a spot in front of the eye and a distinct eyebrow white; round the eye a circlet of whitish feathers; ear-coverts light ashy brown, with narrow whitish shaft-lines; cheeks and throat white, with mesial shaft-lines of brown; breast and flanks pale whity brown, the centre of the body and under tail-coverts white; thighs light brown; under wing-coverts very pale orange-buff, the axillaries whitish, with narrow dark brown shaft-streaks. Total length 8·5 inches, culmen 1·05, wing 4·65, tail 4·7, tarsus 1·3.

Hab. S.E. New Guinea and Northern Australia.

a.	Ad. sk.	Cape York, N. Australia.	J. Gould, Esq.
b.	Ad. sk.	Port Moresby, S.E. New Guinea.	O. C. Stone, Esq. [P.].

4. Collyriocincla rufiventris.

Colluricincla rufiventris, *Gould, P. Z. S.* 1840, p. 164; *id. B. Austr.* fol. pl. 75; *Gray, Gen. B.* i. p. 295; *Bp. Consp.* i. p. 356; *Gould, Handb. B. Austr.* i. p. 222; *Gray, Hand-l. B.* i. p. 385.

Turdus subcinereus, *Sclater, P. Z. S.* 1866, p. 320; *id. P. Z. S.* 1873, p. 555.

Adult. General colour above ashy grey, with a very slight olivaceous tinge; wing-coverts ashy grey like the back; quills brown, externally greyish; lower back, rump, and upper tail-coverts clearer grey than the mantle; tail-feathers grey, rather lighter brown on the inner webs; lores and a spot of feathers in front of the eye white; round the eye a ring of dusky grey feathers, very indistinct; ear-coverts light ashy brown; cheeks greyish white, slightly mottled with obscure brown markings; throat greyish white; fore neck and breast light ashy grey, darker on the sides of the breast, the feathers indistinctly streaked with dusky brown down the centre; rest of under surface of body pale fawn-colour, deeper on the under tail-coverts; thighs dull ashy; under wing-coverts pale fawn-colour, with greyish bases, the quills dull brown below, the inner webs pale yellowish buff towards their bases; "bill blackish brown; feet dark greenish leaden grey; iris dark reddish brown" (*Gould*). Total length 8·5 inches, culmen 0·95, wing 4·8, tail 4, tarsus 1·2.

Hab. Australia.

a.	Ad. st.	Perth, Swan River, March 9, 1839.	J. Gould, Esq.
b.	Ad. sk.	Port Lincoln, S. Australia (*Harvey*).	J. Gould, Esq.
c.	Ad. st.	Port Lincoln, S. Australia (*Harvey*).	J. Gould, Esq.
d.	Ad. sk.	Australia.	Purchased.
e.	Ad. sk.	[Australia.]	P. L. Sclater, Esq. [P.]. (Type of *Turdus subcinereus.*)

5. Collyriocincla pallidirostris.

Adult. General colour above ashy brown, the head rather more grey, with indistinct brown shaft-lines; wing-coverts also a little greyer than the back; quills brown, externally ashy grey, the secondaries rather clearer grey on the outer webs; upper tail-coverts ashy grey; tail brown, all the feathers ashy grey on their margins and slightly paler at the extreme tips; a spot in front of the eye white; round the eye a circlet of whitish feathers; over the eye a very faint eyebrow of dull white feathers; ear-coverts uniform ashy brown, with faint whitish shaft-lines; cheeks and throat dull white, with faint mesial streaks of rufous brown; breast and sides of body light ashy brown, the centre of the body dull white, the entire breast marked with obscure mesial streaks of light rufous brown; thighs ashy brown; under tail-coverts dull white, with light brown shaft-streaks; under wing-coverts light orange-buff, the axillaries ashy brown like the breast; under surface of quills ashy brown, light tawny buff along the inner web; bill pale yellowish or whitish horn-colour; "iris brown" (*Elsey*). Total length 9·2 inches, culmen 1, wing 4·85, tail 4·4, tarsus 1·2.

Hab. Northern Australia.

a. Ad. st.	Port Essington, May 13, 1840.	Capt. Chambers, R.N. [P.].
b. ♀ ad. sk.	Nicholson River, Aug. 24, 1856, lat. 19° 40′, long. 138°.	J. R. Elsey, Esq. [P.].

10. PINAROLESTES*.

Myiolestes, *auct. recent., nec Cabanis,* 1850.

Head of *Pinarolestes megarhynchus.*

Range. New Guinea and adjacent Papuan Islands; Pelew Islands; Northern Australia and islands of Torres Strait; Fiji Islands.

Key to the Species.

a. Bill light yellowish horn-colour or light brown.
  a′. Round the eye a ring of whitish or buff feathers.
    a″. Lores fulvous; throat only a little paler fawn-colour than the breast; ear-coverts rufous brown, with fulvous shaft-streaks .... *megarhynchus*, p. 295.

* Type *P. vitiensis.*

b″. Lores whitish; throat white, in contrast to the breast.
a‴. Over the eye a distinct white eyebrow; under surface very pale buff; under tail-coverts whitish .... *parvulus*, p. 296.
b‴. No distinct eyebrow; under tail-coverts deep fawn-colour, like the rest of the breast .... *rufigaster*, p. 296.
b′. No light ring of feathers round the eye; underneath ashy or olivaceous brown .. *affinis*, p. 297.
b. Bill black, sometimes lighter at the base.
c′. Throat and breast brownish white, yellower towards the abdomen and on the sides of the breast .... *melanorhynchus*, p. 298.
d′. Throat and breast dark brown.
c″. Upper surface brown like the under surface; the head and throat darker .... *tenebrosus*, p. 298.
d″. Upper surface brownish grey, slightly inclining to olive; below grey inclining to brown, the abdomen washed with white .... *obscurus*, p. 299.
e′. Throat and breast grey; flanks tawny brown.
c‴. With white tips to the tail-feathers.
a^{4}. Under tail-coverts and lower abdomen white.................. *heinii*, p. 299.
b^{4}. Under tail-coverts rufous chocolate like the flanks; abdomen grey .. *vitiensis*, p. 299; *buensis*, p. 300; *fortunæ*, p. 300; *compressirostris*, p. 301.
d‴. With no white tips to the tail-feathers .... { *macrorhynchus*, p. 301. *maximus*, p. 301.
f′. Throat black, succeeded by an irregular white band .... *nigrogularis*, p. 301.

This genus is formed for the reception of certain small Shrike-like birds, usually, though erroneously, placed in the genus *Myiolestes* (*vide infrà*, p. 305). These small species are included by Gray in *Collyriocincla*, and in this he is followed by Salvadori. The proportion of the toes, and still more the narrow Laniine bill, appear to me to be characters which justify their separation under a separate generic heading.

Too many species are allowed by ornithologists, in my opinion; and in the case of the Papuan and Australian birds, the series examined has been sufficiently large to enable me to justify the suppression of several supposed species. In the case of the Fijian members of the genus I refrain from expressing an opinion at present, not having seen specimens of four of the species. Two not in the Museum have been examined by me from specimens in the Marquis of Tweeddale's collection. There appear to be only two forms, one with distinct white tips to the tail-feathers, and the other without them. A shade of grey more or less in the plumage appears to be a very uncertain character, dependent on age apparently: thus *M. macrorhynchus* seems to be the only species really distinguishable from *M. vitiensis*, of which all the others are representative forms, differing only in the size of bill.

1. Pinarolestes megarhynchus.

Muscicapa megarhyncha, *Quoy et Gaim. Voy. Astrol. Zool.* i. p. 172, pl. 3. fig. 1.
Myialestes megarhynchus, *Bp. Consp.* i. p. 358; *Gray, Cat. Mamm. &c. N. Guin.* p. 34; *Meyer, SB. Akad. Wien*, lxix. p. 208.
Myiolestes aruensis, *Gray, P. Z. S.* 1868, p. 180.
Colluricincla megarhyncha, *Gray, Hand-l. B.* i. p. 386; *Salvad. Ann. Mus. Civic. Genov.* vii. p. 773.

Adult male (Dorey: *Wallace*). General colour above olivaceous brown, duller on the head; wing-coverts olivaceous brown, the greater series, primary-coverts, and primaries externally edged with tawny brown, the inner secondaries entirely dull tawny brown on their outer webs; upper tail-coverts and tail dull rufous brown, the outermost externally tawny brown; lores and a very indistinct eyebrow dull rufous; feathers in front of the eye dusky greyish; ear-coverts dusky rufous brown; cheeks dull tawny rufous, with black shaft-lines; rest of under surface of body deep tawny buff, washed with brown on the sides of the body, the throat paler and more of an ashy whitish colour, washed with tawny, the latter colour deepening on the fore neck, the latter, as well as the lower throat, being longitudinally streaked with dusky brown; under wing-coverts deep tawny buff like the breast; quills ashy brown below, margined with tawny on the inner webs; "bill dull brown; feet dull lead-colour; iris chestnut-brown" (*D'Albertis*). Total length 8 inches, culmen 0·8, wing 3·65, tail 3·1, tarsus 1.

Obs. The late Mr. G. R. Gray separated the Aru birds as a distinct species, *Myiolestes aruensis*; but from an examination of his types, I must say that I fail to find any valid reasons for so doing. The females appear to be rather smaller, as will be seen by the following measurements:—

	Total length. in.	Culmen. in.	Wing. in.	Tail. in.	Tarsus. in.
a. ♂. Dorey (*Wallace*)	8	0·8	3·65	3·1	1·0
β. ♂. Aru Islands (*Wallace*)..	7·7	0·8	3·6	3·2	1·0
γ. ♀. Aru Islands (*Wallace*)..	7·3	0·8	3·3	3·0	0·95
δ. ♂. Mysol (*Wallace*)	7·5	0·9	3·85	3·15	0·95
ε. ♀. Mysol (*Wallace*)	7·3	0·9	3·45	2·9	1·0

Hab. New Guinea and adjacent islands.

a. Ad. st.	[New Guinea.]	Zoological Society.
b. Ad. st.	New Guinea.	Purchased.
c. Juv. (?) sk.	Salwatti.	A. R. Wallace, Esq. [C.].
d. Ad. st.	Dorey, N. G.	A. R. Wallace, Esq. [C.].
e. ♂ ad. sk.	Dorey, N. G.	A. R. Wallace, Esq. [C.].
f, g. ♂ ♀ ad. sk.	Aru Islands.	A. R. Wallace, Esq. [C.].
h, i, k. ♂ ♀ ad. sk.	Mysol.	A. R. Wallace, Esq. [C.].

2. Pinarolestes parvulus.

Colluricincla parvula, *Gould, P. Z. S.* 1845, p. 62; *Gray, Gen. B.* i. p. 295; *Gould, B. Austr.* ii. pl. 78; *Bp. Consp.* i. p. 366; *Gould, Handb. B. Austr.* i. p. 225; *Gray, Hand-l. B.* i. p. 386.

Adult. General colour above ashy brown, the head slightly darker and more greyish; lores and a streak above the eye dull white, forming a distinct eyebrow; round the eye a ring of dull whitish feathers; ear-coverts fulvous brown; cheeks whitish, washed with brown on the hinder parts; wings brown, the least wing-coverts slightly ashy, the greater series fulvous towards the tips, the quills externally fulvous brown; tail-feathers light brown, washed slightly with ashy on both webs, and tipped narrowly with buffy white; throat dull white; rest of under surface pale fulvous, rather deeper on the flanks and under tail-coverts, the fore neck and chest with narrow streaks of dark brown; under wing-coverts pale fawn-buff, the edge of the wing whitish; lower surface of quills ashy brown, pale fulvous along the base of the inner web; "bill blackish grey; tarsi bluish grey; iris dark brownish red" (*Gould*). Total length 7·6 inches, culmen 0·85, wing 3·7, tail 3·25, tarsus 1·05.

Hab. Northern Australia.

a. Ad. st.	Port Essington.	Dr. Sibbald [P.].

3. Pinarolestes rufigaster.

Colluricincla rufogaster, *Gould, P. Z. S.* 1845, p. 80; *id. B. Austr. Intr.* p. xxxvii; *Bp. Consp.* i. p. 366; *Gould, Handb. B. Austr.* i. p. 226; *Gray, Hand-l. B.* i. p. 386.
Myiolestes gouldi, *Gray, P. Z. S.* 1858, p. 180.
Myiolestes griseatus, *Gray, P. Z. S.* 1858, p. 180.
Colluricincla parvissima, *Gould, Ann. & Mag. Nat. Hist.* (4) x. p. 114.

Adult male (Cape York). General colour above dull olive-brown, the head and hind neck slightly more shaded with greyish, the rump and upper tail-coverts more decidedly olive; round the eye a ring of buffy white feathers; plumes in front of the eye also whitish; ear-coverts greyish brown, with narrow lighter shaft-streaks; least wing-coverts coloured like the back; greater series of coverts and quills brown, washed with olive; tail-feathers brown, slightly washed with olive on the edges of the feathers; cheeks and throat white washed with buff; rest of under surface of body yellowish fawn-colour, deeper on the flanks, the fore neck slightly streaked with ashy brown; under wing-coverts deep fawn-colour; quills ashy brown below, pale fawn-buff from the base upwards along the inner web. Total length 7·2 inches, culmen 0·85, wing 3·55, tail 3, tarsus 1·05.

(*Young* ?). Above olive-greenish, the crown of the head distinctly greyish, in strong contrast to the back; above the eye a broad streak of dull brick-red; round the eye a ring of isabelline feathers; lores isabelline; ear-coverts ashy brown, with lighter shaft-lines;

least wing-coverts olive like the back, the bastard wing, primary-coverts, and greater series dull chestnut; quills dark brown, externally dull chestnut, the inner secondaries somewhat shaded with olive; tail-feathers olive-brown; cheeks and throat buffy whitish, the latter in strong contrast to the breast; rest of under surface of body fawn-colour, rather paler in the centre of the abdomen, the fore neck somewhat streaked with dusky greyish; under wing-coverts and axillaries fawn-colour; lower surface of quills ashy brown, the inner webs internally fawn-buff. Total length 6·5 inches, culmen 0·8, wing 3·4, tail 2·9, tarsus 1·05.

Obs. I regret that I am unable definitely to settle the distinctive plumages of sex and age in this species. There are apparently three stages of plumage represented in the Museum series. Mr. Macgillivray has marked two birds as "males;" and both of these agree in having the wings olivaceous and resembling the back. Two other specimens differ in having the wings externally chestnut, and having likewise a small spot of chestnut behind the eye, like the remains of an eyebrow; these I think may be the old females, though the sex is not determined. From Port Albany comes the bird described in full above; and this I take to be the young. It is evidently in changing plumage, but has the chestnut wings of the foregoing (supposed) females, although it differs from all of the other examples in its very distinct rufous eyebrow. A corresponding specimen from Salwatty, with an equally broad rufous eyebrow, I consider to be the young of *P. megarhynchus*.

Mr. Gould has been so good as to lend me the type of his *Colluricincla parvissima*, which is a chestnut-winged bird, with a distinct rufous eyebrow, wing 3·5 inches. I believe it to be only the young of *Pinarolestes rufigaster*.

Hab. Northern and North-eastern Australia.

a.	[♀] ad. st.	Barnard Isles, June 2, 1848 (*J. Macgillivray*).	Capt. Stanley [P.].
b.	[♀] ad. st.	Brown's River (*J. M.*).	Capt. Stanley [P.]. (Type of *M. gouldi.*)
c.	♂ ad. sk.	Dunk Island (*J. M.*).	Capt. Stanley [P.]. (Type of *M. griseatus.*)
d.	♂ ad. st.	Cape York (*J. M.*).	Capt. Stanley [P.].
e.	Ad. sk.	Port Albany.	Purchased.
f.	[Juv.] sk.	Port Albany.	Purchased.

4. Pinarolestes affinis.

Myiolestes affinis, *Gray, P. Z. S.* 1861, p. 431; *Meyer, Sitz. Akad. Wien*, lxix. p. 209.
Colluricincla affinis, *Gray, Hand-l. B.* i. p. 386; *Salvad. Ann. Mus. Civic. Genov.* vii. p. 773.

Adult male. General colour above olive-brown, the head rather greyer, the whole back more shaded with olive; least wing-coverts olive like the back, the greater and median series externally dull rufous; quills dark brown, dull rufous on their outer aspect, the innermost secondaries slightly washed with olive; tail dull rufescent brown, the feathers slightly washed with olive on their margins; lores dusky brown; ear-coverts and sides of face and of neck dull olive-brown; chin dull whitish; rest of under surface of body light olive-brown, the feathers of the breast with distinct olive-yellow edgings, the lower throat decidedly dusky; thighs olive-brown; under tail-coverts like the breast; under wing-coverts light fawn-colour; quills dark brown below, the feathers rufous along the inner webs. Total length 7·5 inches, culmen 0·8, wing 3·55, tail 3·1, tarsus 0·9.

The Mysol specimen measures—total length 7·2 inches, culmen 0·8, wing 3·55, tail 2·8, tarsus 0·9.

Hab. Islands of Waigiou and Mysol.

a. ♂ ad. sk.	Waigiou.	A. R. Wallace, Esq. [C.].
b. Ad. sk.	Mysol.	A. R. Wallace, Esq. [C.].

5. Pinarolestes melanorhynchus.

Myiolestes melanorhynchus, *Meyer, Sitz. Akad. Wien*, lxix. p. 494.
Colluricincla melanorhyncha, *Salvad. Ann. Mus. Civic. Genov.* ix. p. 931.

Upper surface of body olive-brown, the wing-coverts and lower back reddish brown; hinder part of head somewhat inclining to greenish; region of the nostrils and forehead washed with yellowish brown; under surface brownish white, yellow towards the abdomen and on the sides of the breast; tail above brown, below lighter, and inclining to olive-colour; quills above yellowish brown on the outer webs of the primaries, and reddish brown on that of the secondaries, inner webs blackish, with whitish edges; under surface of quills blackish grey, with clear brownish edgings to the inner webs; under wing-coverts yellowish white, the edge of the wing brownish yellow; bill and claws black; legs clear brown; feet blackish. Total length 6·7 inches, culmen 0·5, wing 3·75, tail 2·85. (*A. B. Meyer.*)

Hab. Island of Mysore, in Geelvink Bay, New Guinea.

6. Pinarolestes tenebrosus.

Rectes tenebrosus, *Hartlaub & Finsch, P. Z. S.* 1868, pp. 6, 118, *iid. P. Z. S.* 1872, p. 99; *Finsch, Journ. Mus. Godeffr.* Heft viii. p. 18, Taf. iii. fig. 1.
Colluricincla tenebrosa, *Gray, Hand-l. B.* i. p. 386.

Adult male. Head above brownish black; chin, throat, and neck somewhat clearer blackish brown; the rest of the upper parts, wings, and tail still clearer umber-brown; under surface of body from the throat lighter earthy brown, the sides of the body dark umber-brown; bill dark brown (*Kubary*), black (*Peters*); feet dark (*Kubary*); iris yellow (*Peters*). Wing 3″ 6‴–3″ 10‴, tail 2″ 6‴–2″ 8‴, culmen 8‴–9‴, tarsus 12‴–13‴. (*O. Finsch.*)

Adult female. Like the male, but somewhat lighter in colour,

especially on the under surface and rump; lower mandible brownish at base. (*O. Finsch.*)
Hab. Pelew Islands.

7. Pinarolestes obscurus.

Rectes obscura, *Meyer, Sitz. Akad. Wien,* lxix. p. 390.
Colluricincla sordida, *Salvad. Ann. Mus. Civic. Genov.* ix. p. 931.

Entire upper surface brownish grey, with a tendency to an olive shade of colour; wings and tail somewhat clearer brownish; under surface grey, with a tendency to a brownish shade, the abdomen washed with whitish; bill and feet black: the claws somewhat paler. Total length 7·7 inches, culmen 0·6, wing 3·7–4, tail 3–3·35, tarsus 0·95. (*A. B. Meyer.*)
Obs. A comparison of the description of the *Colluricincla sordida* of Count Salvadori with that of *Rectes obscura* leaves little doubt of their being identical: the measurements are nearly exactly the same; and the locality is the same. Salvadori notices that the inner webs of the quills are margined with dull rufous—a feature omitted in Meyer's description.
Hab. Island of Jobi, in the Bay of Geelvink.

8. Pinarolestes heinii.

Myiolestes heinei, *Finsch & Hartl. P. Z. S.* 1869, p. 546; *id. J. f. O.* 1870, p. 126, tab. iv.; *Finsch, P. Z. S.* 1876, p. 20.

Adult. Above olivaceous dusky brown, the rump somewhat rufescent; quills obsolete dusky brown, margined with the same colour as the back; eyebrows, sides of the head, throat, and breast dull grey; sides of body ashy; middle of abdomen and under tail-coverts white; under wing-coverts cineraceous; space between the nostrils and the eyes rather broadly blackish; the ear-coverts obscure dusky brown; the four centre tail-feathers uniform with the back, the lateral ones with a white spot at the tips; bill blackish, the edge of the mandibles whitish; feet blackish. Total length 7″–7″ 2‴, bill 9‴, wing 3″ 6‴, tail 2″ 6‴, tarsus 10‴. (*Finsch & Hartlaub.*)
Young. The beak is nearly uniformly dark, with very narrow tomial margins. The end of the tail-feathers is pale, the apical spot not being distinctly discernible. (*F. & H.*)
Hab. Tonga Islands.

9. Pinarolestes vitiensis.

Myiolestes vitiensis, *Hartl. Ibis,* 1866, p. 173; *Layard, Ibis,* 1875, pp. 145, 392; *Finsch & Hartl. Faun. Central-Polyn.* p. 71, Taf. viii. fig. 1.
Colluricincla vitiensis, *Gray, Hand-l. B.* i. p. 386.

Adult male (Wakaia: *E. L. Layard*). General colour above chocolate-brown, slightly more rufous on the back, and especially so on the rump and upper tail-coverts; wings dark brown, with chocolate-brown edgings, the quills externally chocolate-brown; tail dark brown, washed with chocolate near the bases of the feathers, which are also undulated with dusky bars under certain lights; all the tail-feathers obscurely tipped with white at the tips of the inner webs, increasing in extent on the four outermost ones; ear-coverts chocolate-brown like the head; cheeks more ashy brown; throat grey; rest of under surface of body light greyish brown, somewhat washed with olivaceous on the breast, the abdomen rather clearer grey; sides of the body and flanks rufous-chocolate, duller on the under tail-coverts; under wing-coverts ashy grey slightly washed with rufous; lower surface of quills ashy, paler and more grey on the inner webs; "bill blackish brown; legs blue; iris brown" (*E. L. Layard*). Total length 6·7 inches, culmen ·85, wing 3·35, tail 3·1, tarsus ·85 (*Coll. Marquis of Tweeddale*).

Obs. Five other specimens lent to me by Lord Tweeddale measure as follows:—

	Total length.	Culmen.	Wing.	Tail.	Tarsus.
	in.	in.	in.	in.	in.
1. ♀. Wakaia (*E. L. Layard*).	6·5	0·75	3·25	3·0	0·8
2. ♂. Ovalau (*E. L. Layard*).	7·3	0·8	3·55	3·2	0·85
3. ♀. Ovalau (*E. L. Layard*).	6·4	0·75	3·1	2·95	0·9
4. ♀. Levuka (*E. L. Layard*).	6·6	0·9	3·35	3·0	0·9
5. ♀? Rewa (*E. L. Layard*).	7·0	0·85	3·2	3·2	0·9

As in the Papuan species, the female appears to be a little smaller than the male; but there is very little difference in the plumage. Some have a tendency to a greyish eyebrow, and have also the lores grey; and the Rewa bird is clear greyish on the under surface, the sides of the body and under tail-coverts being rufous chocolate.

Subsp. *α*. **Pinarolestes buensis.**

Myiolestes buensis, *Layard, Ibis*, 1876, p. 145.
Myiolestes buaënsis, *Layard, t. c.* p. 392.

Adult male. Similar to *P. vitiensis*, but having the stout bill of *P. macrorhynchus*; bill black; legs ashen grey; iris dark brown. Total length 6·7 inches, culmen ·85, wing 3·55, tail 3·1, tarsus ·85. (*Mus. Marquis of Tweeddale.*)

Adult female. Similar to the male. Total length 7·3 inches, culmen ·85, wing 3·55, tail 3·2, tarsus ·9. (*Mus. Marquis of Tweeddale*).

Hab. Bua, island of Vanua Levu, Fiji.

Subsp. *β*. **Pinarolestes fortunæ.**

Myiolestes fortunæ, *Layard, Ibis*, 1876, p. 145.

Smaller than *M. buensis*, more like *M. vitiensis*; but the throat, chin, and breast are grey; the tails, though damaged, are evidently broadly tipped with white. Total length 6″ 6‴, bill 9‴, wing 3″ 1‴, tail 3″ 3‴, tarsus 11‴ (*E. L. Layard*).

Hab. Fortuna Island, 350 miles N.E. of the Fiji group.

Subsp. γ. **Pinarolestes compressirostris.**

Myiolestes compressirostris, *Layard, Ibis*, 1876, pp. 153, 392.

"This apparently new species is intermediate between my *M. macrorhynchus* and *M. vitiensis*, Hartl.; but the remarkably thin compressed bill, with its very white base, at once distinguishes it. In size it is nearest to *M. vitiensis*, Hartl., being—length 6″ 6‴, wing 3″ 4‴, tail 3″, tarse 10‴, bill 12‴. In coloration it closely resembles *M. macrorhynchus*, but is rather redder in general tint, and has the broad whitish tips to the tail-feathers found in *M. vitiensis*." (*E. L. Layard.*)

Hab. Island of Kandavu, Fiji group.

10. **Pinarolestes macrorhynchus.**

Myiolestes macrorhynchus, *Layard, Ibis*, 1876, pp. 145, 392.

Adult male. Very like *P. vitiensis*, but with a larger bill, and not having the white spot at the tips of the outer tail-feathers, which are only slightly paler at their ends; rufous edgings to the wings also rather darker than in the foregoing bird; "bill black; legs blue; iris brown" (*E. L. Layard*). Total length 7·4 inches, culmen ·95, wing 3·7, tail 3·35, tarsus ·85.

Adult female. Similar to the male. Total length 6·7 inches, culmen ·9, wing 3·5, tail 3·0, tarsus ·85.

Hab. Island of Taviuni, Fiji.

a, b. ♂ ♀ ad. sk.	N'Gila, Taviuni, August 1875.	E. L. Layard, Esq. [C.].

Subsp. α. **Pinarolestes maximus.**

Myiolestes, sp., *Layard, Ibis*, 1876, pp. 153, 392.
Myiolestes maximus, *Layard, Ibis*, 1876, p. 498.

Adult male. General colour throughout warm brown, clearer and paler below, lightest on the upper throat and vent; flanks slightly rufous; no grey tinge about the lower parts as in *M. vitiensis*, Hartl., nor is there the slightest trace of the whitish tips to the tail-feathers, which are very indistinctly edged and tipped with rufous; inner and outer webs of wing-primaries the same, but more distinct; bill very large and strongly hooked, black, with the edges of both mandibles and the tip whitish horn-colour; legs bluish. Length 8″, wing 4″, tail 3″ 6‴, bill 1″ 2‴ (breadth 5‴), tarsus 1″ 1‴. Iris brown. (*Layard.*)

Hab. Island of Kandavu, Fiji group.

11. **Pinarolestes nigrogularis.**

Lalage nigrogularis, *Layard, P. Z. S.* 1875, p. 149.
Myiolestes nigrogularis, *Finsch, P. Z. S.* 1876, p. 20; *id. Journ. Mus. Godeffr.* Heft viii. p. 9.
Pachycephala? nigrogularis, *Layard, Ibis*, 1876, p. 392.

General colour above brown, below cinereous; top of head cinereous—each feather having a pale edge, presents a slightly scaled appearance; wing-feathers brown, inner webs darkest, outer feathers slightly tinged with greenish yellow on the outer edge; tail-feathers in strong light closely barred; covering the nostrils is a patch of black extending into a broadish eyebrow; this coalesces with the black of the throat behind the ear, which is covered by a large white patch; above the black eyebrow a faint whitish streak; chin and upper portion of throat black, this separated from the cinereous of the underparts by a broken irregular white band; vent and under tail-coverts whitish tinged with isabella colour; underside of wing- and tail-feathers paler than the upper, the latter much paler at the tips; upper edge of the wing inside deep black; axillaries whitish; bill and legs blue, changing into black after death; iris brown. Length 8″, wing 4″ 1‴, tail 3″ 9‴, tarse 1″ 1‴, bill 1″ 2‴. (*Layard.*)

Hab. Islands of Ovalau, Taviuni, and Viti Levu, Fiji group.

11. CUPHOPTERUS.

		Type.
Cuphopterus *Hartl. P. Z. S.* 1866, p. 326	....	C. dohrni.

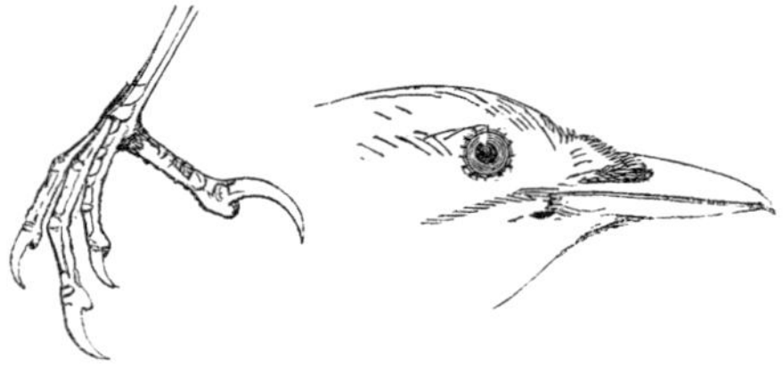

Claw and head of *Cuphopterus dohrni.*

Range. W. Africa, Prince's Island and Gaboon.

Cuphopterus dohrni.

Cuphopterus dohrni, *Hartl. P. Z. S.* 1866, p. 326, pl. xxxiv.; *Sharpe, Cat. Afr. B.* p. 42.

Tephrodornis dohrni, *Gray, Hand-l. B.* i. p. 395.

Adult female. General colour above deep chocolate-brown, the lower back and rump washed with olive; ear-coverts and lores rather darker brown than the head; wings brown like the back, the primaries margined with ashy white; tail deep brown, slightly washed with ashy on the margins of the feathers; cheeks, throat, and underparts white, slightly washed with yellow on the breast; across the fore neck a band of ashy brown; flanks olive-brown, forming a broad lateral stripe; sides of body silky white; thighs olive-brown; under tail-coverts white, with brown bases; wing-coverts white, the latter with a small spot of brown near the outer edge of the greater series; quills brown below, white along the inner web; bill brownish, the lower mandible paler; feet brownish; iris dull chestnut. Total length 5·8 inches, culmen 0·6, wing 2·75, tail 2·7, tarsus 0·95.

Male. Similar to the female (*Dohrn*).

Hab. The same as that of the genus.

a. ♀ ad. sk.	Prince's Island.	Dr. A. Dohrn [C.].
b. Ad. sk.	Prince's Island.	Dr. A. Dohrn [C.].
c. Ad. sk.	Gaboon.	M. Verreaux.

12. FRASERIA.

	Type.
Fraseria, *Bp. C. R.* xxxix. p. 536 (1854)	F. ochreata.
Eucnemidia, *Heine, J. f. O.* 1860, p. 134	F. ochreata.

Claw and head of *Fraseria ochreata.*

Range. Confined to Africa.

Key to the Species.

a. Grey above, with no white spot in front of the eye *ochreata*, p. 303.
b. Grey above; in front of the eye a conspicuous white tuft of plumes over the loral region ...... *cinerascens*, p. 304.

1. Fraseria ochreata.

Tephrodornis ochreatus, *Strickl. P. Z. S.* 1844, p. 102; *Allen, Exp. Niger*, ii. p. 489; *Fraser, Zoologia Typ.* pl. 36; *Bp. Consp.* i. p. 358; *Hartl. J. f. O.* 1855, p. 360; *Gray, Hand-l. B.* i. p. 394.
Fraseria ochreata, *Bp. C. R.* xxxix. p. 536; *Hartl. Orn. W.-Afr.* p. 102; *id. J. f. O.* 1861, p. 171; *Sharpe, Cat. Afr. B.* p. 52; *id. Ibis*, 1872, p. 71; *Bouvier, Cat. Ois. Marche &c.*, p. 20.
Eucnemidia ochreata, *Heine, J. f. O.* 1860, p. 134.

Adult (type of species). Above ashy grey, darker and rather browner on the head, the ear-coverts decidedly browner; wings dusky brown, externally washed with ashy grey like the back, clearer on the primaries; tail also dusky brown, washed on the outer webs with grey; cheeks and underparts white, with a few dusky grey margins to the feathers, very distinct on the sides of the neck and chest, and gradually disappearing on the breast; the flanks rather washed with grey; abdomen and under tail-coverts white; thighs greyish, the feathers margined with white; under wing-coverts grey, broadly edged with white; bill and legs (in skin) dark horn-brown. Total length 6·5 inches, culmen 0·65, wing 3·6, tail 2·85, tarsus 0·9.

Obs. Immature birds generally have some tinge of ochraceous colour on the breast, and remains of fulvous tips to the wing-coverts; the amount of barring on the breast varies very much, some of the individuals being thickly clouded with bars, others having them very thinly distributed; younger birds appear to have the flanks more or less distinctly barred with grey.

Hab. West Africa, from the Gold Coast to Gaboon and Fernando Po.

a. Ad. sk.	Fernando Po.	Louis Fraser, Esq. [C.]. Type of species.
b. Ad., *c.* Imm. sk.	Ashantee.	J. Gould, Esq.
d. Ad. sk.	Fantee (*Governor Ussher*).	R. B. Sharpe, Esq.
e. Ad. sk.	Fantee (*Aubinn*).	R. B. Sharpe, Esq.
f. Ad. sk.	Wassaw, July 1872 (*H. T. Blissett*).	R. B. Sharpe, Esq.
g. Imm. sk.	Accra (*Capt. Haynes*).	R. B. Sharpe, Esq.
h. Ad. sk.	Gaboon (*R. B. N. Walker*).	R. B. Sharpe, Esq.
i. Imm. sk.	Gaboon (*P. B. Du Chaillu*).	R. B. Sharpe, Esq.
k. Ad. sk.	Gaboon.	M. Du Chaillu [C.].
l. Imm. sk.	Fernand Vaz.	M. A. Marche [C.].

2. Fraseria cinerascens.

Fraseria cinerascens, *Hartl. Orn. W.-Afr.* p. 102 (*ex Temm. MS. in Mus. Lugd.*); *Cass. Pr. Philad. Acad.* 1859, p. 51; *Reich. J. f. O.* 1874, p. 27; *Bouvier, Cat. Ois. Marche &c.*, p. 21.

Tephrodornis cinerascens, *Gray, Hand-l. B.* i. p. 394.

Adult. Above deep slaty grey, the head blacker, especially the forehead, lores, and feathers below the eye; in front of the latter, and situated above the lores, a conspicuous little spot of silky white plumes; wings and tail dusky brown, washed with slaty grey; cheeks and under surface of body pure white, thickly mottled with slaty grey margins to the feathers, becoming less distinct towards the abdomen and under tail-coverts, which are pure white; sides of breast and flanks nearly uniform grey; thighs grey, the plumes edged with white; under wing-coverts white, with grey bases to the feathers; bill black (in skin); legs horn-brown. Total length 6·2 inches, culmen 0·7, wing 3·25, tail 2·95, tarsus 0·75.

Young. Brown instead of dark grey, mottled with ochraceous spots on the back, very distinct on the wing-coverts, less so on the crown, the upper tail-coverts mesially streaked with rufous, the quills also edged with ochraceous or pale rufous; underneath white, the breast pale fawn-colour or ochraceous, with greyish edgings to the feathers, producing a mottled appearance; ear-coverts dusky, streaked with pale rufous.

Hab. West Africa, from Guinea to Gaboon.

a. Ad. sk.	West Africa.	Purchased.
b. Ad. sk.	Ogôwe river.	MM. Marche et De Compiègne [C.].
c. ♀ juv. sk.	Gaboon.	M. Du Chaillu [C.].
d. Juv. sk.	Gaboon (*Du Chaillu*).	R. B. Sharpe, Esq.

13. HEMIPUS.

	Type.
Hemipus, *Hodgson, P. Z. S.* 1845, p. 32	H. capitalis.
Myiolestes, *Cabanis, Mus. Hein.* Th. i. p. 67 (1850)	H. obscurus.
Cabanisia, *Bp. C. R.* xxxviii. p. 536 (1854)............	H. obscurus.

Bill and foot of *Hemipus obscurus.*

Range. India and Ceylon; Burmah; Malayan peninsula and islands of Java, Sumatra, and Borneo.

Key to the Species.

a. Without any longitudinal white bar down the wing; tail black, not tipped with white, the outer feather narrowly margined with white along the outer web. *obscurus*, p. 305.
b. With a longitudinal bar down the wing caused by the white edgings to the inner secondaries; tail black, the outer feathers broadly tipped with white.
  a'. White tips to the outer tail-feathers very large, occupying the terminal half of the feather...... *capitalis*, p. 306.
  b'. White tips to the outer tail-feathers very small, occupying not more than one fifth of the feather. *picatus*, p. 307.

1. Hemipus obscurus.

Muscicapa obscura, *Horsf. Tr. Linn. Soc.* xiii. p. 146; *Gray, Gen. B.* i. p. 263.
Lanius no. 12, *Raffl. Tr. Linn. Soc.* xiii. p. 308.
Obscure Flycatcher, *Lath. Gen. Hist.* vi. p. 265.
Muscicapa hirundinacea, *Temm. Pl. Col.* 119; *Horsf. Zool. Researches in Java,* pl. 39.
Hemipus obscurus, *Blyth, J. A. S. B.* xv. p. 305; *id. Cat. B. Mus. A. S. B.* p. 154; *Horsf. & Moore, Cat. B. Mus. E.I. Co.* i. p. 136; *Jerd. B. Ind.* i. p. 413; *Blyth & Walden, B. Burm.* p. 122; *Sharpe, Ibis,* 1877, p. 20.
Tephrodornis obscura, *Gray, Gen. B. App.* p. 13; *Pelz. Reis. Novara, Vög.* pp. 83, 161.
Tephrodornis hirundinaceus, *Swains. An. in Menag.* p. 284; *Bp. Consp.* i. p. 357; *Bernst. J. f. O.* 1859, p. 244; *Scl. P. Z. S.* 1863, p. 217.
Myiolestes obscurus, *Cab. Mus. Hein.* i. p. 67.
Cabanisia hirundinacea, *Bp. C. R.* xxxviii. p. 536.

Adult male. Above glossy greenish black, inclining to steel-green on the head and neck; plumes of lower back black, broadly tipped with pure white where they adjoin the rump, which is pure white;

upper tail-coverts greenish black; tail-feathers black, glossed with greenish, the outermost feather edged on either web with a narrow line of white; wing-coverts greenish black; quills black, washed externally with greenish, more especially on the secondaries; lores, feathers round the eye, and ear-coverts greenish black; cheeks, sides of neck, and entire throat pure white, breast pinkish ashy; rest of under surface with the under wing- and tail-coverts pure white, the flanks very slightly washed with greyish; thighs whitish; "bill and feet shining black; iris brownish black" (*Everett*). Total length 5·4 inches, culmen 0·6, wing 2·6, tail 2·25, tarsus 0·5.

Adult female. Everywhere brownish where the male is black, the head with a slight greenish gloss, ear-coverts darker; wing-coverts with a slight greenish gloss; quills and tail brown, the inner secondaries narrowly edged with ashy brown; underneath dull white, with a pinkish grey shade on the breast. Total length 5·5 inches, culmen 0·6, wing 2·6, tail 2·3, tarsus 0·5.

Hab. Java, Sumatra, Borneo, Malacca, ranging into Tenasserim.

a. ♂ ad. st.	Java.	Purchased.
b, *c*. ♂ ♀ ad. sk.	Java.	Purchased.
d, *e*. ♂ ♀ ad. sk.	Java (*Dr. Horsfield*).	Secretary of State for India [P.].
f, *g*. ♂ ♀ ad. sk.	E. Java.	A. R. Wallace, Esq. [C.].
h, *i*. ♂ ♀ ad. sk.	Malacca.	W. Harvey, Esq. [P.].
k. ♂ juv. sk.	Malacca.	A. R. Wallace, Esq. [C.].
l. ♂ ad. sk.	Sarawak, Borneo.	A. R. Wallace, Esq. [C.].
m, *n*. ♂ ♀ ad., *o*. Juv. sk.	Labuan.	Hugh Low, Esq. [C.].
p. ♀ ad. sk.	Sibu, Sarawak.	Alfred Everett, Esq. [C.].
q. ♂ ad. sk.	Bintulu.	Alfred Everett, Esq. [C.].

2. Hemipus capitalis.

Muscicapa capitalis, *McClell. P. Z. S.* 1839, p. 157.
Hemipus picæcolor, *Hodgs. P. Z. S.* 1845, p. 33.
Hemipus picatus (*nec Sykes*), *Gray, Cat. Mamm. &c. Nepal Hodgs.* p. 93; *Horsf. & Moore, Cat. B. Mus. E.I. Co.* i. p. 136 (pt.); *Gray, Hand-l. B.* i. p. 323; *Blyth & Wald. B. Burm.* p. 122.
Hemipus capitalis, *Blyth, Cat. B. Mus. A. S. B.* p. 154; *Bp. Consp.* i. p. 317; *Godw.-Aust. J. A. S. B.* 1870, p. 99; *Jerd. Ibis*, 1872, p. 116.

Adult male. General colour above brown, the lower back with an indistinct white bar across it, caused by the feathers being white with brown subterminal bars; rump pure white; upper tail-coverts glossy black, some of the outer ones with tiny white tips; head and nape glossy greenish black; chin and cheeks white; throat, breast, and sides of body drab-brown, the abdomen and under tail-coverts white; thighs brown; under wing-coverts white; wing-coverts glossy greenish black, the least series washed with the same brown as the back, the median and greater series broadly edged with white, forming a wing-bar; quills black, the secondaries externally margined with white, in continuation of the bar formed by the coverts; tail glossy black, all but the two centre feathers tipped with white, increasing very much in extent towards the outermost, which are edged with white along the outer web for all but the basal third; bill and legs black; iris sienna-yellow. Total length 5·3 inches, culmen 0·6, wing 2·5, tail 2·55, tarsus 0·55 (*Nepal*).

Adult female. Differs from the male in having the head brown, only a little darker than the back; wings brown; tail blackish brown; the markings on both these exactly as in the male. Total length 5·3 inches, culmen 0·5, wing 2·5, tail 2·6, tarsus 0·55 (*Nepal*).

Range. Himalaya Mountains, ranging into Burmah.

a. ♂ ad. st.	Himalaya Mountains.	
b, c, d, e. Ad. sk.	Nepal.	B. H. Hodgson, Esq. [P.].
f, g. Ad., *h, i.* Jun. sk.	Darjiling.	J. Fotheringham, Esq. [C.].
k. ♂ ad. sk.	Ponsee, Kakhyen hills, March 13, 1868.	Dr. Anderson [C.].

3. Hemipus picatus.

Muscicapa picata, *Sykes*, *P. Z. S.* 1832, p. 85; *Blyth*, *J. A. S. B.* xi. p. 458; *Gray*, *Gen. B.* i. p. 263.
Muscicapa tyrannides, *Tickell*, *J. A. S. B.* ii. p. 574.
Muscicapa hirundinacea, *Jerd. Madr. Journ.* xi. p. 17.
Hemipus picatus, *Blyth*, *J. A. S. B.* xv. p. 305; *Jerd. B. Ind.* i. p. 413; *Blyth*, *Ibis*, 1866, p. 368; *Holdsw. P. Z. S.* 1872, p. 437; *Ball*, *Str. F.* 1873, p. 399; *Hume*, *t. c.* p. 474; *id. S. F.* 1875, p. 93; *Blyth & Walden*, *B. Burm.* p. 122.

Adult male. Above black, with a slight gloss of bronzy green, the head glossed with steel-black; lores, feathers round the eye, and ear-coverts black; cheeks and sides of neck pure white, forming a kind of half-collar round the back of the neck; rump creamy white, some of the plumes tipped with black, forming a transverse bar; wings black, the coverts and quills slightly glossed with greenish black, the median and inner greater coverts tipped with white, and the outer secondaries externally white, forming a longitudinal bar down the wing; upper tail-coverts and tail black, the three outer feathers tipped with white, the outermost also edged externally with the same; under surface of body pinkish ashy, the abdomen and under wing- and tail-coverts pure white, the latter with blackish bases; thighs ashy brown; "bill black; legs and feet black; iris reddish brown" (*Legge*). Total length 5·2 inches, culmen 0·55, wing 2·4, tail 2·3, tarsus 0·5.

Hab. Southern India and Ceylon.

a. Ad. sk.	[India.]	
b. ♂ ad. sk.	Ceylon, June 10, 1870.	Capt. Vincent Legge [P.].

14. BRADYORNIS.

	Type.
Bradornis, *Smith, Ill. Zool. S. Afr.* pl. cxiii. (1847) ..	B. mariquensis.
Melanopepla, *Cab. Mus. Hein.* i. p. 54 (1850)	B. ater.
Sigelus, *Cab. t. c.* p. 68 (1850)	B. silens.
Bradyornis, *Sundev. Öfv. K. Vet. Akad. Förh.* 1860, p. 106 (nom. emend.)	B. mariquensis.

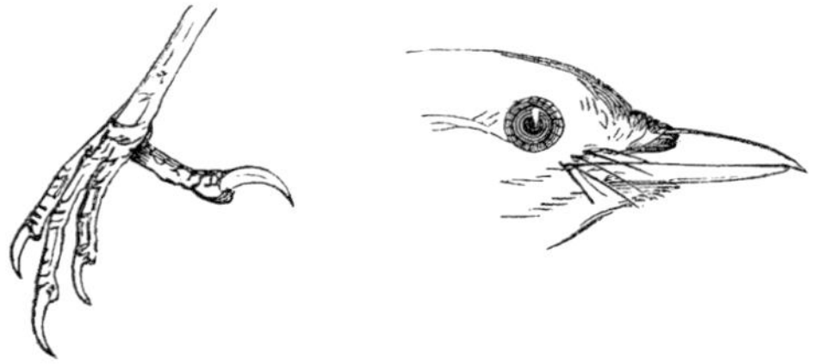

Claw and head of *Bradyornis mariquensis.*

Range. Confined to Africa.

Key to the Species.

a. Upper surface ashy or clear brown, not black.
a'. Above light brown, uniform; under surface dull white; under wing-coverts white .... *mariquensis*, p. 308.
b'. Above light brown, uniform; under wing-coverts light fawn-colour; throat white.
a''. Breast pale fawn-buff; under tail-coverts whitish, washed with light fawn-buff .... *pallidus*, p. 310.
b''. Breast ashy fulvous; under tail-coverts clear fawn-colour .... *modestus*, p. 310.
c'. Above olive-brown; under wing-coverts yellowish buff, the throat and breast also pale yellowish buff: abdomen whitish .... *woodwardi*, p. 311.
d'. Above ashy chocolate; under wing-coverts also ashy brown like the breast and throat, the latter being a little paler.................. *chocolatina*, p. 311.
b. Upper surface black.
e'. Below white.
c''. Lower back and rump black .... *silens*, p. 312.
d''. Lower back and rump white .... *senegalensis*, p. 313.
f'. Below black.
a'''. General colour glossy blue-black above and below; quills deep black below, rather browner on the inner web ...... *ater*, p. 314.
b'''. General colour dull black, rather ashy in appearance; inner linings of quills ashy black, the inner webs light ashy ...... *diabolicus*, p. 314.

1. Bradyornis mariquensis*.

Bradornis mariquensis, *Smith, Ill. Zool. S. Afr.* pl. cxiii.; *Bp. Consp.* i. p. 267.

* A second species, *Bradyornis major*, is named by Bonaparte (Consp. i.

Bradyornis mariquensis, *Sundev. Öfv. K. Vet. Akad. Förh. Stockh.* 1850, p. 106; *Hartl. J. f. O.* 1859, p. 323; *Sharpe, Cat. Afr. B.* p. 28; *Ayres, Ibis,* 1871, p. 156; *Gurney, in Anderss. B. Dam. Ld.* p. 128; *Buckley, Ibis,* 1874, p. 372.
Saxicola mariquensis, *Layard, B. S. Afr.* p. 109.
Melænornis mariquensis, *Gray, Hand-l. B.* i. p. 288.

Adult. General colour above light brown, the scapulars and least wing-coverts like the back; lesser and median coverts darker brown, edged with light brown like the back; primary-coverts and quills dark brown, externally edged with light fawn-colour, paler on the secondaries; upper tail-coverts rather more rufescent than the back; tail brown; lores buffy white; in front of the eye a dusky spot; round the eye a ring of buffy white feathers, and above the ear-coverts a slight streak of whitish feathers; ear-coverts light brown, with obsolete lighter shaft-streaks; cheeks and throat dull white, as also the rest of the under surface, the breast slightly washed with isabelline, the flanks rather browner; under wing-coverts whitish; quills light brown, edged with fulvous along the inner web; "bill and legs black; iris dark hazel" (*T. E. Buckley*). Total length 6·9 inches, culmen 0·5, wing 3·35, tail 3·15, tarsus 0·9.

Young. Brown, mottled with longitudinal streaks down the feathers, widening into a rather broad spot at the tip, very distinct on the wing-coverts; under surface of the body dull whitish, the feathers mottled with brown edges.

Hab. South Africa.

a, b, c. Ad. sk.	South Africa.	Sir A. Smith [C.].	(Types of species.)
d. Juv. sk.	South Africa.	Sir A. Smith [C.].	
e. Ad., *f.* Juv. sk.	South Africa.	Sir A. Smith [P.].	
g. Ad. sk.	South Africa.	Massena Collection.	

---

p. 267) from a statement of Sir A. Smith's in the description of *B. mariquensis*, that he believed in another species "a little larger." The bird so determined by Gray (*Melænornis major*, Gray, Hand-l. B. ii. p. 288) is *Saxicola infuscata*, which resembles the immature of *B. mariquensis* in its mottled plumage.

BRADYORNIS MINOR.

Bradyornis variegatus, *Heugl. Syst. Uebers.* p. 32 (descr. nulla).
Bradyornis minor, *Heugl. Orn. N.-O. Afr.* p. 430.

Young. Above dusky smoky brown, the head streaked with buffy white; neck, interscapulary region, and wing-coverts varied in the same manner and spotted; the white colour on the wing-coverts somewhat washed with rufous buff, marked with a triangular spot of this colour at the tip of the feathers; the outer margin of the first primary pure white; under surface of body whitish, partly washed with fulvous; the cheeks, neck, breast, and flanks streaked with dusky; under wing-coverts, vent, and centre of the abdomen pure white, unspotted; bill and feet dusky plumbeous; iris umber-brown. Total length 5·3 inches, culmen 5¼, wing 2·8¼, tail 2·1, tarsus 8·8. (*Heuglin.*)

Hab. Bari-Negro Land.

Obs. Dr. Von Heuglin only procured a young bird of this species, which is evidently, as he points out, not allied to *B. pallidus*, but to *B. mariquensis*, which it resembles in its white under wing-coverts. It is impossible to give the species a place in the synoptic table from a comparison of the immature plumage.

2. Bradyornis pallidus *.

Muscicapa pallida, *Von Müller, Naum.* 1855, p. 28; *id. Beitr. Orn. Afrik.* pl. 8; *Heugl. Syst. Uebers.* p. 32; *id. J. f. O.* 1864, p. 257.
Bradyornis pallidus, *Finsch & Hartl. Vög. Ostafr.* p. 322, note; *Heugl. Orn. N.O.-Afr.* p. 428.
Melænornis pallida, *Gray, Hand-l. B.* iii. p. 291.
Bradyornis subalaris, *Sharpe, P. Z. S.* 1873, p. 713, pl. lviii. fig. 1.

Adult. General colour above light brown; wing-coverts and quills dark brown, externally margined with paler brown, inclining to whity brown on the secondaries; tail-feathers dark brown, edged with lighter brown; lores isabelline white; round the eye a ring of isabelline feathers; ear-coverts light brown; fore part of cheeks and throat dull white; breast and sides of body light fawn-buff, paler and more of an isabelline whitish on the centre of the body; under wing-coverts, axillaries, and inner lining of quills deeper fawn-buff; bill dark brown in skin, lighter at the base of the lower mandible; feet blackish brown in skin. Total length 6·9 inches, culmen 0·55, wing 3·45, tail 3·25, tarsus 0·8.

Obs. A re-examination of the skins induces me to unite my *Bradyornis subalaris* with *B. pallidus*, as it appears to be a small race only of the latter. *B. modestus* also differs very little from the last-named bird, but may be a distinct species, depending on the characters given in the synoptic table.

Hab. North-eastern and Eastern Africa.

a. Ad. sk.	Bogos Land.	Mr. Esler [C.].
b. Ad. sk.	Mombas (*Wakefield*).	R. B. Sharpe, Esq. [P.]. (Type of *B. subularis.*

3. Bradyornis modestus.

Bradyornis modestus, *Shelley, Ibis,* 1873, p. 140.

Adult (type of species). General colour above clear brown, very

---

BRADYORNIS MURINUS.

Bradyornis murinus, *Hartl. & Finsch, Vög. Ostafr.* p. 866.

Adult male. Above earthy brown or fawn greyish; wings dark brown, slightly edged with rufous brown along the outer web, for the basal half of the inner web clearer and broader reddish fawn; secondary-coverts externally edged with pale reddish fawn; from the base of the nostrils above the lores to the eye an indistinct reddish white streak, the eye surrounded with an ill-defined ring of the same; in front of the eye on the lores an obsolete dusky spot; region of the ear reddish-fawn brown, like the sides of the neck, crop, breast, and sides; under wing-coverts clearer reddish isabelline fawn-colour; chin, throat, lower part of the breast, vent, and under tail-coverts white, with a tinge of fawn-colour; tail-feathers dark brown, externally and at the tips very slightly edged with pale reddish brown; bill and feet blackish horn-colour; "iris chestnut-brown" (*Anchieta*). Total length about 7 inches, wing 3″ 9‴, tail 2″ 10‴, bill 6‴, tarsus 10‴. (*Hartlaub & Finsch.*)

Hab. Benguela (*Anchieta*).

Obs. This species is apparently the same as my *B. subalaris*, which I now consider to be only a race of *B. pallidus*. It is most likely that *B. modestus* will also hereafter be united; and then there will be one form, *Bradyornis pallidus*, differing in size from different localities.

faintly washed with rufous on the upper tail-coverts; lores conspicuously dull white; round the eye a ring of isabelline feathers; ear-coverts light brown, with paler shafts to the feathers; cheeks brown, the feathers below the eye mottled with white; throat dull white, shaded with brown, especially on the sides; fore neck and breast pale isabelline brown; rest of under surface dull whitish, inclining to fawn-colour on the vent and under tail-coverts; flanks also pale fawn-colour; under wing-coverts light fawn, the inner edges of the quills rufous fawn-colour; wings above brown, the coverts with paler whity-brown edges to the feathers, the primaries externally edged with rufous fawn-colour, shading into fulvous towards the tips of the quills, the secondaries externally margined with whity brown; tail brown, the feathers edged with paler brown near the base; "bill brown, paler towards the base of the lower mandible; tarsi and feet brown; irides brown" (*Shelley*). Total length 6·4 inches, culmen 0·55, wing 3·35, tail 2·85, tarsus 0·85. (*Mus. G. E. Shelley.*)

Hab. Gold Coast.

4. Bradyornis woodwardi. (Plate XIV.)

Adult. General colour above olive-brown, the wing-coverts resembling the back; greater wing-coverts and quills ashy brown, edged with the olivaceous colour of the back, the secondaries resembling the latter, and having paler tips; tail-feathers ashy brown, margined with olivaceous; lores and an indistinct eyebrow buffy whitish; round the eye a ring of buffy white feathers; ear-coverts light fulvous brown; sides of neck washed with clear grey; cheeks and under surface of body yellowish buff, whiter on the centre of the abdomen, the flanks darker and more ochraceous brown; under tail-coverts white, with dusky brown centres; under wing-coverts yellowish buff; quills below ashy brown, the inner webs ashy whitish along their inner edge; bill brown, lighter at base of lower mandible; feet brown (in skin). Total length 5·8 inches, culmen 0·45, wing 3·25, tail 2·4, tarsus 0·85.

Hab. Natal.

a. Ad. sk. Berea Hills, near Durban, Natal. Messrs. Woodward [P.].

5. Bradyornis chocolatina.

Muscicapa chocolatina, *Rüpp. Neue Wirb.* p. 107; *id. Syst. Uebers.* Taf. 20; *Guérin et De Lafresn. in Ferr. et Galin. Voy. Abyss.* ii. p. 211, pl. 7; *Heugl. Syst. Uebers.* p. 31.
Muscicapa fumigata, *Guérin, Rev. Zool.* 1843, p. 161.
Curruca chocolatina, *Rüpp. Syst. Uebers.* p. 37, Taf. 14; *Bp. Consp.* i. p. 294.
Butalis chocolatina, *Bp. Consp.* i. p. 318.
Sylvia chocolatina, *Heugl. Syst. Uebers.* p. 25.
Bradyornis chocolatina, *Finsch & Hartl. Vög. Ostafr.* p. 323, note; *Heugl. Orn. N.-O. Afr.* p. 429; *Blanf. Geol. & Zool. Abyss.* p. 346.
Melænornis chocolatina, *Gray, Hand-l. B.* i. p. 288.

Adult. General colour above ashy chocolate-brown, the wing-coverts resembling the back; primary coverts and quills dark brown, edged with the lighter chocolate-brown of the back; upper tail-coverts with obsolete lighter tips; tail dark brown, washed with chocolate-brown near the base, the feathers obscurely washed with dusky under certain lights; sides of face chocolate-brown like the head; throat and breast light ashy chocolate, paler on the throat; remainder of under surface dull white, the lower flanks and under tail-coverts washed with isabelline; thighs ashy brown; under wing-coverts pale ashy chocolate; quills light brown below, with ashy margins to the inner webs; "bill and feet bluish black" (*Heuglin*); "bill bluish grey" (*Blanford*); "iris dusky" (*Heuglin*), "of a peculiar greyish yellow" (*Blanford*). Total length 6·6 inches, culmen 0·55, wing 3·35, tail 3·1, tarsus 0·85.

Hab. North-eastern Africa.

a. Ad. sk. Adigrat, 8500 ft., March 23, 1868. W. T. Blanford, Esq. [C.].

6. Bradyornis silens.

La Pie-Grièche Silencieuse, *Le Vaill. Ois. d'Afr.* ii. pl. 74; *Sundev. Crit. om Levaill.* p. 32.
Lanius silens, *Shaw, Gen. Zool.* vii. p. 330 (*ex Levaill.*); *Vieill. N. Dict. d'Hist. Nat.* xxvi. p. 153.
Saxicola incomta, *Licht. Verz. Säug. u. Vög. Kafferl.* p. 13.
Dromolæa incomta, *Hartl. Rev. Zool.* 1842, p. 13.
Laniarius silens, *Gray, Gen. B.* i. p. 299; *Layard, B. S. Afr.* p. 162.
Tephrodornis silens, *Bp. Consp.* i. p. 357; *id. C. R.* xxxix. p. 356.
Bradyornis vittatus, *Sundev. Oefv. K. Vet. Akad. Förh. Stockh.* 1850, p. 106.
Bradyornis leucomelas, *Sundev. t. c.* p. 106.
Sigelus silens, *Cab. Mus. Hein.* i. p. 68; *Sharpe, Cat. Afr. B.* p. 28; *Buckley, Ibis*, 1874, p. 372.
Melanosoma tænioptera, *Sundev. MS. in Mus. Holm.*, *undè*
Melanornis tænioptera, *Bp. Consp.* i. p. 350; *Gray, Hand-l. B.* i. p. 288.
Melanosoma simplex, *Sundev. MS. in Mus. Holm.*, *undè*
Melanornis simplex, *Bp. Consp.* i. p. 350; *Gray, Hand-l. B.* i. p. 288.
Bradyornis silens, *Hartl. J. f. O.* 1859, p. 324.
Melænornis silens, *Gray, Hand-l. B.* i. p. 288.

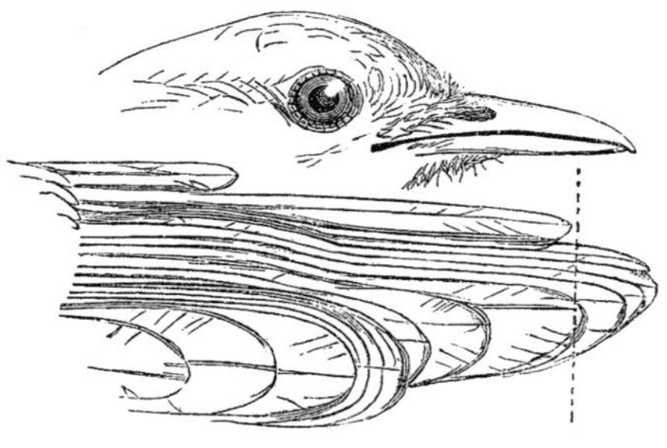

Head and tail-feathers of *Bradyornis silens*.

Young. Chocolate-brown above, all the feathers mottled with buffy whitish, and terminally edged with dark brown; upper wing- and tail-coverts edged with rufous; ear-coverts brown, longitudinally streaked with fulvous; under surface of body dull white, the feathers mottled with narrow edgings of dark brown, the sides of the upper breast rather more rufous; under wing-coverts white, with dusky bases; quills brown below, with white bases to the feathers, the alar speculum and the white margins to the secondaries very distinct; tail-feathers blackish brown, with the same white bases to the outer feathers as in the adult, but not so extended; the tips of all the feathers also white.

Adult male. General colour above glossy black, all the feathers with ashy-grey bases; wing-coverts black like the back; quills black, the inner primaries white at the base of both webs, forming a large alar speculum, the outer secondaries white along their outer margin, forming a longitudinal bar down the wing; two centre tail-feathers black, all the other feathers white with a black terminal band, the base of the inner web also dusky blackish; lores, sides of face, and sides of neck black; cheeks and under surface of body dull white, the flanks greyish; under wing-coverts white, the quills black below, white at the base of the inner webs; "bill and feet black; iris dark hazel" (*T. E. Buckley*). Total length 7 inches, culmen 0·55, wing 3·6, tail 3·2, tarsus 0·95.

Adult female. Brown where the male is black, the alar speculum not so distinct, and the outer edges of the secondaries also not so broadly white, the white on the outer tail-feathers also not so extended; under surface of body dull ashy white, the under wing-coverts whitish with brown bases; greater series of under wing-coverts brown, edged with white. Total length 6·5 inches, wing 4·5, tail 3·2, tarsus 0·95.

Hab. South Africa.

a. [♂] ad. st.	South Africa.	Purchased.
b. [♀] ad. sk.	South Africa.	Sir A. Smith [P.].
c, *d*. ♀ imm. sk.	South Africa.	Sir A. Smith [C.].
e. Juv. sk.	South Africa.	Sir A. Smith [C.].
f. ♂ ad. sk.	Natal.	Dr. Gueinzius [C.].
g. Ad. sk.	Potchefstroom, Transvaal, June 20, 1876 (*T. Ayres*).	R. B. Sharpe, Esq. [P.].

7. Bradyornis senegalensis.

Sigelus senegalensis, *Hartl. Orn. W.-Afr.* p. 112.
Bradyornis senegalensis, *Hartl. J. f. O.* 1859, p. 325.
Melænornis senegalensis, *Gray, Hand-l. B.* i. p. 288.

Above black, with scarcely any metallic gloss; wings and tail rather more dusky; a little streak between the nostrils and the feathers of the upper part of the eyebrow white; back and rump, entire under surface of body, and under wing- and tail-coverts, white; quills margined with white at the base of the inner web;

lores black; bill horn-colour; feet lead-colour. Total length 6½ inches, bill 8‴, wing 1″ 10‴, tail 2″ 2‴, tarsus 10‴. (*Hartlaub.*)

Hab. Senegal (*Mus. Lugd.*).

8. Bradyornis ater.

Bradyornis ater, *Sundev. Öfv. K. Vet. Akad. Förh.* 1850, p. 105.
Melanornis atra, *Bp. Consp.* i. p. 350; *Gurney, Ibis*, 1866, p. 268; *Layard, B. S. Afr.* p. 156; *Shelley, Ibis*, 1875, p. 76.
Melanopepla atronitens, *Cab. Mus. Hein.* Th. i. p. 54.
Melanornis atronitens, *Licht. Nomencl.* p. 20; *Kirk, Ibis*, 1864, p. 319.

Adult male. General colour, both above and below, glossy blue-black; wing-coverts like the back; quills and tail black, glossed externally with blue like the back; under wing-coverts blue-black like the breast; quills ashy black below; "bill, tarsi, and feet black; iris very dark brown" (*Ayres*). Total length 7·3 inches, culmen 0·55, wing 4·15, tail 3·7, tarsus 0·85.

Hab. South-eastern Africa, from Natal to the Zambesi.

a. ♂ ad. sk.	Durban, Natal, March 5, 1874.	Capt. G. E. Shelley [P.].
b. Ad. st.	Shupanga, Zambesi.	Dr. Kirk [C.].
c. Ad. st.	South Africa.	Sir A. Smith [C.].

9. Bradyornis diabolicus.

Melanornis pammelæna, *Gray, Hand-l. B.* i. p. 288 (*nec Stanley*).
Bradyornis pammelæna (pt.), *Finsch & Hartl. Vög. Ostafr.* p. 320.
Melanopepla pammelæna, *Gurney in Anderss. B. Dam. Ld.* p. 128.

Adult male. General colour above and below blue-black, with a dull gloss; quills and tail dark brown, the secondaries externally glossy black, the innermost entirely like the back; tail-feathers blackish brown, externally glossed with black; under wing-coverts like the breast; quills ashy below, decidedly lighter grey towards the base of the inner webs. Total length 7·9 inches, culmen 0·7, wing 4·5, tail 3·8, tarsus 0·9.

Obs. An examination of the types may prove that some of the names referred by me to *Melænornis edolioides* may really belong to this species, which I at present separate as a dull black representative of *B. ater*, and *not* a *Melænornis* at all. The Abyssinian bird is somewhat smaller than the typical Damara form, and measures as follows—total length 6·8 inches, culmen 0·55, wing 4, tail 3·7, tarsus 0·95. I may add that Von Müller's figure of *Muscicapa lugubris* appears to be drawn from an example of the true *Melænornis*, though the measurements given by him show a slight superiority of the length of wing over that of the tail. Altogether these black Wood-Shrikes of Africa are in a state of great confusion, from which only an extensive series of specimens can extricate them.

Hab. North-eastern and South-western Africa.

a. ♂ ad. sk.	Abyssinia.	Purchased.
b. Ad. sk.	Elephant Vley, Damara Land.	C. J. Andersson, Esq. [C.].

15. MELÆNORNIS.

	Type.
Melasoma, *Swains. B. W. Afr.* i. p. 257 (1837, *nec Lafresn.*)	M. edolioides.
Melænornis, *Gray, List Gen. B.* 1840, p. 36	M. edolioides.

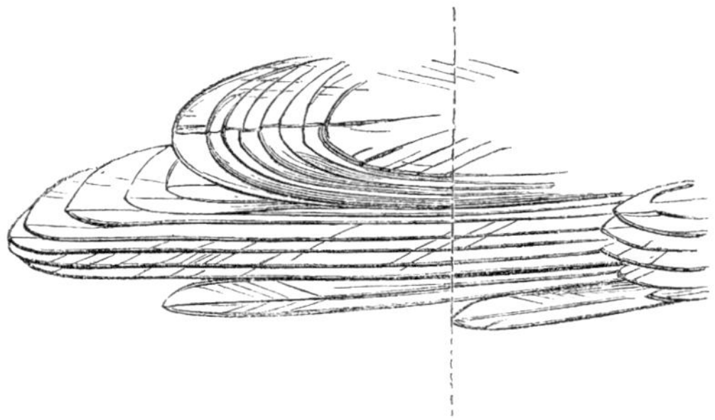

Wing of *Melænornis edolioides.*

Range. Confined to Africa.

1. Melænornis edolioides.

Melasoma edolioides, *Swains. B. W. Afr.* i. p. 257, pl. 29; *Allen & Thoms. Exp. Niger*, ii. p. 42.
Melænornis edolioides, *Gray, List Gen. B.* 1840, p. 36; *id. Gen. B.* i. p. 288; *Bp. Consp.* i. p. 350; *Cab. Mus. Hein.* Th. i. p. 54; *Heugl. Syst. Uebers.* p. 32; *Hartl. Orn. W.-Afr.* p. 102; *id. J. f. O.* 1861, p. 70; *Gray, Hand-l. B.* i. p. 288; *Heugl. Orn. N.O.-Afr.* p. 427; *Sharpe, Cat. B.* i. p. 28.
Argya edolioides, *Lafresn. in D'Orb. Dict. d'Hist. Nat.* ii. p. 126.
Muscicapa lugubris, *Von Müller, Beitr. Orn. Afr.* Taf. ii.
Melænornis melas, *Heugl. Syst. Uebers.* p. 32.
Melasoma nigerrima, *Pr. Würt. Naum.* 1857, p. 433; *Heugl. J. f. O.* 1867, p. 296.
Melanopepla lugubris, *Hartl. J. f. O.* 1859, p. 323.
Melænornis intermedia, *Heugl. J. f. O.* 1862, p. 303.

Adult male. General colour glossy black above, with somewhat of an ashy lustre; quills and tail black, glossy on the edges of the feathers; under surface of body rather more ashy black than the upper, the quills clearer ash-colour towards the bases of the inner webs; "bill and feet black; iris brown, in some yellow" (*Heuglin*).

Total length 7·5 inches, culmen 0·55, wing 3·7, tail 4, tarsus 9.

Adult female. Greyer and somewhat smaller (*Hartlaub*).

Obs. This species has probably a Robin-like young plumage, as a specimen in the Museum still preserves remains of fulvous tips to the wing-coverts.

Hab. Western and North-eastern Africa.

a. Ad. st.	West Africa.	
b, c. Ad. sk.	River Gambia.	Purchased.
d. Ad. sk.	River Gambia, Dec. 24, 1864.	R. B. Sharpe, Esq.
e. Ad. sk.	Accra (*J. Smith*).	R. B. Sharpe, Esq.

16. HYPOCOLIUS.

Type.

Hypocolius, *Bp. Consp.* i. p. 420 (1850) .... H. ampelinus.

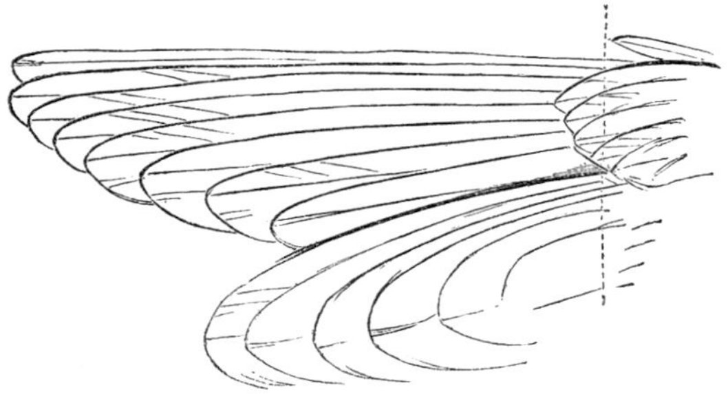

Wing of *Hypocolius ampelinus.*

Range. North-eastern Africa. Sindh.

1. Hypocolius ampelinus.

Hypocolius ampelinus, *Bp. Consp. Av.* i. p. 420; *Heugl. Ibis,* 1868, p. 181, pl. v.; *id. Orn. N.O.-Afr.* p. 420. no. 364, cxi.; *Salvad. Ibis,* 1870, p. 539; *Blanf. Ibis,* 1876, p. 388; *Scl. P. Z. S.* 1875, p. 633; *Blanf. Str. F.* 1875, p. 358; *Hume, t. c.* p. 416.
Ceblepyris isabellina, *Heugl. Syst. Uebers.* p. 32. no. 308.
Lanius ampelinus, *Gray, Hand-l. B.* i. p. 392. no. 5962.

Adult. General colour above ashy grey, slightly washed with creamy buff on the rump; head crested, deep isabelline, greyer on the crown; a narrow line of black running from the nostrils along the base of the bill and joining the feathers in front of the eye, which are also black; ear-coverts and a nuchal band black, shaded with grey, the hinder neck rather darker grey; wing-coverts drab-grey, like the back, bastard wing and primary-coverts ashy brown, the latter washed with clearer grey near the base and on the outer web; quills black, white towards their tips, shaded with brown at their extreme ends, the secondaries ashy grey like the back, the outer ones blackish on the inner webs; tail grey, with a broad black tip; cheeks and throat creamy isabelline, breast and sides of body ashy grey, the abdomen, thighs, and under tail-coverts pale creamy isabelline, the vent whitish; under wing-coverts isabelline buff; "bill horn-coloured, dusky towards the tip; legs flesh-coloured" (*Blanford*). Total length 8·7 inches, culmen 0·6, wing 3·95, tail 4·45, tarsus 0·95.

Adult female. Slightly smaller, olivascent greyish isabelline; the head, the eyebrows, and the lores duller; no black on the face or nape; the primaries externally margined with pale dusky grey, with an abrupt and conspicuous white apical margin; the black colour at the tip of the tail less distinct; under wing-coverts pale and dull isabelline. (*Von Heuglin.*)

Young. Like the female, but duller in tint; the quills nearly uniform, only slightly blackish at the tips, and not margined with white. (*Von Heuglin.*)

Hab. The same as that of the genus.

a. Ad. st.	White Nile.	M. Verreaux.

17. PLATYLOPHUS.

	Type.
Platylophus, *Swains. in Faun. Bor.-Am.* p. 482 (1831).	P. galericulatus.
Lophocitta, *Gray, List Gen. B.* p. 50 (1841)........	P. galericulatus.

Range. Malayan peninsula, from Province Wellesley southwards; Sumatra; Java; Borneo.

Key to the Species.

a. Jet-black all over, the mantle and lower parts included .... *galericulatus*, p. 317.
b. Slaty black, the back inclining to olive-brown; breast and abdomen slaty grey .... *ardesiacus*, p. 318.
c. Rufous brown in colour, both above and below .. *coronatus*, p. 318.

1. Platylophus galericulatus.

Le Geai noir à collier blanc, *Levaill. Ois. de Paradis*, i. p. 124, pl. 42.
Garrulus galericulatus, *Cuvier, Règne An.* 1817, vol. i. p. 399; *Schl. Mus. P.-B.* Coraces, p. 66.
Lanius scapulatus, *Licht. Verz. Doubl.* p. 49.
Pica galericulata, *Wagler, Syst. Av.* Pica, sp. 22.
Vanga cristata, *Griffith, An. Kingd. Birds*, pl. to p. 486.
Platylophus galericulatus, *Swains. in Faun. Bor.-Am. Birds*, p. 482; *Horsf. & Moore, Cat. B.* ii. p. 574 (excl. syn.); *Cab. J. f. O.* 1866, p. 309; *Gray, Hand-l. B.* ii. p. 2.
Lophocitta galericulata, *Gray, List Gen. B.* 1841, p. 50; *id. Gen. B.* ii. p. 305, pl. 74. fig. 5; *Bp. P. Z. S.* 1850, p. 79; *id. Consp. Gen. Av.* i. p. 374; *Cab. Mus. Hein.* Th. i. p. 218.
Lophocitta scapulata, *Licht. Nomencl. Av.* p. 9.

Adult. General colour above and below black, the wings somewhat browner; the tail black, with wavy cross markings under certain lights; head with an enormous crest, nearly three inches long; above and below the hinder part of the eye a small white spot; on each side of the neck a broad white patch, forming a demi-collar, interrupted by the black of the neck; under wing-coverts black, some of the feathers margined with white; bill and legs black. Total length 11 inches, culmen 1·15, wing 5·6, tail 5·3, tarsus 1·3.

Hab. Java.

a. Ad. sk.	West Java, 1863.	A. R. Wallace, Esq. [C.]

2. Platylophus ardesiacus.

Lophocitta galericulata, *Blyth, J. A. S. B.* xiv. p. 571 (*nec Cuv.*); *id. Cat. B. Mus. A. S. B.* p. 94; *Moore, P. Z. S.* 1859, p. 446; *Stoliczka, J. A. S. B.* 1870, p. 325; *Walden, Ibis*, 1871, p. 175.

Lophocitta ardesiaca, *Cab. Mus. Hein.* Th. i. p. 219; *Bp. Consp. Av.* i. p. 374.
Platylophus ardesiacus, *Cab. J. f. O.* 1866, p. 309; *Gray, Hand-l. B.* ii. p. 2.
Platylophus malaccensis, *Cab. J. f. O.* 1866, p. 310; *Gray, Hand-l. B.* ii. p. 2.

Adult male. General colour dark slaty grey, the back and wings distinctly washed with olive-brown; head and neck all round black, including the long crest-feathers; a small spot of feathers above and below the hinder part of the eye white; a longitudinal patch down the sides of the neck also white, the feathers of the neck immediately behind the white patch also black; throat black; the rest of the under surface slaty grey, as also the under wing-coverts, some of the latter being margined with white; lower surface of quills ashy brown, much paler near the base of the inner webs, which are also narrowly edged with rufous towards the tips; bill and legs black. Total length 10·5 inches, culmen 1·15, wing 5·6, tail 4·95, tarsus 1·3.

Young. Much paler and browner than the adult, the forehead greyish; lores whitish; the white feathers behind the eye and on the neck-patch narrowly margined with blackish; wings reddish brown, the coverts tipped with fulvous, the secondaries also lighter-tipped; under surface of body slaty grey, all the feathers broadly barred and tipped with white; the under tail-coverts more reddish brown, and having obsolete white tips. N.B. The disappearance of the whitish margins to the breast-feathers, and of the fulvous tips to the wing-feathers, seems to be effected gradually without a moult.

Hab. Malayan peninsula.

a. Ad. st.	?	Purchased.
b. ♂ juv. sk.	Malacca.	Dr. Cantor [C.].
c. Jun. sk.	Malacca.	Dr. Cantor [C.].
d. Ad. sk.	Malacca.	W. Harvey, Esq. [P.].
e. Juv. sk.	Malacca.	W. Harvey, Esq. [P.].
f. ♂ ad. sk.	Malacca.	A. R. Wallace, Esq. [P.].

3. **Platylophus coronatus.**

Lanius coronatus, *Raffl. Tr. Linn. Soc.* xiii. p. 306.
Vanga coronata, *Vigors, Mem. Raffl.* p. 658.
Lophocitta histrionica, *Bp. P. Z. S.* 1850, p. 79 (*ex Müll. MS. in Mus. Lugd.*); *id. Consp. Gen. Av.* i. p. 374.
Lophocitta coronata, *Cab. Mus. Hein.* i. p. 219.
Garrulus histrionicus, *Schl. Handl. Dierk.* i. p. 327.
Garrulus rufulus, *Schl. Handl. Dierk.* i. p. 327 (*ex Temm. MS. in Mus. Lugd.*); *id. Mus. P.-B.* Coraces, p. 66.
Platylophus coronatus, *Cab. J. f. O.* 1866, p. 309; *Gray, Hand-l. B.* ii. p. 2; *Salvad. Ucc. Born.* p. 280; *Sharpe, P. Z. S.* 1875, p. 107.

Adult. General colour above rufous brown, the head also of this colour, but rather lighter than the back; two long feathers in the centre of the crest elongated and spatulate, black; side of the face rufous brown, rather washed with blackish; above and below the hinder part of the eye a small white spot; a large patch of white on the sides of the neck, the feathers immediately behind it black; wings rufous brown like the back, all the feathers, both of the coverts and quills, blackish; upper tail-coverts slaty black washed with rufous brown; tail slaty black; entire under surface of body bright rufous brown; vent, thighs, under wing- and tail-coverts slaty grey; bill and legs blackish. Total length 11 inches, culmen 1·15, wing 5·25, tail 5·15, tarsus 1·25.

Young. Similar to the adult, but rather more dingy; wing-coverts and secondaries tipped with pale rufous buff; sides of face and throat greyish, barred with dull rufous, as also the upper breast; the rest of the under surface slaty grey washed with rufous brown; outer tail-feathers tipped with white; bill light yellowish horn-colour.

Hab. Sumatra and Southern Borneo.

a, b. Ad. st.	Sumatra.	Purchased.
c. ♀ jun. sk.	Sumatra.	A. R. Wallace, Esq. [C.].
d, e. Ad. sk.	Sumatra.	A. R. Wallace, Esq. [C.].

18. PRIONOPS.

Type.
Prionops, *Vieill. N. Dict. d'Hist. Nat.* iii. p. 145 (1816).. P. plumatus.

Head of *Prionops plumatus*

Range. Confined to Africa.

Key to the Species.

a. Head grey, surmounted by a crest of white over-shadowing the crown; throat white.
  a'. Crest enormous, composed of long plumes, reaching, when depressed, as far as the mantle .... *plumatus*, p. 320.

b'. Crest smaller and composed of hair-like plumes not reaching beyond the nape .... *poliocephalus*, p. 31.
b. Head grey, the frontal plumes rigid but not elongated back to any extent; chin white, with grey edgings to the feathers................ *talacoma*, p. 321.

1. Prionops plumatus.

Le Geoffroy, *Levaill. Ois. d'Afr.* pls. 80, 81; *Sund. Crit. om Levaill.* p. 33.
Lanius plumatus, *Shaw, Gen. Zool.* vii. pt. 2. p. 292 (*ex Levaill.*).
Prionops geoffroyi, *Vieill. N. Dict.* iii. p. 145 (*ex Levaill.*); *id. Gal. Ois.* pl. 142.
Prionops plumatus, *Swains. B. W. Afr.* i. p. 246, pl. xxvi.; *Gray, Gen. B.* i. p. 292; *Cab. Mus. Hein.* i. p. 68; *Bp. Consp.* i. p. 359; *Hartl. Orn. W.-Afr.* p. 107; *Bocage, Jorn. Lisb.* i. p. 138; *Gray, Hand-l. B.* i. p. 395; *Sharpe, Cat. Afr. B.* p. 49; *Bouvier, Cat. Ois. Marche &c.* p. 21.

Nestling. Head all round, a very slight crest, and under parts snowy white; back light brown, the feathers margined with greyish white; wing-coverts also pale brown, the inner, median, and greater coverts white with brown bases, quills black, with a slight greenish gloss; all the primaries with a broad bar of white on the inner web, the inner secondaries narrowly edged with white towards the tip; tail greenish black, tipped with white, more broadly towards the outermost, which is entirely pure white.

Adult. Head white, with an enormous crest, the loral and frontal plumes erect and recurved over the nostrils, and behind this smaller crest a large one of long white plumes, extending, when at rest, as far as the mantle; above and below the eye a large wattle, yellow in skin but probably red in life; ear-coverts, sides of neck, and nape inclining to slaty grey; hind neck white, forming a collar; entire back glossy black, some of the outer upper tail-coverts tipped with white on their outer web; wings black, the median and inner greater coverts pure white, as also the outer webs of the middle secondaries, forming a bar down the wing, the rest of the secondaries tipped with white, the primaries broadly barred with white on their inner web; tail greenish black, the feathers tipped with white, the tips being gradually broader towards the outer feathers, the outermost being entirely pure white, excepting a small patch of brown at the base of the inner web; entire under surface of body pure white, the flanks somewhat greyish streaked with white; under wing-coverts black. Total length 9 inches, culmen 0·85, wing 4·65, tail 4·05, tarsus 0·95.

Hab. West Africa.

a. Ad. st.	Sierra Leone.	
b. Juv. st.	River Gambia.	Governor Rendall [C.].
c. Ad., *d*. Juv. sk.	River Gambia.	R. B. Sharpe, Esq.
e. ♂ ad. sk.	Senegal.	R. B. Sharpe, Esq.
f. Pull. st.	Senegal.	Purchased.
g. Ad. sk.	W. Africa.	Purchased.

2. Prionops poliocephalus.

Lanius poliocephalus, *Stanley, in Salt's Trav. Abyss.* Appen. p. 50.
Prionops cristatus, *Rüpp. Faun. Abyss.* p. 30, pl. 12. fig. 2; *Vierth. Naum.* 1852, p. 48; *Heugl. Syst. Uebers.* p. 34.
Prionops geoffroyi, *Des Murs, in Lefebvre Voy. Abyss. Zool.* p. 98 (*nec Vieill.*).
Prionops concinnatus, *Sundev. Oefv. K. Vet. Akad. Förh. Stockh.* 1850, p. 130.
Prionops poliocephalus, *Gray, Gen. B.* i. p. 292; *Rüpp. Syst. Uebers.* p. 62; *Bp. Consp.* i. p. 359; *Cab. Mus. Hein.* i. p. 68; *Hartl. P. Z. S.* 1863, p. 105; *Scl. P. Z. S.* 1864, p. 109; *Antin. Cat. descritt. Ucc.* p. 53; *Heugl. Orn. N.O.-Afr.* p. 454; *Gray, Hand-l. B.* i. p. 395; *Antin. & Salvad. Ucc. Bogos*, p. 81.

Adult male. Back black, with a greenish gloss, the outer upper tail-coverts white on their outer web; wings black like the back, with a broad longitudinal bar of white formed by the inner median and greater coverts and the outer webs of the middle secondaries, the rest of the secondaries tipped with white, the primaries with a broad white band across the inner web; tail greenish black, tipped with white, which increases gradually towards the outermost, this being entirely white with a black base to its inner web; lores and frontal plumes recurved and hiding the nostrils; fore part of head crested, but the plumage stiff, and not extending beyond the nape; sides of face white; crown of head dark pearly grey; hind neck white, forming a collar; entire under surface of body white, the flanks mixed with grey; under wing-coverts black, as also are a few feathers at the sides of the upper breast; "bill black; eye-wattle sulphur-yellow; feet vermilion; iris pearly" (*Heuglin*). Total length 9 inches, culmen 0·85, wing 4·55, tail 3·9, tarsus 0·95.

Hab. North-eastern Africa.

a.	Ad. st.	Abyssinia.	Dr. Rüppell [C.].
b.	♂ ad. sk.	Abyssinia.	R. B. Sharpe, Esq.
c.	Ad. sk.	White Nile (*Mus. Holm.*).	R. B. Sharpe, Esq.

3. Prionops talacoma.

Prionops talacoma, *Smith, Report Exp. Expl. C. Afr.* App. p. 45; *id. Ill. Zool. S. Afr.* Aves, pl. v.; *Gray, Gen. B.* i. p. 292; *Bp. Consp.* i. p. 359; *Gray, Ann. N. H.* (3) xiv. p. 374; *Kirk, Ibis,* 1864, p. 319; *Gurney, P. Z. S.* 1864, p. 3; *Anderss. t. c.* p. 5; *Layard, B. S. Afr.* p. 160; *Gurney, Ibis,* 1868, p. 267; *Gray, Hand-l. B.* i. p. 395; *Ayres, Ibis,* 1869, p. 293; *Sharpe, P. Z. S.* 1870, p. 148; *Finsch & Hartl. Vög. Ostaf.* p. 365; *Sharpe, Cat. Afr. B.* p. 49; *Gurney in Anderss. B. Dam. Ld.* p. 141; *Bocage, Jorn. Lisb.* 1874, pp. 7, 24; *Buckley, Ibis,* 1874, p. 376.

Adult male. Above black, with a greenish gloss, some of the outer upper tail-coverts broadly tipped with white; wings black like the back, with a broad white bar down the wing, formed by the inner median and greater coverts and the outer web of the middle secondaries, all of which are tipped with white, the primaries with a broad bar of white across the inner web; tail greenish black, broadly tipped with white, which increases gradually towards the outermost feather, this being entirely white, with the exception of the base of the inner web, which is black; head deep pearly grey, darker in front of the eye, the frontal feathers whiter, and recurved over the nostrils; round the hind neck a white collar; sides of face whitish washed with grey, with a crescentic ring of black behind the ear-coverts; entire under surface of body white, the chin mottled with greyish tips to the feathers; flanks greyish streaked with white; under wing-coverts black; "bill black; eye-wattle orange; legs flesh-colour; iris bright lemon-yellow" (*Andersson*). Total length 7·8 inches, culmen 0·85, wing 4·25, tail 3·5, tarsus 0·95.

Adult female. Exactly similar to the male. Total length 7·4 inches, wing 4·25, tail 3·55, tarsus 0·9.

Young. Similar to the adults but browner, especially on the crown, which is ashy brown, deepening towards the nape, the feathers of the hind neck edged with ashy brown. A specimen with remains of nestling-plumage, shows that, as in the other species, this is brown above, with whitish edgings to the feathers.

Hab. South Africa.

a. Ad. st.	South Africa (*Sir A. Smith*).	S. African Museum. (Type of species.)
b. Ad. st.	South Africa.	Sir A. Smith [C.].
c. Ad. sk.	South Africa (*Sir A. Smith*).	Purchased.
d, e. ♂ ♀ ad. sk.	Transvaal (*T. Ayres*).	R. B. Sharpe, Esq.
f, g, h. Imm. sk.	Tete, Zambesi.	Dr. Kirk [C.].
i. Ad. st.	Tete, Zambesi.	Dr. Kirk [C.].
k. ♂ ad. sk.	Ovaquenyama, Damara Land, June 6, 1867 (*C. J. Andersson*).	R. B. Sharpe, Esq.
l. ♀ ad. sk.	Ovaquenyama, May 23, 1867 (*C. J. A.*).	R. B. Sharpe, Esq.
m. Juv. sk.	Ondonga, Ovampo Land, Jan. 22, 1867 (*C. J. A.*).	R. B. Sharpe, Esq.
n, o. ♂ ♀ ad. sk.	Galungo, Angola, Aug. 30 (*Sala*).	R. B. Sharpe, Esq. (*Cf.* Sharpe, P. Z. S. 1870, p. 148.)
p. Ad. sk.	Angola.	J. J. Monteiro, Esq. [C.]. (*Cf.* Ibis, 1862, p. 327.)

19. SIGMODUS.

Type.

Sigmodus, *Finsch & Hartl. Vög. Ostafr.* p. 364, 1870 (*ex Bp.**) .... S. caniceps.

Range. Confined to Africa.

Key to the Species.

a. Back black, the head grey; frontal crest grey or orange, not black like the crown.
  a'. Frontal tuft whitish or whitish grey.
    a''. Chin and ear-coverts pearly grey; abdomen deep tawny or chestnut .... *rufiventris*, p. 323.

* In none of the places where Prince Bonaparte mentions *S. caniceps* does he give the characters of the genus; the above-named authors are the first to do so, as far as I can determine.

b''. Chin and ear-coverts black like the rest of the throat; abdomen light tawny buff .... *caniceps*, p. 323.
b'. Frontal tuft orange buff .... *scopifrons*, p. 324.
b. Back ashy or drab-brown; the head black, as well as the frontal crest; hind neck black like the head; under surface black.
c'. With a white bar on the inner web of the primaries.
c''. Size very much larger: wing 5·25 inches; back ashy brown .... *retzi*, p. 324.
d''. Size much smaller: wing 5 inches; back drab-brown .... *tricolor*, p. 325.
d'. With no white bar on the primaries .... *graculinus*, p. 325.

1. Sigmodus rufiventris.

Sigmodus rufiventris, *Bp. Rev. et Mag. de Zool.* 1853, p. 441; *Hartl. Orn. Westafr.* p. 105; *id. J. f. O.* 1861, p. 171; *Cass. Pr. Acad. N. Sci. Philad.* 1859, p. 13; *Sharpe, Cat. Afr. B.* p. 50.
Prionops rufiventris, *Gray, Hand-l. B.* i. p. 395; *Finsch & Hartl. Vög. Ostaf.* p. 364; *Heugl. Orn. N.O.-Afr.* p. 455, note.

Adult. General colour above black with a slight greenish gloss, the quills and tail uniform with the back, all the primaries with a broad bar of white across the inner web; head pearly grey, with a slight gloss of lilac, the forehead and lores white, the grey colour occupying the chin and entire sides of face as well as the head; throat black, forming a collar which runs round the neck and separates the head from the greenish black of the back; breast greyish white; rest of under surface light chestnut or deep tawny; under wing-coverts black; bill and feet red. Total length 7·5 inches, culmen 0·85, wing 4·45, tail 3·2, tarsus 0·85.
Hab. Gaboon.

a. Ad. sk.	Gaboon.	Purchased.
b. Ad. st.	Gaboon (*F. B. N. Walker*).	R. B. Sharpe, Esq.
c. Ad. st.	Moonda River, Gaboon.	M. B. Du Chaillu [C.].

2. Sigmodus caniceps.

Sigmodus caniceps, *Bp. Consp.* i. p. 365 (*ex Temm. MS. in Mus. Lugd.*); *Jard. Contr. Orn.* 1852, p. 140, pl. 94; *Schl. Naum.* ii. pt. 2. p. 27; *id. Handl. Dierk.* i. p. 286, *Atl. Vog.* fig. 17; *Bp. Rev. et Mag. de Zool.* 1853, p. 441; *Hartl. Orn. Westafr.* p. 105; *Sharpe, Cat. Afr. B.* i. p. 50.
Prionops caniceps, *Gray, Hand-l. B.* i. p. 395; *Finsch & Hartl. Vög. Ostafr.* p. 364, note; *Heugl. Orn. N.O.-Afr.* p. 455, note.

Adult. Entire surface above black with a greenish gloss, the primaries barred across with white on the inner web of all except the first one; crown of head, lores, and feathers in front of the eye pearl-grey; the entire throat, sides of face, and neck all round deep black, without any metallic gloss like the back; entire breast white, the abdomen and under tail-coverts light tawny buff; under wing-coverts black; bill and feet red, fading to yellow in skin. Total length 8·3 inches, culmen 0·85, wing 4·65, tail 3·45, tarsus 0·85.

Young. Differs from the adult in being greyish black above, with whitish lores, and streaks on the forehead; throat and sides of face light tawny buff, with greyish edgings to the ear-coverts; outer tail-coverts rather pointed, and having a fulvous tip to the outer web; on the inner web of the quills the same white bar as in the adult. This latter character seems to vary, insomuch that it is sometimes absent on the first primary.

Hab. Gold-Coast district of Western Africa, ranging up to Sierra Leone.

a. Ad. st.	Gold Coast.	Purchased.
b. Juv. sk.	Fantee.	A. Swanzy, Esq. [P.].
c, *d*. Ad., *e*. Juv. sk.	Fantee (*Gov. Ussher*).	R. B. Sharpe, Esq.
f. Imm. sk.	Denkera (*Gov. Ussher*).	R. B. Sharpe, Esq.
g. Ad. sk.	Sierra Leone (*Dr. Clark*).	R. B. Sharpe, Esq.

3. **Sigmodus scopifrons.**

Sigmodus scopifrons, *Peters, J. f. O.* 1854, p. 422; *Hartl. Orn. W.-Afr.* p. 105, note.
Prionops scopifrons, *Gray, Hand-l. B.* i. p. 395; *Finsch & Hartl. Vög. Ostafr.* p. 368; *Heugl. Orn. N.O.-Afr.* p. 455, note.

Adult. Above slaty blackish; underneath slate-colour; a crest on the forehead orange-buff; two centre tail-feathers uniform, the outer ones white at the tip and on the outer edge, the rest only white at the tip; vent and under tail-coverts pure white; lower abdomen faintly varied with slate-colour and white; quills marked on the inner web with rather a large spot of white; under wing-coverts ashy grey; bill and feet coral-red. Total length 6·5 inches, wing 3·8, tail 2·5, tarsus 0·8. (*Finsch & Hartlaub.*)

Hab. Mozambique.

4. **Sigmodus retzii.**

Prionops retzii, *Wahlb. Oefv. K. Vet. Akad. Förh. Stockh.* 1856, p. 174; *id. J. f. O.* 1857, p. 1; *Anderss. P. Z. S.* 1864, p. 5; *Gurney, t. c.* p. 3; *id. in Anderss. B. Dam. Ld.* p. 142; *Sharpe, Cat. Afr. B.* p. 50; *Bocage, Jorn. Lisb.* i. p. 344, 1874, pp. 7, 24, 1876, p. 7.

Adult male. Above dull ashy brown, the wing-coverts similar to the back, the greater series rather duller brown; primary-coverts and quills black with a slight greenish gloss, the inner secondaries externally ashy brown like the back; upper tail-coverts and tail black, the latter tipped with white, more broadly on the outer feathers; entire head and neck all round, as well as the whole of the under surface of the body, blue-black, the vent and under tail-coverts white; under wing-coverts black, the quills also greyish black below, the primaries barred with white on the inner web; round the eye a "wattle of orange-red; bill orange-red, paling off towards the extremity; legs pale orange, toes yellowish; iris light orange" (*Andersson*). Total length 9 inches, culmen about 0·9, wing 5·25, tail 3·85, tarsus 0·95.

Young. Differs from the adult in being uniformly ashy brown both above and below, deepening into black on the upper tail-coverts; wings and tail as in adult, the vent and under tail-coverts white, as in the old bird also, but washed with brown; the white on the quills also more extended; "soft parts similar to those of the adult, but paler" (*Andersson*).

Hab. South-western Africa.

a. Ad. sk.	Elephant Vley, Damara Land (*C. J. Andersson*).	R. B. Sharpe, Esq.
b. ♂ juv. sk.	Elephant Vley, Aug. 19, 1859 (*Andersson*).	R. B. Sharpe, Esq.
c. ♂ ad. sk.	Maconjo, July 1870 (*Anchieta*).	R. B. Sharpe, Esq.

5. Sigmodus tricolor.

Prionops tricolor, *Gray, Ann. N. H.* (3), xiv. p. 379; *id. P. Z. S.* 1864, p. 45; *Kirk, Ibis,* 1864, p. 319.

Prionops retzii, *Gray, Hand-l. B.* i. p. 366; *Finsch & Hartl. Vög. Ostafr.* p. 366.

Adult. Back and scapulars drab-brown, the wing-coverts uniform with the back, the greater series and inner secondaries more ashy brown; primary-coverts and remainder of quills blackish with a slight greenish gloss; upper tail-coverts blackish; tail greenish black, the two centre feathers very slightly, all the others broadly tipped with white, extending some little way up the outer web of the outer feather; head and neck all round, as well as the under surface of the body, glossy black, slightly washed with grey on the lower flanks; vent and under tail-coverts white, the latter with broad black bases; under wing-coverts blackish, the primaries with a broad bar of white across the inner web. Total length 8 inches, culmen 0·8, wing 5, tail 3·9, tarsus 0·85.

Hab. Zambesi district in S.E. Africa.

a, b, c. Ad. sk.	Tete, Zambesi.	Dr. Kirk [C.]. (Types of species.)
d. Ad. st.	Lower Shiré valley, July 1861.	Dr. Kirk [C.].

6. Sigmodus graculinus.

Prionops graculinus, *Cab. in Von der Decken, Reis.* iii. p. 24, Taf. iii.; *Gray, Hand-l. B.* i. p. 395; *Finsch & Hartl. Vög. Ostafr.* p. 368; *Sharpe, P. Z. S.* 1873, p. 714.

Adult. Above ashy brown, including the scapulars and upper wing-coverts; primary-coverts and primaries black, slightly washed with greenish externally, the secondaries browner, the outer web of the inner secondaries ashy brown like the back; head all round, hind neck, and upper part of mantle, as well as the sides of the neck and entire under surface of body purplish black, including the under wing-coverts; vent and under tail-coverts white; upper tail-coverts black; tail also black, with a greenish gloss, all the feathers tipped with white, the outer ones more broadly; round the eye a red caruncle; bill red; feet also red, both yellow in skin. Total length 8·9 inches, culmen 0·9, wing 4·85, tail 3·85, tarsus 0·95.

Hab. Eastern Africa.

a. Ad. sk.	Mombas (*Wakefield*).	R. B. Sharpe, Esq. [P.] (*Cf.* P. Z. S. *l. c.*)

Subfam. EURYCEROTINÆ.

20. EURYCEROS.

Type.

Euryceros, *Less. Cent. Zool.* pl. 74 .... E. prevosti.

Range. Confined to Madagascar.

1. Euryceros prevosti.

Naracarac Hornbill, *Lath. Gen. Hist. B.* ii. p. 326.

Euryceros prevosti, *Less. Cent. Zool.* pl. 74; *id. Illustr. Zool.* pl. 13; *id. Bull. Sc. Nat.* xxv. p. 243; *id. Compl. Buff.* viii. p. 460; *Hartl. Faun. Madag.* p. 58; *Schl. & Poll. Faun. Madag. Ois.* p. 159; *id. Dierk.* p. 188; *Grand. Rev. et Mag. Zool.* 1867, p. 388; *Gray, Hand-l. B.* ii. p. 23; *Sharpe, P. Z. S.* 1875, p. 77.

Adult male. Head and neck velvety black, with a purplish-blue gloss; entire back, scapulars, and wing-coverts bright bay, the bastard wing, primary-coverts, and quills black; tail black, with the exception of the two centre feathers, which are bright bay, with a small terminal spot of black; entire under surface of body black, with a few indistinct remains of narrow fulvous bars on the abdomen. Total length 11·8 inches, culmen 1·9, depth of bill 1·1, wing 5·8, tail 5, tarsus 1·25.

Adult female. The bill lighter-coloured than in the male, with black tip; secondaries externally olive-brown; belly thickly crossed with narrow fulvous bars, the under wing- and tail-coverts rufescent, with irregular zigzag bars of blackish; thighs brownish. Total length 12 inches, culmen 2, depth of bill 1·05, wing 5·7, tail 4·8, tarsus 1·3.

Young. Similar to the female, the general colour of the upper surface brown, gradually assuming the bright bay of the adult; the head dull brown like the mantle, with here and there indications of the approaching black hood; cross lines on abdomen very distinct.

Hab. Madagascar.

a. [♂] ad. st.	Madagascar.	Purchased.
b. [♀] ad. st.	Madagascar.	Purchased.
c. ♂ ad. sk.	S.E. Madagascar (*Crossley*).	R. B. Sharpe, Esq. [P.].
d. ♀ ad. sk.	S.E. Madagascar.	Mr. A. Crossley [C.].
e. ♀ juv. sk.	S.E. Madagascar.	Mr. A. Crossley [C.].

APPENDIX.

A FEW alterations and additions have become necessary during the progress of the present volume through the press.

Page 14. **Corvus corax.**

Dr. Taczanowski (Bull. Soc. Zool. France, vol. i.) considers the Raven of Eastern Siberia distinct, and names it *Corvus coronivox* on Dr. Dybowski's authority. A specimen from the Onon river is included in our series; and I cannot see that it is in any way specifically distinct from the common Raven.

Page 32. **Corone capellana,** Scl.

Corvus capellanus, *Sclater, P. Z. S.* 1876, p. 694, pl. lxvi.

Page 44. **Corone orru.**

Since the account of this bird was written the Museum has obtained, by exchange, from the Darmstadt Museum, one of the specimens described by Dr. Brüggemann as *C. annectens.* It is a large Crow of the *C. enca* type, but distinguished by its greater size and by the more pronounced lanceolations on the throat, produced by the splitting up of the ends of the feathers. The Marquis of Tweeddale, who has seen the specimen, identifies it as true *C. valida*; but I consider that it is undistinguishable from *C. orru*, to which its measurements (total length 17·2 inches, culmen 2·75, wing 13·05, tail 6·7, tarsus 2·2) ally it. Dr. Brüggemann has established the existence of a second species of Crow in the island of Celebes, a very interesting fact. The following specimen is therefore to be added:—

a. ♀ ad. sk. Gorontalo (*Schneider*). Darmstadt Museum [E.].

Page 66. **Pica leucoptera.**

Add the following:—

a. ♀ ad. sk. Kashgar, Feb. 16, 1874. Capt. J. Biddulph [C.].

Page 148.

For **Pyrrhocorax alpinus** *read* **Pyrrhocorax pyrrhocorax.**

Page 151. **Podoces biddulphi.**

Add:—

a. ♀ ad. sk. Maralbashi, Jan. 10, 1874. Capt. T. Biddulph [C.]. (Type of species.)

Add:— Page 183. **Manucodia comrii.**

Sclater, *P. Z. S.* 1877, p. 43.

Page 224. **Sphecotheres maxillaris.**

In description of adult female, for *S. flavigaster* read *S. flaviventris*.

Page 298. **Pinarolestes melanorhynchus.**

The following is a description of the type specimen recently lent to me by Dr. Meyer:—

Adult male (type of species). General colour above of an olive-brown, with somewhat of a greenish cast, more especially on the head and nape; wings reddish brown, the least wing-coverts slightly washed with olive, the primary-coverts dusky, externally olivaceous; quills dark brown, the primaries tawny on the outer web, the secondaries externally chestnut, the innermost olivaceous on their inner webs; tail rufous olive; lores and feathers in front of the eye fulvous, as also a ring of plumes round the latter; over the eye an indistinct shade of buff; ear-coverts fulvous brown, with narrow shaft-lines of a light colour; cheeks and under surface of body yellowish buff, more distinctly yellow on the lower abdomen and under tail-coverts, the flanks rather more tawny, and the sides of the upper breast shaded with brown; under wing-coverts yellowish buff, the quills brown below, with paler edges of rufous buff along the inner web; bill black, brown towards the base of the lower mandible. Total length 6·5 inches, culmen 0·8, wing 3·6, tail 2·5, tarsus 1. (*Mus. Dresden.*)

Page 299. **Pinarolestes obscurus.**

Dr. Meyer having kindly forwarded me the typical specimen for examination, I give a detailed description:—

Adult male (type of species). General colour above dull ashy brown with a distinct olivaceous shade, the head and neck greyer; least wing-coverts like the back; remainder of the wing-coverts brown shaded with olive, especially on the primary-coverts, which are darker; quills dark brown, externally olivaceous brown, the innermost entirely of the latter colour; tail clearer brown, slightly washed with olive; lores ashy; sides of face brown, like the head; under surface of body dingy grey, somewhat shaded with olive, especially on the chest, the flanks decidedly olivaceous; centre of the breast clearer grey, dull whitish on the centre of the abdomen; under wing-coverts pale fulvous, the quills dusky brown below, pale rufous buff along their inner webs; bill black. Total length 7·4 inches, culmen 0·9, wing 3·55, tail 3, tarsus 0·95. (*Mus. Dresden.*)

ALPHABETICAL INDEX.

END OF THE THIRD VOLUME.

PRINTED BY TAYLOR AND FRANCIS, RED LION COURT, FLEET STREET.

LIST OF PLATES.

Pl. I.

J. G. Keulemans del. Mintern Bros. imp.

Physocorax moneduloides.

J. G. Keulemans del. Mintern Bros. imp.

Strepera crissalis.

J. G. Keulemans del. Mintern Bros. imp.

Dendrocitta occipitalis.

J. G. Keulemans del. Mintern Bros. imp.

Garrulus leucotis.

J. G. Keulemans del

Mintern Bros. imp.

1, *Perisoreus capitalis.*
2, *Perisoreus obscurus.*

Pl. VI.

J. G. Keulemans del. Mintern Bros. imp.

Xanthura melanocyanea.

Pl. VII.

J.G. Keulemans del. Mintern Bros. imp.

Xanthura turcosa.

J. G. Keulemans del. Mintern Bros. imp.

Xanthura quindiuna.

Pl. IX.

J. G. Keulemans del.

Mintern Bros. imp.

Psilorhinus cyanogenys.

Pl. X.

J. G. Keulemans del.

Mintern Bros. imp.

Oriolus steerii.

Pl. XI.

J. G. Keulemans del.

Mintern Bros. imp.

Oriolus viridifuscus.

J. G. Keulemans del.

Mintern Bros. imp.

Sphecotheres salvadorii.

Pl. XIII.

J. G. Keulemans del.

Mintern Bros. imp.

Chaetorhynchus papuensis.

J. G. Keulemans del.

Mintern Bros. imp.

Bradyornis woodwardi.

Made in the USA
Las Vegas, NV
03 November 2026

57424064R00233